ADVANCES IN GENETICS

VOLUME 11

ADVANCES IN GENETICS

VOLUME 11

Edited by

E. W. CASPARI

*Biological Laboratories
The University of Rochester
Rochester, New York*

and

J. M. THODAY

*Department of Genetics
University of Cambridge
Cambridge, England*

Editorial Board

G. W. BEADLE   MERLE T. JENKINS
WILLIAM C. BOYD   JAY L. LUSH
M. DEMEREC   ALFRED MIRSKY
TH. DOBZHANSKY   J. T. PATTERSON
L. C. DUNN   M. M. RHOADES
           CURT STERN

1962

ACADEMIC PRESS • NEW YORK AND LONDON

Copyright© 1962 by Academic Press Inc.

All Rights Reserved

No part of this book may be reproduced in any form by photostat, microfilm, or any other means, without written permission from the publishers.

ACADEMIC PRESS INC.
111 Fifth Avenue
New York 3, N. Y.

United Kingdom Edition
PUBLISHED BY
ACADEMIC PRESS INC. (London) Ltd.
17 Old Queen Street, London S.W. 1

QH
431
A1
A3
v.11

Library of Congress Catalog Card Number 47–30313

PRINTED IN THE UNITED STATES OF AMERICA

CONTRIBUTORS TO VOLUME 11

W. F. BODMER,* Department of Genetics, University of Cambridge, Cambridge, England

ALLAN M. CAMPBELL, Department of Biology, University of Rochester, Rochester, New York

RALPH E. CLELAND, Department of Botany, Indiana University, Bloomington, Indiana

A. W. F. EDWARDS†, Department of Genetics, University of Cambridge, Cambridge, England

D. A. HOPWOOD,‡ Botany School, University of Cambridge, Cambridge, England

P. LISSOUBA, Laboratoire de Génétique Physiologique du Centre National de la Recherche Scientifique, Gif-sur-Yvette (Seine et Oise), France

J. MOUSSEAU, Laboratoire de Génétique Physiologique du Centre National de la Recherche Scientifique, Gif-sur-Yvette (Seine et Oise), France

P. A. PARSONS,§ Department of Genetics, University of Cambridge, Cambridge, England

G. RIZET, Laboratoire de Génétique Physiologique du Centre National de la Recherche Scientifique, Gif-sur-Yvette (Seine et Oise), France

J. L. ROSSIGNOL, Laboratoire de Génétique Physiologique du Centre National de la Recherche Scientifique, Gif-sur-Yvette (Seine et Oise), France

G. SERMONTI, Istituto Superiore di Sanità, Rome, Italy

* Present address: Department of Genetics, Stanford University School of Medicine, Palo Alto, California.
† Present address: Istituto di Genetica, Universita di Pavia, Pavia, Italy.
‡ Present address: Department of Genetics, University of Glasgow, Glasgow, Scotland.
§ Present address: Melbourne University, Melbourne, Australia.

CONTENTS

Contributors to Volume 11 v

Linkage and Recombination in Evolution
W. F. Bodmer and P. A. Parsons

I. Introduction. 2
II. Polygenic Balance 4
III. Balanced Polymorphisms and Linkage 32
IV. Chiasma Frequencies and Cytological Evidence 40
V. Genotypic and Environmental Variables Affecting Recombination . . 47
VI. Linkage and Recombination in Microorganisms 52
VII. Theoretical Genetics of the Linked Gene Complex. 65
VIII. Discussion 84
      Acknowledgments 87
      References 87

Episomes
Allan M. Campbell

I. Introduction. 101
II. The Autonomous State. 107
III. The Integrated State 109
IV. Immunity 114
V. Lysogenization. Transition from the Vegetative to the Integrated State. 121
VI. Transition from the Integrated to the Autonomous State. 124
VII. Curing 125
VIII. "Gene Pick-Up" by Episomes. 126
IX. Partial Diploidy in *Escherichia coli* 132
X. Transfer of Episomes by Other Episomes 134
XI. Relationship to Cellular Regulatory Mechanisms 135
XII. General Discussion. 135
      Acknowledgments 137
      References 137

The Cytogenetics of *Oenothera*
Ralph E. Cleland

I. Introduction. 147
II. Outline of Genetic Behavior 148
III. Early Cytological Work 152

IV. The Relation of Cytological to Genetic Behavior	154
V. The Cause of Circle Formation in *Oenothera*	156
VI. Chromosome Structure	160
VII. Early Meiotic Stages	164
VIII. Factors Influencing Genetic Analyses	167
IX. Factor Analysis and the Location of Genes	170
X. Crossing-Over	172
XI. Position Effect	180
XII. Gene Conversion	183
XIII. The Nature of de Vries' Mutants	188
XIV. Induced Mutations	202
XV. Incompatibility	204
XVI. The Plastid Behavior	208
XVII. Evolutionary Considerations	215
References	229

Genetics and the Human Sex Ratio

A. W. F. Edwards

I. Introduction	239
II. The Importance and Origin of Sexual Differentiation	240
III. The Determination of Sex	241
IV. The Genetic Problem of the Human Sex Ratio	244
V. Evidence on the Variability and Heritability of the Human Sex Ratio	246
VI. Evidence on the Variability and Heritability of the Sex Ratio in Other Species	258
VII. Natural Selection and the Sex Ratio	259
VIII. Conclusion	265
References	265

The Genetics of *Streptomyces coelicolor*

D. A. Hopwood and G. Sermonti

I. Introduction	273
II. Strains and Methods of Culture	276
III. Morphology of the Organism	285
IV. Studies of Recombination by Selective Analysis	290
V. Heteroclones	304
VI. Analysis of Short Chromosome Regions	332
VII. Speculations on Nuclear Behavior during the Life Cycle	335
Acknowledgments	339
References	339

Fine Structure of Genes in the Ascomycete *Ascobulus immersus*

P. Lissouba, J. Mousseau, G. Rizet, and J. L. Rossignol

I. Introduction	343
II. Reasons for Choosing the Ascomycete *Ascobulus immersus*	345
III. Occurrence, within the Same Series, of Both Tetrads with Reciprocal and Tetrads with Non-Reciprocal Recombinants: Series 75	346

IV.	Localization of Crossing-Over and Conversion in Series 19	349
V.	Analysis of Series 46; Evidence for a Polarized Genetic Unit: the Polaron	355
VI.	Structure of Series 19 and the Concept of the Polaron.	361
VII.	Tentative Interpretation of the Properties of the Polaron.	362
VIII.	Linkage Structures and Relationships between Polarons	364
IX.	Occurrence of Odd Segregations	374
X.	Conclusions.	375
	Acknowledgments	378
	References	379

Author Index 381

Subject Index 389

LINKAGE AND RECOMBINATION IN EVOLUTION

W. F. Bodmer* and P. A. Parsons†

Department of Genetics, University of Cambridge, Cambridge, England

	Page
I. Introduction	2
II. Polygenic Balance	4
A. General Considerations	4
B. Evidence from Artificial Selection Experiments	7
C. Evidence from Studies on Natural Populations	21
III. Balanced Polymorphisms and Linkage	32
A. Introduction	32
B. Incompatibility Systems	32
C. Linked Complexes in Animal Populations	35
D. A Linked Complex Built Up by Artificial Selection	37
E. Possible Linked Complexes in Man	38
IV. Chiasma Frequencies and Cytological Evidence	40
A. Natural Selection and the Chiasma Frequency	40
B. Supernumerary Chromosomes	44
C. The Recombination Index	44
V. Genotypic and Environmental Variables Affecting Recombination	47
A. Factors Affecting Recombination	47
B. Genotypic Variation	50
C. Changing Recombination by Selection	50
D. Conclusions	52
VI. Linkage and Recombination in Microorganisms	52
A. Genetic Systems	52
B. Polygenic Variation	55
C. Linkage between Related Biochemical Loci	58
VII. Theoretical Genetics of the Linked Gene Complex	65
A. Introduction	65
B. Formulation of the Two Locus Theory	67
C. Initial Progress of New Gametic Combinations	70
D. Equilibrium Conditions for Viabilities Which Correspond to Mather's Concept of Balance	78
E. The Evolution of the Balanced Complex	81
VIII. Discussion	84
Acknowledgments	87
References	87

* Present address: Department of Genetics, Stanford University School of Medicine, Palo Alto, California.

† Present address: Reader in Human Genetics, Melbourne University, Melbourne, Australia.

I. Introduction

The discovery of large amounts of naturally occurring genetic variability is one of the most striking features of modern evolutionary studies. The explanation of the continued persistence of such variability under the action of natural selection is a central problem of genetic evolutionary theory.

The classic theoretical studies of population genetics did not consider the relevance of linkage. They were due mainly to Fisher, Haldane, and Wright, and referred principally to the simple situation of two alleles segregating at a single locus leading to the concept that the basic evolutionary process was the successive replacement of one allele by another. On this view, only a small proportion of an organism's genetic material would be expected to be heterogeneous at a given time. The basic store of variability immediately available would be largely the direct product of mutation. It was, however, emphasized by Fisher (1930a) that the available genetic variance would determine the rate of increase in fitness of an organism and so the rate of evolution by natural selection. In the specific context in which he showed this statement to hold, Fisher referred to it as the "fundamental theorem of Natural Selection."

The recognition of the importance of linkage in population genetics is due to Fisher (1930a). He pointed out that if two factors, with alleles A, a and B, b, occurring on the same chromosome, were such that A was advantageous in the presence of B but disadvantageous in the presence of b and vice versa, then natural selection would favor closer linkage between the two factors A and B. This is because, although natural selection will favor the gametes AB and ab, recombination in heterozygotes AB/ab will produce the less favored gametes Ab and aB and the consequent reduction in fitness will not be balanced by recombination in the rarer heterozygotes Ab/aB producing the favored gametes AB and ab. Having found an agency which would consistently favor tighter linkage, Fisher emphasized the existence of a further agency which would have the opposite effect. This is the constant spread of advantageous mutations which, unless they occur so seldom that each has become predominant before the next appears, can only come together in the same gamete by means of recombination. Here we first meet the conflict between the need for immediate fitness and the need for long-term flexibility to permit adaptation to new environments which is analyzed in the classic work of Darlington (1939, 1958) and Mather (1943a) and which most undoubtedly lies at the basis of any explanation of the continued existence in natural populations of large amounts of genetic variability.

The immediate fitness of a species depends on a certain uniformity of the genotype whereas long-term survival depends on the production of adequate genetic variability. Genetic variants departing from the optimum needed for present adaptation will be relatively unfit. Darlington (1939) discussed the antagonism between these two needs, in terms of a number of examples showing how they had been to some extent resolved by different types of "genetic systems." He emphasized that the over-all amount of crossing-over must be adjusted to an optimum, and coined the term "recombination index" for the sum of the haploid chromosome number and the average total meiotic chiasma frequency for all the chromosomes. Mather (1943a), on the other hand, developed the concept of the "balanced polygenic combination." On the assumption that natural selection favors intermediate phenotypes, gametic combinations which are well "balanced" with respect to high and low factors will be continually selected at the expense of poorly balanced combinations which give rise to extreme values. The poorly balanced gametes will be continually produced from the well balanced gametes by recombination, so that natural selection will favor closer linkage between factors occurring in a well balanced combination or combinations of factors which are already more closely linked. In outbreeding species, selection may thus be expected to cause the accumulation of chromosomes heterozygous for linked balanced polygenic combinations which, although phenotypically uniform, release variation by recombination.

The experimental evidence for the existence of such balanced complexes is now overwhelming. It comes principally from the analysis of artificial selection experiments in *Drosophila melanogaster* following the lead of Mather (1941, 1942), and from the classic work by Dobzhansky and others on naturally occurring inversion polymorphisms in various *Drosophila* species. Theoretical work on the way in which such balanced complexes could evolve (Wright, 1952; Kimura, 1956; Lewontin and Kojima, 1961) is by no means so convincing. A degree of linkage tighter than that which has usually been observed is generally indicated by the theory. We must however beware, as Mather (1953) warns us, "not to let mathematical argument blind us to biological reality." In fact, it appears likely that the intensity of linkage between the loci of a balanced complex can vary widely, from very loose linkage, to very tight linkage, such as occurs in complex loci making up a balanced polymorphism.

From an evolutionary point of view, therefore, we would expect recombination to be a highly variable phenomenon. Evidence has been accumulating for many years which shows that diverse environmental agencies alter recombination and interference. Maternal age and sex also affect recombination, as do different genetic backgrounds. Furthermore, artificial alteration of the genetic background by selection or inbreeding frequently may change recombination.

In microorganisms diverse alternatives to the sexual system exist for the promotion of genetic recombination, such as transduction, transformation, the quasi-sexual mating system of *Escherichia coli* and the parasexual cycle of the filamentous fungi. These systems provide ample opportunity for recombination.

Most microorganisms are haploid, so that potential genetic variability can only be stored as a balanced polygenic complex. However, in filamentous fungi, heterokaryosis provides a mechanism for storing variability equivalent to dominance in a diploid organism.

In some microorganisms, cases of linked blocks of genes controlling closely related biochemical reactions occur. The loci often occur in the same order as the steps in the biochemical synthesis. Such systems must surely have an evolutionary basis.

It is the purpose of this review to try to integrate these different lines of evidence which point to the importance of linkage and recombination in evolution.

II. Polygenic Balance

A. General Considerations

It was first pointed out by Fisher (1930a) that natural selection will favor closer linkage between two factors whose only effect is on a metrical character with respect to which intermediate values give optimum fitness. The existence of a genic balance for factors affecting quantitative characters was suggested by Bridges (1922, 1939) with reference to his work on sex determination and the autosome sex-chromosome balance in *Drosophila*. It was further pointed out by Goodale (1937, 1938) that certain of the results obtained with selection experiments for quantitative characters could only be explained by the assumption that "a series of alternating, but sometimes staggered, plus and minus (or zero) genes were located in one pair (possibly more) of chromosomes and that the minus genes almost balanced the plus genes." These balanced chromosomes are carried in the unselected stock and cause but limited observable variation. Selection rearranges plus and minus modifiers on one or more pairs of chromosomes to produce genotypes that far transcend the original unselected stocks. The integration of these ideas on genic balance and the importance of linkage in maintaining the integrity of balanced polygenic combinations, is contained in a series of classic papers by Mather (1941, 1942, 1943a).

Consider the simple case of two segregating loci with alleles A-a, B-b, where A, B affect the metrical character in one direction and a, b in the other, and there is no interaction between the effects of the various alleles. The coupling and repulsion heterozygotes AB/ab and Ab/aB, giving intermediate phenotypes and so optimum fitness, are the only genotypes whose gametic output is influenced by linkage between the two loci. If the linkage is complete, the repulsion heterozygote Ab/aB will segregate only intermediate gametes, whereas from the coupling heterozygote AB/ab only extreme gametes will be formed. The average fitness of the offspring from the repulsion heterozygotes will thus, in general, always be more than that from the coupling heterozygotes. Recombination between the loci leads to the production by Ab/aB of "unbalanced" gametes AB and ab, and the production by AB/ab of the "balanced" gametes Ab and aB. It is thus clear that natural selection will, in the short run, favor close linkage between the two loci, and so the preservation of the balanced gametes Ab and aB. Too close a linkage would, however, be a disadvantage in the long run if conditions were ever to change in such a way as to favor the gametes AB and ab instead of Ab and aB. These opposing tendencies reflect the need to maintain intact existing optimal polygenic combinations and yet allow the production of adequate genetic variability for future adaptation to new environments.

When there are more than two loci the number of possible heterozygotes increases rapidly with the number of loci, and the relationship between heterozygous genotypes and the fitness of their offspring is more complicated. Thus with three loci, A-a, B-b, and C-c, there are four possible genotypes heterozygous at all loci, one of which, the tri-coupling arrangement ABC/abc, contains only gametes giving extreme phenotypes when homozygous, whereas the other three, ABc/abC, AbC/aBc, and aBC/Abc, each contains pairs of gametes which, when homozygous, give phenotypes intermediate between the optimum and the extreme. The average fitness of the offspring of the ABC/abc genotype is clearly less than that of the others because at least half the gametes they carry are unbalanced with respect to genes of opposite effects. Genotypes heterozygous for only two of the loci such as AbC/aBC are intermediate in value between the two categories of triple heterozygotes. Recombination in a single interval in the tri-coupling heterozygote ABC/abc gives partially balanced gametes such as ABc. Recombination in the heterozygotes ABc/abC or aBC/Abc gives mainly either partially balanced gametes such as aBc and AbC, or the unbalanced gametes ABC and abc, whereas recombination in the heterozygote AbC/aBc only gives unbalanced gametes if crossing-over occurs simultaneously in both the regions A-B and B-C. Thus once again, natural selection will favor the preservation of the partially balanced gametes AbC and aBc.

When four loci are segregating there are eight possible genotypes heterozygous at all loci. With complete linkage three of these, $ABcd/abCD$, $AbcD/aBCd$, and $AbCd/aBcD$ produce only fully balanced gametes. Unbalanced gametes $ABCD$ or $abcd$ can arise by a single crossover from $ABcd/abCD$, but only by a double crossover from $AbcD/aBCd$ and by a triple crossover from $AbCd/aBcD$. Genotypes which are not heterozygous at all loci will usually be at a disadvantage as compared with those that are. In this case the most favored gametes are, therefore, the arrangements $AbCd$ and $aBcD$. It seems worth noting that only combinations heterozygous for an even number of loci can produce the majority of their offspring at the optimum phenotypic value so that natural selection may, in fact, favor combinations consisting of an even number of factors. This, of course, assumes approximately equal values for each locus. However, the effective unbalance produced by recombination in a single interval will decrease with increasing numbers of loci, although the greater length of chromosome concerned will increase the probability of multiple crossovers. In general, under the simple condition of independent cumulative effects of the factors, the "alternating" gametes $AbCdE\ldots$ and $aBcDe\ldots$ will be those favored by natural selection.

Interactions between the factors will clearly alter the details of the arguments presented above although not the general qualitative conclusion that natural selection will favor the preservation, by close linkage, of balanced polygenic combinations. In Mather's (1943a) words: "An arrangement which maintains the optimum, or near optimum, balance of fitness and flexibility will be referred to as a balanced polygenic combination, and is characterized by the twin properties of having a phenotypic effect near to the optimum for the constituent polygenes, and of releasing its variability only slowly by recombination with other homologous combinations of the same chromosome."

In outbreeding species where a large proportion of individuals may be heterozygous for any given section of the genetic material, balanced polygenic combinations will be exposed to natural selection in a predominantly heterozygous condition (Mather, 1953, 1955a; Parsons and Bodmer, 1961). Natural selection will therefore adjust polygenic combinations so as to give their optimum when heterozygous, and Mather (1943a) refers to the optimal combinations as showing "relational" balance. The evolution of relational balance is really one facet of the general phenomenon of the "evolution of overdominance" discussed by Parsons and Bodmer (1961).

The reverse situation will occur in inbreeding species where polygenic combinations will be exposed in a predominantly homozygous condition. Natural selection will adjust the combinations to give their optimum when homozygous, and Mather (1943a) refers to this as "internal" balance. Mather has pointed out that the association of good relational with poor internal balance and vice versa requires dominance of the factors concerned, and has drawn the analogy with theories of heterosis which depend on dominance.

The poise between the complementary needs for preserving balanced polygenic combinations and releasing adequate variability by recombination must be achieved by the control of crossing-over within the chromosome. The significance of inversion heterozygotes, which are of widespread occurrence in many species, in mediating such control of recombination was first discussed by Sturtevant and Mather (1938). The experimental demonstration of such control and of the relationship between different inversion sequences is due to the classic work of Dobzhansky and others on various species of *Drosophila*. This work will be discussed in Section II,C.

In his discussion on polygenic inheritance and natural selection, Mather (1943a) emphasized the role of balanced polygenic combinations in maintaining the store of potential, as opposed to free, genetic variability. Free variability is that producing a direct effect on the phenotype, whereas potential genetic variability is exhibited by different genotypes giving the same phenotype. Free variability is bound to produce individuals departing from the optimum, whereas potential variability need not do so. A simple means of storing potential variability is by heterozygosity for a dominant gene. For example, a population consisting entirely of homozygotes AA or aa has all its variability with respect to this locus free, but if Aa individuals occur which have the same phenotype as AA, potential genetic variability exists which is released on intercrossing or on crossing to the homozygous recessive aa. Potential variability may also be carried in homozygotes when, for example, there are two loci with non-interacting genes A, a and B, b such that homozygotes $AAbb$ and $aaBB$ produce the same phenotype.

The storage and release of potential variability is one way of maintaining the compromise between immediate fitness and future flexibility. The balanced polygenic combination is clearly a most important mechanism for the storage of potential genetic variability. As pointed out above the recombination frequency must control the rate of release of this variability.

A more detailed consideration of these problems in terms of specific mathematical models will be deferred to Section VII.

B. Evidence from Artificial Selection Experiments

1. *General Evidence*

Much of the specific evidence for the existence of balanced polygenic complexes comes from the analysis of selection experiments for various quantitative traits in *Drosophila*. Early selection experiments such as those of Payne (1918, 1920) and Sturtevant (1918) demonstrated unequivocably that the response to selection for a quantitative character was due to the action of many genes. They were able moreover to detect linkage of genes causing the response to selection with specific genetic markers. It was partly the results of Payne's (1918) experiments which prompted Goodale (1937) to postulate the existence of balanced polygenic combinations. Payne (1918) selected for increased numbers of scutellar chaetae in *Drosophila*. He was able to increase the mean chaeta number from just over 4 to 9.85 in 38 generations of selection.

Mather (1941) selected for high and low abdominal chaeta number in *Drosophila melanogaster*. He produced stable high and low selection lines whose means differed by about 17 chaetae. Using marked balancers which inhibit recombination in the relevant chromosome and are lethal when homozygous, it is possible to combine in all possible combinations chromosomes extracted from a selection line with control chromosomes and so study their individual effects and interactions on a quantitative character, such as chaeta number. Since there is no recombination in the male in *Drosophila*, whole chromosomes may be maintained indefinitely for such purposes, by backcrossing males which carry them to females homozygous for a recessive marker on each chromosome. An analysis in this way of whole chromosome effects in Mather's (1941) high and low selection lines revealed significant genetic activity affecting chaeta number on each of the chromosomes studied (Mather, 1942). Whereas the parental stocks from which the selection lines originated differed only slightly in the chaeta producing strength of the two large autosomes II and III, the two selected lines differed markedly and significantly from each other in both these chromosomes. A comparison of each with their parents showed clearly that effective recombination of polygenes must have occurred during the course of selection.

If a suitable multiply marked chromosome is available, differences in chaeta number between the various classes of recombinant genotypes produced by backcrossing flies heterozygous for the marked and selected chromosomes to a multiple recessive stock provide information as to the polygenic activity of different parts of the same chromosome. Using such a technique Mather (1942) was able to show that all the effects of chromosome III in the selected lines were located in a section of the chromosome near the *th* locus. Thus recombination of polygenes must have occurred in this restricted part of the chromosome and the recombinants must have been selected in both the high and low lines. Analysis of the X-chromosome showed no major significant differences, but provided suggestive evidence for a redistribution by recombination of $+$ and $-$ genes along the chromosome.

A simple direct demonstration of the general need for recombination between heterogeneous chromosomes in order to produce responses to selection is given by Harrison and Mather (1950). Stocks were built up which were homozygous for chromosome II's extracted from flies caught in the wild, but which were homogeneous for chromosomes X and III from a standard Oregon inbred line. Four such stocks were crossed in all possible pair-wise combinations to give six "heterozygous" lines. Selection was practiced for increased and decreased numbers of abdominal

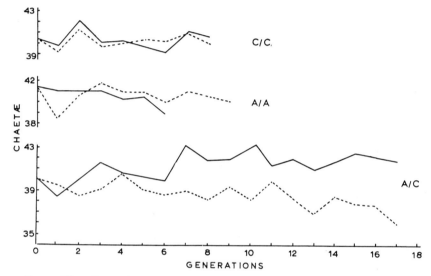

FIG. 1. The effects of selection for chaeta number in two homozygous lines A/A and C/C, and in the "heterozygous" line A/C formed by crossing A/A and C/C. Selection for high chaeta number is shown by solid lines and for low chaeta number by dotted lines (after Harrison and Mather, 1950).

chaetae in these six lines and in the four homozygous controls. Generally little or no response occurred in the lines which were homozygous for an extracted chromosome II although one line showed a limited response which could be attributed to nonrecurrent mutation in the original stock. However, the heterozygous lines all responded to selection and mostly transgressed the parental differences, demonstrating that the response must be due to recombination of balanced polygenic complexes. An example of the results they obtained for two of the homozygous lines and the heterozygote formed from them is illustrated in Fig. 1.

2. *Accelerated Responses to Selection*

The observed response to selection for chaeta number in Payne's (1918) selection experiment showed two periods of rapid increase which followed periods of comparative stability. These were explained by Goodale (1937) as the result of recombination between plus and minus modifiers on the same chromosome. Sismanidis (1942) confirmed Payne's results and was able to show that such rapid responses could be attributed to changes in one or more particular chromosomes. His various selection lines fell into a limited number of classes according to their response to selection. The production of such variability from initially uniform stocks can only be due either to mutation or to recombination between linked balanced polygenic combinations. The repeatability of the results indicates clearly that recombination is the only plausible mechanism which would account for these "accelerated" responses to selection.

Similar irregular responses to selection were observed by Mather and Wigan (1942). They selected in separate lines for increased and decreased numbers of sterno-pleural and abdominal chaetae starting from the highly inbred Oregon stocks. Over twenty-one generations of selection, sterno-pleural chaeta number showed a slow steady response in the low line whereas the high line remained stable. The response to selection for abdominal chaetae, however, occurred in a number of short bursts separated by periods of comparative stability. A detailed consideration of their results showed that if mutations of major effect were involved, at least five or six separate mutations must have occurred. At this level the distinction between major and polygenic mutation breaks down, more especially in the light of the analyses of the genetic control of chaeta number such as that due to Mather (1942). Mather and Wigan consider that their results can most plausibly be explained by recombination of polygene mutations. These mutations must have accumulated without producing significant visible effects, until recombinants having more extreme effects were produced.

The most striking example of the phenomenon of repeated accelerated responses to selection is probably the work of Thoday and Boam (1961) on a series of lines of *Drosophila melanogaster* which were selected for increased sterno-pleural chaeta number. Three lines of the same origin, but which had varying initial periods of mass culture or back-selection, showed remarkably similar patterns of response (lines dp 1, dp 2, and dp 6—Fig. 2). Each of the lines showed an accelerated response from about 24 to 28 chaetae, which occurred over a period of five to ten generations and after which the lines reached a plateau at about 30 chaetae. A fourth line, related to these three, showed a similar accelerated response which, however, occurred after a rather longer initial period of comparative stability. This response continued beyond the level of 30 chaetae reached by the first three lines and reached a plateau at about 37 chaetae (line vg 4—Fig. 2). A fifth line was obtained from a cross between this fourth line before its accelerated response had occurred, and one of the lines which had reached a plateau at 30 chaetae. This line responded to selection extremely rapidly and ultimately reached a level of 46 chaetae. The patterns of response of the first four lines are illustrated in Fig. 2.

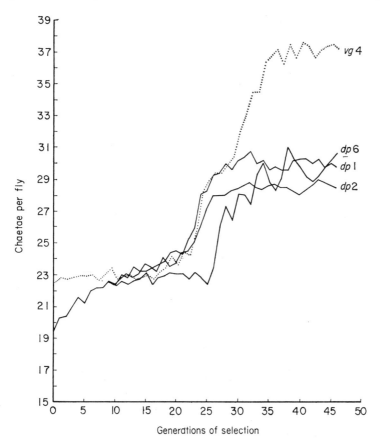

FIG. 2. Accelerated responses to selection in *Drosophila*. The solid curves show the similarity of response of 3 lines $dp1$, $dp2$, and $dp6$. The dotted curve shows the response of a fourth line, $vg4$, plotted so that its accelerated response coincides with those of the dp lines (after Thoday and Boam, 1961).

Only one of the lines which had reached a plateau of 30 chaetae proved to be unstable on mass culture.

Ordinary "random" mutation could hardly account for these repeated responses, all of which occurred when the lines had reached about 24 chaetae. In fact it has recently been possible to show that the response from 24 to 28 chaetae depends on two loci at about 28 and 32 cM (centi-

Morgans) on chromosome III (Thoday, 1961). The only plausible hypothesis for such repeated accelerated responses is that they are caused by recombination within a balanced polygenic complex perhaps of the simple repulsion type, Ab/aB. Thoday and Boam (1961) suggest two possible mechanisms which may account for the delay in the production and selection of recombinants. One is that selection may cause developmental instability in the same way that inbreeding does and this is known also to affect chromosome pairing and chiasma formation and so perhaps cause a general or a localized increase in recombination (see Section IV). The other, and perhaps more plausible, suggestion is based on the postulate that one of the gametes, say aB, is initially rare so that the frequency of heterozygotes and so the probability of producing recombinants from them, is low. If this gamete aB has a slight positive effect on chaeta number, it will be selected for and so increase in frequency. This will increase the frequency of heterozygotes and so the probability of producing recombinants. Appreciable delay will only occur with this mechanism if the markers are very closely linked. In either case the maintenance of a polymorphism for the two balanced repulsion gametes over considerable initial periods of mass culture and even back-selection implies the existence of some associated viability effects.

The tremendous response of the fifth line which occurred after crossing a high line with a lower one suggests that there existed in the lower line one or more genetic factors which could only be exploited in the presence of a factor which had already been selected in the initial high line. The later response of the lower line to a level about 7 chaetae higher than the initial high line, which reached 30 chaetae, is evidence of the same phenomenon. Recently (Thoday, personal communication) specific evidence has been obtained for the occurrence of such a hypostatic factor on chromosome II.

The accumulation of successively interacting factors was postulated by Cavalli and Maccacaro (1952) in their striking but much neglected studies on drug resistance in the bacterium *E. coli*. As suggested by Thoday and Boam (1961), such effects are a clear indication to the practical plant and animal breeder that their selection programs should be flexible and should involve backcrossing improved varieties to unimproved stocks.

3. *Correlated Responses to Selection*

It is well known that response to selection for quantitative characters is generally associated with decreased fitness and fertility. This phenomenon was studied by Wigan and Mather (1942) who showed that selective response in either direction for abdominal and sterno-pleural chaeta number in *Drosophila* was accompanied by a correlated fall in fertility. There are two possible explanations for such a correlated response. The first depends on the assumption that the selected polygenes are pleiotropic in action, and affect simultaneously chaeta number and fertility. The second, developed by Wigan and Mather (1942), explains the correlated response by an intermingling of chaeta and fertility genes along the chromosome. A given polygene affecting one character may be linked not only with like genes affecting the same character, but also with unlike genes affecting other characters. Thus a particular recombination may simultaneously unbalance several combinations affecting different characters. Appropriate further recombination may, however, in time re-order the various polygenes in such a way that response to selection for one character need not necessarily be associated with a correlated response of a different kind. This was in fact observed by Wigan and Mather (1942) in one of their lines which showed an initial drop in fertility with response to selection for increased sterno-pleural chaeta number until the tenth generation, when it recovered sharply. Such a result is very difficult to explain on the basis of pleiotropic action of the polygenes.

Many of the effects of selection discussed so far are illustrated by an extensive selection experiment for increased and decreased abdominal chaeta number in *Drosophila melanogaster*, carried out by Mather and Harrison (1949). Their low line rapidly developed a balanced sterility system so that attention was mainly confined to the high line. This showed both delayed responses to selection and correlated reduction in fertility. Sometimes chaeta number would be stable for periods of up to fifty generations after which there would be a rapid response for a few generations and then further stability. Mather and Harrison suggest that these responses are most probably due to the release of variability by recombination following crossing-over in new positions in the chromosome, resulting possibly from changed environmental conditions. They showed that selection had built up significant effects on chromosomes X, II, and III. High lines were established whose chaeta number was stabilized at three levels. The line at the upper level differed from that at the middle in chromosome II only, while that at the lower level differed from the middle one in chromosomes X and III only. Crosses between the levels indicated that the three chromosomes behaved as units of segregation. That selection lines which had reached a plateau and were stable under mass culture were not homogeneous was demonstrated by the efficiency of back-selection in these lines.

The pattern of association of fertility with response to selection clearly indicated that any correlated responses were due to the mechanical relationship of polygenes affecting fertility and chaeta number as suggested by Wigan and Mather (1942). Correlated responses for a number of other characters were also observed, some of which could be satisfactorily explained by physiological relationships between chaeta number and the characters concerned. In plants and other organisms, the evidence is limited, but Haskell (1954, 1959) favors an interpretation of correlated responses similar to that advanced for *Drosophila*.

4. *Location of Polygenes*

So far little mention has been made of the specific location and distribution of genetic material affecting quantitative characters. Early work using marker genes to locate polygenic activity is summarized by Wigan (1949a). He himself established regions of polygenic activity affecting sterno-pleural chaeta number in *D. melanogaster* in a number of positions on the X-chromosome.

More extensive investigations have been carried out by Breese and Mather (1957, 1960) on the polygenic activity affecting abdominal and sterno-pleural chaeta number and also viability. Their starting material consisted of chromosome III's extracted from two lines H and L, selected, respectively, for high and low numbers of abdominal chaetae. From these they constructed a series of recombinant chromosomes which included known segments from H and L chromosome III's and were otherwise homogeneous. The recombinant segments (A, B, C, D—Fig. 3a) were built up by using the *rucuca* chromosome, which contains markers spread along the third chromosome. The recombinant segments of course contained no marker genes. The effects of the recombinant chromosomes were tested by a series of diallel crosses which were used to estimate parameters representing the additive and dominance effects of the various sections of chromosome. All segments showed significant genetic activity affecting chaeta number and most showed dominance. Results using overlapping segments were in agreement with each other. There was, however, a major effect on abdominal chaeta number associated with the region *st–cu*, which extends over 2 cM and is close to the centromere. Breese and Mather pointed out that near the centromere, the genetic map length greatly underestimates the physical length of the chromosome so that a major effect in this region may simply reflect a concentration of the more diffuse distribution of polygenic activity over the rest of the chromosome. However, Mather (1942) also located a major effect near this same region, using the marker *th*. Furthermore, Wigan (1949a) could find no simple correlations between polygenic activity and either crossover length, physical length, or distribution of major gene loci on a chromosome. The pattern of results for sterno-pleural chaetae was similar to that for abdominal

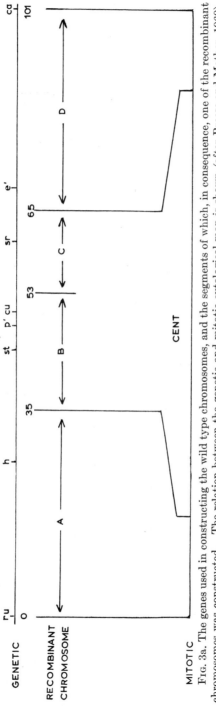

FIG. 3a. The genes used in constructing the wild type chromosomes, and the segments of which, in consequence, one of the recombinant chromosomes was constructed. The relation between the genetic and mitotic cytological map is shown (after Breese and Mather, 1960).

chaetae although the two systems of polygenes were at least in part distinct from each other. Interactions were generally small.

Viability showed a completely different polygenic architecture to that of chaeta number. Simple pleiotropy could account for at most a minor

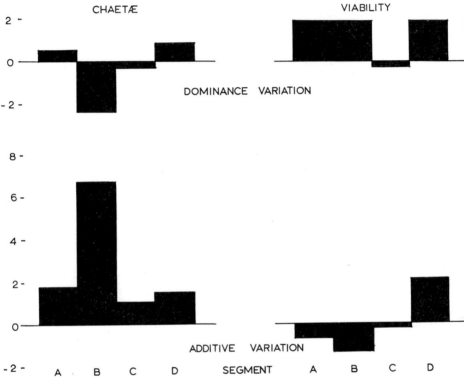

Fig. 3b. The contributions made by the four segments A–D see (Fig. 3a) to the additive and dominance variation in abdominal chaeta number and viability. The sign indicates the difference between parents (high − low) for additive variation, and the direction of the enhancement of expression of chaeta number and viability for dominance variation. The units of measurement are chaetae and angularly transformed percentage viabilities, respectively (after Breese and Mather, 1960).

part of the viability differences. Any interactions that did occur could only be explained by linkage, within the chromosome, of different polygenic systems. Whereas polygenes for chaeta number showed ambi-directional dominance, those for viability only showed dominance in the direction of higher viability. This was predicted by Fisher (1930a), on the basis of his theory of the evolution of dominance. Unidirectional dominance would be expected when selection is directional, as it must be for viability, but dominance in either direction would be evolved for genes affecting a quantitative trait such as chaeta number, which gives optimum fitness at an intermediate value. The difference in dominance effects between viability and chaeta number polygenes thus reflects the different histories of the action of natural selection on these characters. The larger interactions of polygenes affecting viability are explained in the same way. The over-all results obtained by Breese and Mather are illustrated in Fig. 3.

Recent work by Thoday (1958, 1959, 1960), Thoday and Boam (1959), Gibson and Thoday (1959, 1962), and Millicent and Thoday (1960, 1961) has arrived at a more specific location of polygenic differences which have been built up in selection experiments. The aims and results of this work, in respect to the location of polygenic differences, have been summarized by Thoday (1961). For the more specific location of polygenes, the analysis needs to be taken one generation further than the type of recombination analysis carried out by Breese and Mather (1957, 1960). The extra principle involved is that of progeny testing recombinant marker chromosomes to determine the number of classes which can be distinguished with respect to the relevant quantitative character. Suppose, for example, a "High" effect has been associated with recessive linked markers a and b, where the double recessive ab/ab represents a "Low" effect. We want to consider the outcome of backcrossing "High"/ab heterozygotes to "Low" homozygotes ab/ab. If we assume one locus between a and b with "High" allele H and "Low" allele L, then single crossover gametes $a+$, $+b$ will fall into two genotypic categories L or H, depending on where the crossover occurred. However, if there are two loci between a and b single crossover gametes will fall into three classes, the parental types HH and LL, and one or other of the recombinant types HL and LH. In general, with n factors between a and b, single crossover gametes will give the two parental categories and one or other of two sets of $n-1$ complementary recombinant chromosomes. Loci outside the interval a–b will not affect the distribution of types obtained from single crossovers between a and b. Recombinant chromosomes are assayed when heterozygous with the low chromosome. Genotypic categories will, of course, not necessarily correspond to observed phenotypic categories so that these can only really provide estimates of the minimum number of loci involved. Further extensive progeny testing of these recombinants would be needed to be sure of the maximum number of loci.

Apart from the location of the factors on chromosome III of *D. melanogaster* responsible for the first repeated response in the selection experiments of Thoday and Boam (1961), all the other differences investigated by Thoday and his colleagues have been produced by disruptive selection for sterno-pleural chaeta number. This involves mating extremes in each generation of selection. It was predicted by Mather (1953, 1955b) that this system of selection could produce a polymorphism in spite of the occurrence of gene flow between selected parts of the population. Thoday and Boam (1959) were able to produce a polymorphism in this way. They showed that the low chaeta number flies were heterozygous for low and intermediate second chromosomes, which have been analyzed in more detail by Gibson and Thoday (1959, 1962). The low second chromosomes were lethal when homozygous, and the intermediate second chromosomes could be separated by appropriate progeny testing into two categories which may be designated as $+-$ and $-+$, where $+$ has a positive, and $-$ a negative effect on chaeta number. Low second chromosomes were produced by recombination in $+-/-+$ heterozygotes and so must be $--$. Reciprocal $++$ recombinants expected from these heterozygotes were not obtained so that this gametic combination must be dominant lethal. The two loci involved in this system were shown to lie between dp and cn at about 27 and 47 cM on the second chromosome.

It is most significant that both the $-+$ and $+-$ second chromosomes have been found in the wild stock which provided the starting material for the selection experiments. Out of 48 chromosomes assayed, 46 were $+-$ and 2 were $-+$. This system thus provides a model demonstration of the simplest form of balanced combination postulated by Mather (1943a). It is not insignificant that the $+-/-+$ repulsion linkage must be maintained in the wild even though the factors are separated by as much as 20 cM.

The high chaeta number flies produced by Thoday and Boam's (1959) disruptive selection line have been analyzed by Wolstenholme (unpublished results). He has shown that they are polymorphic for two loci on the third chromosome, which are separated by less than 2 cM. There is, as yet, no evidence for a balanced system or any lethality. The tighter the linkage, the smaller the interaction needed between two loci to maintain a balanced polymorphism (see Section VII). Hence for very closely linked loci, it may be difficult to detect interactions between them.

Further disruptive selection lines with varying amounts of gene flow between the two parts of the population have been produced by Millicent and Thoday (1960, 1961). These have in all cases given rise to divergence and polymorphism. Preliminary analysis of second chromosomes has revealed two factors at 47 and 64 cM, the first of which may be the same as one of the factors analyzed by Gibson and Thoday (1962). The low chromosomes are not lethal when homozygous or when tested against Gibson and Thoday's low chromosomes. All the seven loci which have so far been located by Thoday and his colleagues are illustrated on the map in Fig. 4. It is remarkable how much variation has been produced from a stock which started with just one female collected from the wild.

Direct evidence of the way in which stabilizing and disruptive selection can maintain particular linked combinations of polygenes is given by Thoday (1960). He started stabilizing and disruptive selection lines for sterno-pleural chaeta number, using the F_2 derived from a cross between a stock homozygous for the third chromosome markers *se cp e* and the

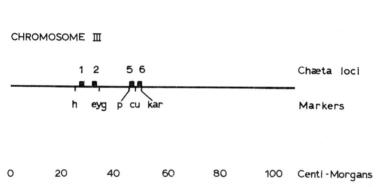

FIG. 4. The location of the chaeta loci, 1–7, found by Thoday and his colleagues. In most cases, the locations are not precise; this is indicated by showing the likely region of the chromosome containing the loci. The markers used in locating the loci are also given (after Thoday, 1961).

very high selection line reported by Thoday and Boam (1961), which carried the marker *vg* on chromosome II. This high line was then known to carry "high" factors in the *se-cp* region of chromosome III, and is now known to carry two high factors between the loci *h* and *eyg*. The second chromosome carries high factors hypostatic to the chromosome III factors (Thoday, unpublished). Stabilizing selection reduced the frequency of *vg* on chromosome II to a low level and established the recombinant *se cp*⁺ chromosome at a high frequency. On the other hand, disruptive selection, which favors more extreme phenotypes, made the *vg* marker homozygous and maintained + + and *se cp* coupling chromosomes in high frequency despite considerable recombination pressure.

5. Discussion and Conclusions

There is an apparent contrast between the approach of Thoday (1961) and that of Breese and Mather (1957, 1960). The former aims at location of specific polygenes and a refinement of technique which makes it possible to treat the polygene almost as if it were a major factor. The latter consider that "the activity of any region of chromosome is then not an attribute of a single gene but of that constellation of genes which the region happens to contain." It seems possible to resolve this contrast by a consideration of the complexity of the character being studied and the biochemical or physiological level of action of the polygenes we wish to isolate. Viability will clearly be influenced by an enormous number of basic biochemical differences so that polygenic activity affecting this character may be expected to be ubiquitous. Chaeta number is probably influenced by fewer variables at the basic biochemical level, so that we might expect to find more localized polygenic activity having a profound effect on this character. The factors found by Mather (1942) and Breese and Mather (1957) in chromosome III and all those analyzed by Thoday and his colleagues are probably of this sort. However, there must exist characters which have an indirect pleiotropic effect on chaeta number. For example Gibson et al. (1961) and Parsons (1961a) have shown that there is a significant correlation between sterno-pleural chaeta number, sterno-pleural plate area, and body weight. Body weight is a character of great complexity comparable to that of viability, and the ubiquity of small but significant effects as found by Breese and Mather (1957) may be explained perhaps partly in terms of such factors affecting body weight. Ultimately all polygenes must be resolvable into different discrete segregating units, even if it does turn out that some polygenes are not just minor mutations at major loci affecting protein structure, but are a different type of genetic material which may affect the production of a number of structurally different proteins. Evidence will be presented in Section VI of a polygenic system in *Neurospora* which is concerned with the modification of a single enzyme deficiency. Even at this most elementary biochemical level, there is considerable complexity.

Apart from this reference to a polygenic system in *Neurospora*, almost all the work so far described in this section was done with *D. melanogaster*. Direct evidence of such systems in other species is almost nonexistent at present, but will undoubtedly turn up in the future. Cooper (1959, 1960), however, infers the existence of balanced polygenic systems in his selection experiments in *Lolium*.

In mice, results of a selection experiment for white head spotting reported by Goodale (1937) suggested the existence of balanced polygenic combinations with respect to this character. Bodmer (1960a) has provided specific evidence that the recessive genes for pallid (*pa*) and fidget (*fi*) affect the incidence of polydactyly both when they are heterozygous and homozygous. Polydactyly is caused by a major recessive gene *py* in linkage group XIII. When first discovered *py* was poorly manifested and incompletely penetrant, but its expression has been radically altered by selection after initial outcrosses to a number of different stocks (Holt, 1945). It is remarkable that both the modifiers *pa* and *fi*, and a specific suppressor of polydactyly (Fisher, 1950, and later unpublished observations) are located in the fifth linkage group. There is most probably on this chromosome a linked system of modifiers affecting polydactyly.

Falconer (1955) has reported extensive selection experiments in mice for body weight, lactation, and litter size. Responses, when they occurred, were more or less uniform and showed no obvious periods of accelerated response which might be due to recombinant products. However, the selection line for low 6-week body weight showed a sudden and remarkable increase in its variability between the seventh and ninth generations which was maintained until at least the twenty-fourth generation. Falconer considers that this is most likely the result of "an irregular scale effect." A more plausible explanation is surely the production of variation by recombination, which has perhaps established a cryptic polymorphism or a balanced sterility system as occurred in one of Mather and Harrison's (1949) low lines.

There can be little doubt that there exists a great store of naturally occurring cryptic variability which is maintained in the form of linked balanced polygenic complexes. This variability is released for the action of selection by recombination. The lesson to be learned by the practical plant and animal breeder is brought home most clearly by the results of Thoday and Boam (1961). He should select up and down, and by disruptive selection in order to produce variability by recombination, and he should not hesitate to cross back to unimproved stocks in order to pick up "hypostatic" variability. His problem is to achieve that rearrangement of linked balanced complexes which best suits his selection aims. It is disconcerting that standard textbooks on quantitative genetics make practically no mention of the significance of linkage and interacting polygenic complexes, and assess the results of selection by a measure of response based on the heritability, which is completely ineffective in the face of the disruption of polygenic complexes by recombination.

C. Evidence from Studies on Natural Populations

1. *Inversion Polymorphism*

Heterogeneity in wild populations for inversion sequences is well known in many species. The first detailed analysis of such variation, occurring in wild populations of *D. pseudoobscura*, was made by Dobzhansky and Sturtevant (1938). The classic work of Dobzhansky and others on different inversion sequences in various species of *Drosophila* (see Dobzhansky, 1951, 1957 and da Cunha, 1955 for reviews) provides a most thorough analysis of such systems. Dobzhansky and Sturtevant (1938) suggested that heterozygosity for inversions may be of advantage because it restricts crossing-over. The way in which such restrictions of crossing-over may be advantageous was considered in more detail by Sturtevant and Mather (1938). The explanation lies in the advantage of maintaining intact balanced polygenic combinations which have developed relational balance, as discussed in Section II,A. Inversion heterozygosity restricts crossing-over in the chromosome on which the inversion occurs and so prevents recombination in a balanced complex which occurs within and just outside (Dobzhansky and Epling, 1948) the bounds of an inverted sequence. This serves to prevent the disruption of such balanced complexes by recombination. However, as Sturtevant and Mather (1938) point out, inversion heterogeneity in *D. pseudoobscura* is in general only found on one of the four chromosomes, which means that there must also be some advantage in recombination. The advantage lies in the fact that recombination increases the flexibility of the species to adapt to new environments. Inversion heterozygosity for just one chromosome is an example of the way in which natural selection achieves the balance between immediate fitness and future flexibility.

The way in which the amount of recombination controls the balance between fitness and flexibility is nicely illustrated by studies on the differences between populations of a species occurring at the margin of its distributional range and those at the center. In certain species of *Drosophila*, notably *D. robusta* and *D. willistoni*, the amount of inversion heterozygosity rapidly decreases as the margin of the species range is approached. Dobzhansky et al. (1950) and da Cunha and Dobzhansky (1954) have suggested that this is because the species are better adapted to their diverse environments at the center of their range and require more flexibility to adapt to new conditions at the margin. The larger amount of inversion heterozygosity at the center serves to maintain the more highly adapted polygenic complexes which occur there. The greater chromosomal uniformity at the margin allows freer recombination, which is necessary in order to maintain the flexibility needed to adapt to the new conditions encountered at the margin. Carson (1955) has confirmed this in *D. robusta*. He has also shown that the response to selection in strains originating from the marginal populations tends to be greater than that in strains from the central populations (Carson, 1958). This undoubtedly reflects the occurrence of more recombination in the marginal than in the central populations.

Extensive work by Dubinin and Tiniakov (1946, and other papers) has shown a remarkable association between frequency of inversion heterozygotes and ecology of habitat in *D. funebris*. This species, which generally lives in man-made environments, shows maximum heterozygosity in urban industrial concentrations and a steady decline in heterozygosity, reaching its lowest point in rural districts. For example, the frequency of heterozygotes in the center of Moscow is 88.54%, and decreases gradually to 12.1% at the boundaries of the city, whereas the frequency in the surrounding rural villages is as low as 1.5%. Dubinin and Tiniakov also found clines in the frequency of particular inversions going from the north to the south of Russia. They have been able to correlate these clines with differences in the ability of flies carrying different inversions to survive at low temperatures. They have also demonstrated seasonal changes in the frequency of inversions in the city populations. Their work is a clear testimony to the role of natural selection in maintaining the pattern of inversion heterozygosity.

There are other lines of evidence which confirm that the frequency of different chromosome arrangements occurring in natural populations must be selectively controlled. Cyclic temporal changes and large directional changes in the frequencies of particular arrangements have often been observed (for review see da Cunha, 1955). Perhaps the most striking demonstration of selective control comes from observations on the changes in frequency of various arrangements after introduction into "population cages." It was shown by Wright and Dobzhansky (1946) that two arrangements coming from the same population would reach stable equilibrium with both arrangements present. This, in general, implies that heterozygotes for different inversions must be "fitter" than homozygotes. The superiority of the heterozygotes has also been demonstrated more directly in other ways. However, Dobzhansky (1950) has shown that this heterozygote superiority no longer exists when pairs of arrangements from *different* populations are compared. Clearly the polygenic complexes "locked" in the inversion have, within a population, developed relational balance leading to heterozygote superiority, but polygenic complexes coming from different populations will not have had this same opportunity for mutual adjustment of their genic contents. Heterozygosity for polygenic complexes contained within inversion sequences coming from different populations will not, therefore, be expected to result in increased fitness. Dobzhansky (1950) calls this mutual adjustment of polygenic complexes occurring together in the same population "co-adaptation." The concept of co-adaptation is synonymous with Mather's (1943a) concept of the evolution in outbreeding species of relational balance between polygenic complexes.

The development of heterozygote superiority for pairs of inversion sequences from different populations, which did not at first show any such advantage, has been demonstrated by Dobzhansky and Levene (1951). They found that in certain of their experiments with population cages, two such inversions would be maintained ultimately in a stable equilibrium, even though viability tests at the start of the experiment showed that one of the homozygotes had a higher mean fitness than the heterozygote. It is clear that some sort of selective modification of the heterozygote viabilities must have occurred during the course of the experiment. Further evidence for alteration of viabilities by selection is given by Levene et al. (1954), who showed that their results for experimental populations containing three different arrangements could not be explained by a model with constant viabilities. Haldane (1957) has examined theoretically the conditions necessary for the establishment of altered inversion sequences and concludes that there must be a general trend toward cumulative overdominance. Similar calculations have been done by Bodmer and Parsons (1960) who also investigated the nature of the changes in viability needed to explain the results of Levene and associates (1954). The necessary selective alteration of viabilities may occur through selection of modifiers which interact with the genic contents of the inversion, or by selection of inversions carrying altered genic contents which have, perhaps, arisen by recombination. Suggestive evidence for a lowering of heterozygote fitness because of disruption of balanced complexes by recombination is given by Dobzhansky and Pavlovsky (1958). They found that whereas population cages started with heterozygotes for pairs of inversions from the same locality maintained high frequencies of heterozygotes, cages started with flies heterozygous for inversions from different localities soon showed a considerable preponderance of homozygotes.

An ingenious theory to explain the incidence of different arrangements maintained in the same population has been put forward by Wallace (1953). He suggests that certain "triads" of overlapping inversions will be selected against because they allow recombination within the overlapping parts of the inversions and so facilitate serial transfer of genes from one arrangement to another. This will cause a breakdown of the relational balance or co-adaptation of the polygenic complexes maintained within the inversions. An analysis of observations on the frequencies of such triads occurring within populations shows that in most cases, though not all, only two members of the triad occur in appreciable frequencies within the same population. However, sets of three inversions which do not constitute an overlapping triad are often all maintained in appreciable frequencies within a population. Wallace also points out that in most cases of triads one of the three possible pairs formed from a set of three inversions is rare or absent from all populations. These situations are, perhaps, best understood by a consideration of the conditions which will allow a new arrangement to gain a hold in a population. A theoretical discussion along these lines (see Bodmer and Parsons, 1960) will be presented in Section VII.

2. Interactions between Polygenic Complexes

A striking demonstration of the relationship between linkage and interaction of polygenic complexes is provided by studies on interactions between certain linked inversions in *D. robusta* and other species (for review see Levitan, 1958a). Levitan (1955) has studied inversions on each arm of the second chromosome of *D. robusta*. The frequencies of the structural karyotypes in each arm fit expectations calculated from the Hardy-Weinberg law on the assumption of random mating and no viability disturbances. However, the associations of the left and right arm arrangements in "double" heterozygotes are nonrandom. In the left arm of chromosome II, there are three possible inversions *2L-1*, *2L-2* and *2L-3*, and in the right arm there is one inversion *2R-1*. Levitan (1958a) gives numbers of heterozygotes found in natural populations as follows:

$$49 \frac{S}{2L\text{-}3}\frac{2R\text{-}1}{S} : 27 \frac{S}{2L\text{-}3}\frac{S}{2R\text{-}1} \qquad \chi_1^2 = 6.4 \quad (P < 0.02)$$

$$40 \frac{2L\text{-}1}{2L\text{-}3}\frac{2R\text{-}1}{S} : 16 \frac{2L\text{-}1}{2L\text{-}3}\frac{S}{2R\text{-}1} \qquad \chi_1^2 = 10.3 \quad (P < 0.01)$$

where S refers to the standard arrangement in each arm and the χ^2's test deviations from a 1:1 ratio. Studies of the amount of crossing-over between these linked inversions shows it to be very variable (Levitan, 1958b), ranging from 1 to almost 50%. The amount of crossing-over is influenced by inversion heterozygosity in the other chromosomes. Similar, but less marked, nonrandom associations between inversions on different arms of the X-chromosome have been observed. Crossing-over between these inversions is generally less than 1% (Levitan, 1958b) and is less influenced by the situation on other chromosomes. The amount of recombination between balanced complexes maintained by the inversions is clearly related to the nonrandomness of the heterozygous linkage types. This must be maintained by natural selection, in an analogous way to the control of recombination within balanced complexes.

In *Drosophila guaramunu*, Levitan and Salzano (1959) have found extreme nonrandom association between two linked inversions H/h and E/e. Out of 1151 larvae examined only 20 (1.7%) were heterozygous for one arrangement and homozygous for the other. Of the other 98.3%, 73.8% were double homozygotes and 24.5% were double heterozygotes, *Ee Hh*. Levitan and Salzano (1959) suggest that the arrangements *EH* and *eh* are favored, and the arrangements *eH* and *Eh* are selected against. Further cases of nonrandom associations between linked inversions, such as that studied by Acton (1957) in *Chironomus*, are reviewed by Levitan (1958a), and a recent case found in *D. pavani* is analyzed by Brncic (1961). Interactions between inversions really represent second-order interactions, the primary interactions occurring between the genic contents, occurring in the form of balanced complexes, of single inversions.

TABLE 1
Estimates of Relative Viabilities of Inversion Arrangements in Different Chromosomes of the Grasshopper *Moraba scurra**

Chromosome EF	Chromosome CD		
	S/S†	S/Bl	Bl/Bl
S/S	1.002	1.000	0.927
S/Td	0.646	0.849	1.044
Td/Td	0.000	1.054	0.626

* After Lewontin and White (1960).
† S represents the standard arrangement on each chromosome.

Interactions between polymorphic inversion systems occurring on different chromosomes have been studied by White in the grasshopper *Moraba scurra* (White, 1957, 1958; Lewontin and White, 1960). Estimates of viability show that there are epistatic interactions between the viabilities of the arrangements occurring on different chromosomes. An example of estimates obtained by Lewontin and White (1960) for the arrangement *Td* on chromosome *EF* and *Bl* on chromosome *CD* is given in Table 1. The *Td* arrangement on chromosome *EF* causes severe viability deficiencies when in combination with the homozygous standard arrangement on chromosome *CD*, but when suitably combined with *S/Bl* and *Bl/Bl*, it gives rise to an increase in viability. The equilibrium configuration given by such a set of viabilities is not stable. Lewontin and White (1960) suggest that equilibrium in natural populations is maintained by a combination of selection depending on gene frequencies and annual fluctuations in the environment. We have here the complement of the balance of the linked polygenic combination, which is the balance between complete chromosomes. Tentative evidence for such a balance in the control of chaeta number is given by the results of Thoday and Boam's (1961) selection experiments described in Section II,B. These interactions between chromosomes again represent second-order interactions, the primary interactions occurring between the gene contents of balanced complexes on single chromosomes.

3. *Storage of Genetic Variability in Balanced Complexes*

In Section II,A it was pointed out that Mather (1943a) emphasized the importance of polygenic complexes in maintaining the store of concealed or potential genetic variability occurring in natural populations. The need to look for concealed genetic variability in natural populations was first realized by Tschetwerikoff (1926). He suggested that such variability could be carried by individuals heterozygous for recessive lethals. His work and that of Dubinin, Dobzhansky, and others (see Dobzhansky, 1951 for review) has shown that a high proportion of chromosomes extracted from wild populations carry recessive mutations having a wide range of deleterious effects on the homozygote, but virtually no effect on the heterozygote.

Such studies, however, reveal only the net effect of homozygosis for whole chromosomes. The effect of homozygosis for recombinant chromosomes derived from crossing stocks in which different chromosomes extracted from the wild were maintained by "balancers" was studied by Dobzhansky (1946). He revealed the remarkable fact that such recombinant chromosomes were sometimes lethal when homozygous although both constituent parent chromosomes were viable when homozygous. An example of the results obtained with two of the chromosomes A and B is illustrated in Fig. 5. This shows the distribution of the viability of homozygotes for recombinant chromosomes, assessed at both $16\frac{1}{2}°C$ and $25\frac{1}{2}°C$. Both homozygotes A and B were viable at $16\frac{1}{2}°C$ but A was lethal when homozygous at $25\frac{1}{2}°C$. A considerable range of viabilities was produced by recombination. Particularly noteworthy are three chromosomes which were lethal at both $16\frac{1}{2}°C$ and $25\frac{1}{2}°C$ in contrast to A which was lethal only at $25\frac{1}{2}°C$. Although Dobzhansky did not run simultaneous controls for chromosomes obtained from A × A and B × B crosses, the probability of three such lethals occurring independently by mutation is remote. These are therefore most probably examples of what Dobzhansky (1946) called "synthetic lethals," produced by recombination between chromosomes which are viable when homozygous. They provide an extreme example of the disruption by recombination of relationally balanced polygenic complexes.

The occurrence of such lethals was studied by Misro (1949) in second chromosomes of *D. melanogaster* which were extracted from wild populations. He found that females heterozygous for a pair of wild chromo-

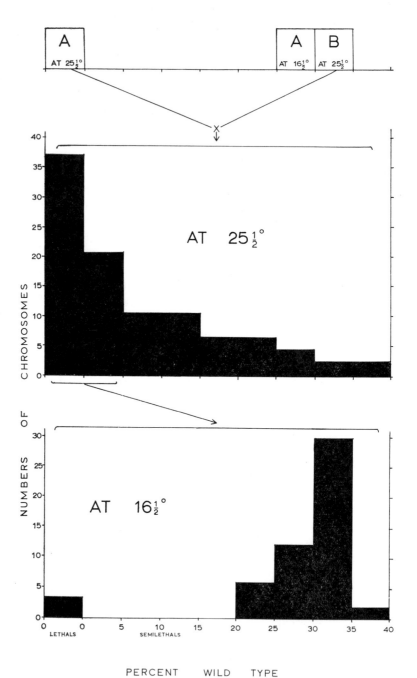

FIG. 5. Results of recombination of the viability genes carried in two wild second chromosomes A and B of *D. pseudoobscura* at two temperatures $25\frac{1}{2}$°C and $16\frac{1}{2}$°C. Under the mating system used, cultures containing between 30 and 35% wild type arise when the viability of homozygotes for a given chromosome is about equal to the heterozygotes. Failure of wild type individuals to appear in the cultures indicates that the chromosome involved is lethal to homozygotes. The lethals and near lethals at $25\frac{1}{2}$°C were retested at $16\frac{1}{2}$°C (after Dobzhansky, 1946).

somes produced a significantly higher proportion of recessive lethals than females homozygous for a wild chromosome or males heterozygous for a wild chromosome. Since there is no recombination in male *D. melanogaster* this provides clear evidence that the observed excess of lethals from heterozygous females must be due to recombinant products. Wigan (1949b) localized lethals produced by Misro (1949) and by himself. He was able to distinguish "synthetic lethals" and "point mutation" lethals, and found that they differed in their distribution along the chromosome. Evidence for the production of "synthetic lethals" by recombination in second chromosomes of *D. melanogaster* was also obtained by Wallace *et al.* (1953). Hildreth (1955), however, found no evidence for the production of synthetic lethals in the X-chromosome. This may well be expected, for the organization of polygenic complexes is not likely to lead to the same degree of relational balance between X-chromosomes as between autosomes, since the X-chromosome is always carried in hemizygous condition in the male, although Tebb and Thoday (1954) have presented evidence showing that both internal and relational balance is a feature of the X-chromosome in *D. melanogaster*. Hildreth's (1956) inability to obtain evidence for the production of synthetic lethals from chromosome III is more surprising. Breese and Mather (1960) have presented positive evidence for the occurrence of synthetic lethals from this chromosome, so that Hildreth's (1956) results must be a product of the particular stocks he used. Certainly the best studied recombinational lethal system is that analyzed by Gibson and Thoday (1959, 1962) on chromosome II of *D. melanogaster* and derived from Thoday and Boam's (1959) disruptive selection experiment, as described in Section II,B.

More extensive investigations, following the lines developed by Dobzhansky (1946), have been reported by Wallace *et al.* (1953). They studied the homozygous effects on chaeta number and viability, of crossover chromosomes obtained from crossing in all combinations two sets of five non-lethal chromosomes derived from two experimental populations of *D. melanogaster*. As mentioned above they obtained synthetic lethals. The pattern of results produced by a particular pair of chromosomes was, in general, very different from that produced by either of the pair in combination with other chromosomes. The effects of the homozygous recombinant chromosomes on chaeta number were much less striking than their effects on viability. The general conclusion is that there is a remarkable amount of variability stored in the form of balanced polygenic combinations and which can be released by recombination.

Further investigations of the production by recombination of wide variations in viability have been reported more recently by Spassky *et al.* (1958), Spiess (1958, 1959), Dobzhansky *et al.* (1959), and Levene (1959) for *D. pseudoobscura, D. prosaltans,* and *D. persimilis.* They extracted second chromosomes for each species from two natural populations separated by distance and in some cases also distinguished by ecological habitat. The chromosomes were chosen to have fairly high viabilities when homozygous. All possible intercrosses were made between the extracted chromosomes, and from each intercross ten chromosomes were tested for their effect on viability when homozygous. In general, viabilities of chromosomes obtained from heterozygous crosses were lower than those obtained from homozygous crosses. The range of observed viabilities extended from lethality to supervitality. Most of this release of genetic variability must be due to recombination between unlike chromosomes.

There were significant differences between the three species studied. In *D. persimilis*, the frequency of lethals was no greater than could be accounted for by mutation, whereas in *D. pseudoobscura* the frequency was twice that expected by mutation. A remarkably high proportion of the total variance in viability present in the populations from which the chromosomes were extracted could be attributed to recombination in all three species. In *D. pseudoobscura* it was about 43%, and in *D. persimilis* and *D. prosaltans* about 24 and 25%, respectively. From these studies it can be inferred that a large amount of genetic variability can be released by recombination disrupting balanced polygenic complexes. This variability is concealed in the balanced complex, as discussed by Mather (1943a), and is only revealed by recombination occurring within the complex.

There is therefore suggestive evidence that the release of variability by recombination in *D. pseudoobscura* is greater than that in the other two species. *D. pseudoobscura* is a widespread, common, and ecologically versatile species, whereas *D. persimilis* and *D. prosaltans* are more specialized species adapted to a narrower range of habitats. Dobzhansky *et al.* (1959) suggest that the difference in the amount of variability released by recombination reflects a difference in the amount of concealed genetic variability in these species which can be explained by their different distributions. They conclude that "natural selection will in every population select those 'supergenes' which interact favourably in heterozygous combinations with other 'supergenes' present in the same population." This is really a restatement, in a less specific form, of Mather's (1943a) concept of the evolution of relational balance between polygenic complexes.

4. *Relational Balance in Natural Populations*

Vetukhiv (1956, 1957) studied the fecundity and longevity of flies obtained from crosses between different geographic races of *D. pseudo-*

obscura. He found heterosis in the F_1 in most cases, but in the F_2 the superiority was lost, and in some cases the F_2 was inferior to the parents. In the F_2, the heterosis in the F_1 caused by certain combinations of polygenes is broken down due to recombination disrupting balanced combinations. Vetukhiv (1954) has done tests on other species, and in some but not all cases finds heterosis in the F_1 and commonly a breakdown in the F_2. Thus various situations can arise in the F_1 and F_2 obtained from crossing two races, depending on the exact genetic constitution of the races. An eventual breakdown of fitness usually occurs if the races are sufficiently dissimilar. Similar studies by Wallace (1955) and Wallace and Vetukhiv (1955) showed how such a breakdown of heterosis was directly correlated with the amount of the genome which was heterozygous.

The effect of chromosomes extracted from wild populations of *D. melanogaster* on chaeta number was studied by Wigan (1944). He compared the effects on sterno-pleural chaeta number of various combinations of pairs of extracted chromosomes with those of marked tester chromosomes. He found that in females the variability was less between families where the homologous chromosomes were derived from different wild individuals than between families where the homologs were derived from the same individual. This provides evidence for the relational balance of polygenic complexes controlling sterno-pleural chaeta number. Pairs of homologs derived from individuals from the same locality showed less variability than ones derived from different localities. This shows that the species is divided into small populations in which the balanced systems have, to some extent, diverged.

Dobzhansky and Wallace (1953) obtained similar results from studies of the rates of survival in crowded cultures of homozygous and heterozygous populations derived from four *Drosophila* species. They showed that the heterozygotes gave relatively uniform survival despite the environmental variations, whereas the homozygotes varied much more. The uniformity of the heterozygotes was postulated to be due to co-adapted or well-balanced heterozygous complexes. In the homozygotes, relational balance will be broken down, with a resultant deterioration in fitness, as is shown by their poorer viability and greater variability in response to environmental change. Similar results have been presented by Dobzhansky and Levene (1955).

Once again the evidence for the existence of balanced polygenic complexes in natural populations comes almost exclusively from work with various species of *Drosophila*. However, the existence of such complexes in other organisms can be inferred from the widespread occurrence, for example, of inversion heterozygosity and in an extreme form, the balanced lethal system in *Oenothera*. There is some further suggestive evidence for balance in plants (see, for example, Mather and Vines, 1951; Stebbins, 1958a) but there are no specific analyses.

III. Balanced Polymorphisms and Linkage

A. Introduction

The association between balanced polymorphism and linkage was first pointed out by Fisher (1930b) in the grouse locust (*Apotettix*) and in the land snail (*Cepaea*). Such polymorphisms often concern sets of closely linked genes controlling characters whose effects interact in respect to fitness, but which are not necessarily related physiologically. This is in contrast to some cases of close linkage between loci having related biochemical effects, such as will be discussed in Section VI.

The degree of overdominance necessary to maintain the polymorphism is probably determined by interactions between the linked loci of the polymorphic complex. The way in which such linked interacting complexes may evolve was clearly shown by Fisher (1930a) when discussing a model, involving interactions between two loci, which could lead to tightening of linkage between them, as described in Section I. There is an analogy between such interacting loci and the balanced polygenic complex since it is likely that clusters of linked polymorphic genes represent the final outcome of reducing recombination within a balanced complex.

In the fish *Lebistes*, various sex-linked color mutants have been found, the color variants being apparently advantageous in males and disadvantageous in females. Fisher (1930b) suggested that the interaction between sexes and the genes has led to a system of closely linked loci on the Y-chromosome controlling color.

B. Incompatibility Systems

Complexes of linked polymorphic genes are now known to be widespread. One of the first examples was the supergene in *Primula* (Ernst, 1933), which controls the different features of the heterostyly system. One complex gene S-s controls the difference between pins and thrums, thrum being the heterozygote Ss, and pin the recessive homozygote ss. Although crossing-over within the S complex has not been proven, there is evidence suggesting that the heterostyly system does break down to give self-fertile homostyles by crossing-over. Homostyle variants in normally heterostyled species are much commoner when the species are polyploid, which, following Darlington (1941), suggests a relation between the occurrence of homostyly and increased chiasma formation (see Lewis, 1954). Thus in *P. obconica*, Dowrick (1956) used polyploidy as a method of increasing crossing-over, and breaking down the S-complex.

The S-complex is made up of three genetically distinct groups of characters (Ernst, 1933, 1936). These are (1) length of style and stigmatic papillae, and stylar incompatibility reaction (G), (2) anther height (A), and (3) pollen grain size and incompatibility reaction (P). G, P, and A refer to the thrum form of these characters and the corresponding genes in the heterozygote are all dominant. Of the eight possible phenotypes, Ernst (1933, 1936) has found all but the self-fertile thrum (GpA).

The eight possible gametic genotypes and corresponding phenotypes are shown in the tabulation (Ernst, 1933).

Self-compatible		Self-compatible	
Thrum	GpA	Short homostyle	Gpa
Pin	gPa	Long homostyle	gPA
Self-incompatible		Self-incompatible	
Thrum	GPA	Short homostyle	GPa
Pin	gpa	Long homostyle	gpA

There are, of course, three possible orders for the genes, G, P, and A, and Dowrick in $P.$ $obconica$ considers the above order the most likely. The order PGA is unlikely, as the self-compatible homostyle, which is the most common aberrant type, would require a double crossover for its formation. Dowrick prefers the order GPA over GAP, since only one of the possible double crossover types from GPA has been found.

Mather and De Winton (1941) suggested that initially pin arose as a deleterious mutant. When outbreeding became advantageous, the initial disadvantage of the pin was compensated by its promotion of crossing. Consequently, it then spread, and at the same time linked modifiers strengthening the incompatibility reaction were accumulated by selection until the heterostyly system was fully developed as we find it today.

During this process the S-s chromosome will always be heterozygous so that, in Mather's (1943a) terminology, there should evolve relational balance between the S and s chromosomes. Homozygous SS thrums, produced by forced selfing of Ss thrums, were shown by Mather and De Winton (1941) to have a viability of 72.8% relative to Ss thrums and pins (ss). This is to be expected as the SS thrums have not been produced in natural populations since the evolution of an efficient outbreeding system. Appreciable proportions of self-fertile long homostyles in natural populations of the primrose $P.$ $vulgaris$ have been reported by Crosby (1940, 1949). The homostyly is controlled by the complex $gPA \equiv S^h$. Homozygous homostyles (S^hS^h) will be homozygous for a part of the S-carrying chromosome and so may be expected to show a similar viability deficiency to the SS thrums. Crosby (1949) attempted to explain the observed distribution of homostyly entirely in terms of such a viability deficiency and on the assumption that homostyles exclusively self-fertilize. However, Bodmer (1958) showed that homostyles may cross-fertilize to a considerable extent under natural conditions and that this cross-fertilization probably provided the basic controlling mechanism that prevented an immediate rapid increase in the proportion of homostyles as soon as they occurred (Bodmer, 1960b). This tendency to cross-fertilize undoubtedly reflects the residual genotype which has been selected to bolster up the efficiency of the basic switch-gene mechanism and which is still partially effective in its absence. It seems most likely that the incompatibility system has evolved from an initial difference of the sort Mather and De Winton (1941) ascribe to the archetype pin, and subsequent accumulation of linked modifiers increasing the efficiency of the incompatibility mechanisms.

Heterostyly systems presumably controlled by similar complexes of tightly linked genes to that in *Primula* are very common among the flowering plants (see, e.g., Lewis, 1954). Such complexes are probably relationally well balanced. There is abundant evidence of naturally occurring polygenic variability affecting polymorphic systems, but there is, as yet, little evidence that it is organized into linked balanced complexes. More specific studies of such genetic variability should provide data of considerable interest.

In many species of plants, the incompatibility reaction is not associated with variation in floral morphology, but only with a reaction between the genetic constitution of the zygote producing the female gamete and in most cases the genetic constitution of the male gamete. Mather (1943b) presented evidence in *Petunia* strongly suggesting that this type of incompatibility reaction, due to a series of "S" alleles, is built up slowly, each allele gradually acquiring a stronger incompatibility reaction when outbreeding is advantageous. This process occurs by the selection of non-allelic, possibly linked, genes modifying the strength of the incompatibility reaction.

Work done by Lewis (1952) shows that the S alleles produce specific substances in the pollen which differ antigenically, and presumably these substances exist in the style also. Thus incompatibility is probably due to an antigen-antibody reaction. Lewis favors a few simple mutational steps as the mechanism whereby the incompatibility system is built up. However, the available evidence seems to favor the gradual strengthening of an initially incomplete incompatibility reaction, by the accumulation of favorable modifiers.

Some evidence of the complexity of loci controlling mating type reactions is available from microorganisms. Leupold (1958) has shown that rare homothallic recombinants are obtained in *Schizosaccharomyces pombe* from crosses between the two mating types + and −. The use of linked markers has demonstrated that these recombinants are produced by conventional crossing-over and therefore that the mating type locus is complex. The production of homothallic *Schizosaccharomyces* by crossing-over at the mating type locus closely parallels the postulated origin of homostylic *Primula* by crossing-over within the incompatibility locus. More complex genetic mechanisms are involved in the control of mating type reaction in some Basidiomycetes, and some of these have also been shown to be associated both with loose linkage and allelic recombination (Day, 1961; Crowe, 1960; Parag and Raper, 1960).

C. Linked Complexes in Animal Populations

A review by Sheppard (1953) summarizes much of the evidence on the evolution of linked systems of genes (see also Ford, 1955) with special reference to animals and man.

In *Cepaea nemoralis*, the genes for pink shell and unbanded are linked (Fisher and Diver, 1934). Further closely linked loci have been found affecting both shell color and banding. The proportions of different varieties vary greatly from one colony to the next, and it appears that the specific polymorphism in each colony is maintained by a balance between predation by birds (visual selection) and the exact background in which the colony lives. The polymorphism must be very old as it can be detected in the fossil record (Diver, 1929). At least three to four loci control color and banding. The selective values associated with the polymorphism vary considerably between populations. Furthermore, the degree of linkage varies between colonies from a situation where separate loci behave as multiple alleles, to recombination values of the order of 20%. This variation is perhaps in part due to an association with inversions (Cain and Sheppard, 1950, 1954). We have here interactions between loci adjusting linkage values so as to provide a balance between selective advantage in a specific environment and genetic flexibility in a changing environment.

Mather (1955b) stressed the importance of interacting modifiers for the building up of a polymorphic supergene in a paper discussing polymorphism as a possible outcome of disruptive selection. Once such a polymorphism has arisen, it will not remain static. Modifiers will be incorporated to build up a supergene. The artificial production of a polymorphic population under disruptive selection in *Drosophila*, and its genetic basis, has been studied by Thoday (see Section II,B). Ecological studies in the mimetic butterfly *Papilio dardanus* demonstrate polymorphisms under disruptive selection in naturally occurring populations (Clarke and Sheppard, 1960b).

Perhaps the most interesting recent study on the build-up of a "supergene" controlling a polymorphism is the work of Clarke and Sheppard (1960a,b,c) on mimicry in the females of the butterfly *Papilio dardanus*. The mimetic patterns are controlled by a supergene which they suggest has evolved step by step due to selection for increased linkage between loci, whose effects interact. For example, a mutant in the presumed primitive non-mimetic Madagascar stock may modify the costal bar of the wing so as to produce a pattern rather like that of race *hippocoonides* (which mimics *Amauris niavius*), except that the pale area of the wing would be yellow, and not white as in the model. A second mutant, which replaced the yellow by white, would, however, enhance the resemblance between mimic and model, but unless the second mutant were on the same chromosome as the first, it may not be able to establish itself, since the replacement of yellow by white in the non-mimetic pattern is likely to be disadvantageous. If, however, the two new mutants are together in the same individual on the same chromosome, so that both are at an advantage, then there will be strong selection for increased linkage between them. This would be followed by other mutants altering the exact distribution of pigment, which would all accumulate on one chromosome, so producing a supergene controlling the range of patterns found (Sheppard, 1959).

Sheppard (1955) proposed a specific model of a genetic system leading to closer linkage by natural selection which is based on that of Fisher (1930a). Assume a locus with a pair of alleles A and a kept in a state of balanced polymorphism by overdominance of the heterozygote Aa. Another pair of alleles B and b are at another locus in the same chromosome, and interact with the genes at the first locus in such a way that A is advantageous with B and disadvantageous with b, while gene a is disadvantageous with B and advantageous with b. This in fact is the same as Fisher's (1930a) model, but with the added condition of a balanced polymorphism at the A,a locus. Kimura (1956) investigated this model theoretically, and showed that provided linkage is sufficiently close, the second locus will remain polymorphic. Under such conditions, reduced crossing-over between the loci would be favored by natural selection. Such a specific reduction in crossing-over could, for example, be caused by an appropriate inversion. The model was applied to the colorations of the land snail *Cepaea*, and shown to provide a good fit with the selective values assumed for the relevant genotypes (see also Section VII). As mentioned earlier, crossing-over between the loci varies according to the background in which the snails live; sometimes linkage is very strong, and it is postulated that this may be due to an inversion having been incorporated into the system. The model given by Kimura shows how such an incorporation could occur.

The situation at the T locus of the mouse has been considerably amplified by recent work of Dunn and his collaborators (1958; Dunn, 1956, 1957). The first allele T found at this locus is lethal when homozygous and produces the short tailed condition, brachyury, when heterozygous $(T/+)$. Various t alleles have been found in natural populations. These are recessive lethals, apart from two exceptions, and heterozygotes T/t are tailless but viable. Given two alleles t_0 and t_1 of independent origin, the cross $T/t_0 \times T/t_1$ will only produce normal offspring if t_0 and t_1 are "non-allelic" for otherwise the genotype t_0/t_1 would be lethal. A series of sites which are "non-allelic" by such a test have been found. It has been suggested (Dunn, 1956) that these sites are connected with different components of a chain of processes with inductive relations between parts of the early embryo, but we must await the results of further investigations for confirmation of this hypothesis. Other closely linked genes (Ki kinky tail, and Fu fused tail) have developmental effects similar to T (Gluecksohn-Waelsch, 1954). Thus it is likely that we have in the region of the T locus a linked complex of interacting genes.

In wild populations of mice, high frequencies of various t alleles have been found indicating a polymorphism, thought to be maintained by an abnormal male segregation ratio whereby the t carrying gametes from the heterozygote $t/+$ fertilize far more frequently than those carrying the $+$ alleles. A selective advantage for the male heterozygotes has also been postulated (Dunn et al., 1958). Bateman (1960) has suggested that the selective fertilization of genetically different eggs may be a further factor maintaining the polymorphism.

Whatever the cause of a polymorphic gene complex, it seems likely to have been built up slowly by the accumulation of modifiers, accompanied by selective modification of linkage in the way already discussed, so leading to the evolution of a supergene. In this case, it is possible to identify sites within the supergene, but we are unable, using the techniques currently available, to study recombination within the supergene. This is a necessary preliminary to ascertaining the relationship between the structure of the supergene, its function, and its ultimate role in maintaining the balanced polymorphism.

D. A Linked Complex Built Up by Artificial Selection

In the ever-sporting stocks, *Matthiola annua*, Winge (1931) showed the pure-breeding single flower to be SV/SV, and the ever-sporting single flower Sv/sV where S is a dominant gene for singleness, such that ss represents doubleness, and v is a gene lethal to pollen. The locus of v is linked to the S-s locus with a recombination fraction of rather less than 1%. Selfing the ever-sporting single Sv/sV gives mainly progeny Sv/sV and sV/sV, the latter being sterile doubles favored by horticulturalists but which leave no offspring. Since the *ss* combination is sterile, the only way it can be produced is by selfing the eversporting single type. Now, it is presumed that the first step in the building up of this system was a mutation from $S \to s$, which at first can only have been perpetuated by continual selection by growers. By selecting from families with the highest proportion of doubles, as soon as a pollen lethal occurred in the *S* chromosome of the heterozygote, it would be favored because it gives a greater proportion of doubles, and this would make the ever-sporting single more prevalent. Fisher (1933) therefore suggested that selection for ever-sporting singles would lead to a progressive tightening of linkage between the two loci, so explaining the somewhat remarkable coincidence that the lethal needed to balance the doubleness occurred exactly where it was required. Fisher (1933) cites frequent batches of impure seed in data reported by Saunders (1911), which suggest that recombination in most strains is higher than in the very reliable ever-sporting strain used by Winge. The intensity of the linkage is, therefore, to a certain extent correlated with the intensity of selection.

Thus we have a polymorphism, maintained by artificial selection, producing a system analogous to a naturally occurring complex polymorphism.

E. Possible Linked Complexes in Man

Many of the blood groups in man are polymorphic. Ford (1945) suggested that the stability of the polymorphic *ABO* blood group ratios in populations indicates that the gene frequencies are maintained by natural selection. This was, to some extent, substantiated by Aird *et al.* (1953) who showed that cancer of the stomach caused a slight selection against blood group *A*. Since then further evidence on the association of blood groups and disease has been presented (Roberts, 1959). Whether the selective forces so far found are the true ones determining the polymorphism seems questionable, since these polymorphisms are presumably very old and evolved under conditions very different from those existing today. Having once evolved, it is unlikely that they would be lost by drift if the genes became neutral, since population sizes are excessively large. However, differences of the sort associated with *ABO* and other blood groups, even if they are not the operative selective forces involved, show that there do exist physiological forces connected with blood groups of a type that may well cause important selective differences. Recent work of Chung and Morton (1961) suggests that fetal incompatibility may be a major selective factor in maintaining the *ABO* blood group frequencies.

There is at present no direct evidence that the *ABO* blood groups are not simply controlled by multiple alleles. However, more refined techniques than at present available may well show that the *ABO* locus is complex as are so many of the other loci which are known to be polymorphic in natural populations.

In the case of the *Rh* blood groups, evidence has been presented suggesting that at least three very closely linked loci are involved (Fisher, 1947; Race and Sanger, 1958). This interpretation has been disputed by Wiener (see Wiener and Wexler, 1958), who regards the system as a series of multiple alleles. Fisher, on the basis of the results available at the end of 1943, postulated that three closely linked gene pairs were responsible, which he called *C-c*, *D-d*, and *E-e*. A possible further gene pair *F-f* has been found more recently. Fisher's conclusion was based on his observation that the reactions of anti-*C* and anti-*c* serum were antithetical (Table 2). He drew up a table, based on existing data, from which he made predictions as to combinations not discovered at that stage, almost all of which have since been found (Table 2). Wiener's notation, R_1, R_2, ... etc., is given in the table for comparison.

TABLE 2
Serological Confirmation of Fisher's Theory for the Rhesus System*†

Antibodies	English % positive	Genes and antigens							
		$\dfrac{CDe}{R_1}$	$\dfrac{cDE}{R_2}$	$\dfrac{cde}{r}$	$\dfrac{cDe}{R_0}$	$\dfrac{cdE}{R''}$	$\dfrac{Cde}{R'}$	$\dfrac{CDE}{R_2}$	$\dfrac{CdE}{R_y}$
anti-*C*	70	+	−	−	−	−	+	+	+
anti-*D*	85	+	+	−	+	−	−	+	−
anti-*E*	30	−	+	−	−	+	−	+	+
anti-*c*	80	−	+	+	+	+	−	−	−
anti-*d*	65	−	−	+	−	+	+	−	+
anti-*e*	96	+	−	+	+	−	+	−	−

* From Race and Sanger (1958).

† Only the reactions within the enclosure were known before Fisher formulated his theory. The predictions made by the theory are shown outside the enclosure: all of them have been confirmed serologically save that no anti-*d* serum has yet been identified with certainty.

In a classic paper Fisher (1947) analyzed the evidence in detail. The frequencies of the various genotypes in the population fit the hypothesis that the likely gene arrangement is *D-C-E*. In some people, the *C* and *E* antigens are totally absent which suggests a possible deletion of the *C* and *E* loci (Race and Sanger, 1958). This agrees with the order *D-C-E* as *C* and *E* are adjacent. Thus the evidence seems, on the whole, to favor Fisher's interpretation. There is also some recent evidence of interaction between at least two of the loci (Race and Sanger, 1959). It would be reasonable for interactions to occur if the *Rh* locus turns out to be a complex cistron, as may be likely (see Carlson, 1958 for further discussion).

The other blood groups are perhaps less well known, but linkage has been suggested to account for the *MNSs* system (Race and Sanger, 1959).

No doubt the question of whether the polymorphic blood groups are due to extremely closely linked interacting loci, or due to multiple alleles will eventually be resolved. The evidence (see also Parsons and Bodmer, 1961) already presented, pointing to the importance of linkage, seems to argue against overdominance leading to a polymorphism associated with a genetic difference at a single site [muton or recon in Benzer's (1957) terminology]. As in microorganisms and *Drosophila*, when methods of analysis gain progressively in their powers of genetic resolution, we will be able to see whether in fact one or several sites and also perhaps several "cistrons" are involved. Similar analyses in animal blood groups, which also give balanced polymorphisms, will be of interest. In plants too, incompatibility systems governed by a series of *S* alleles may well turn out to consist of a series of linked interacting cistrons each of which controls different facets of the incompatibility reaction. Evidence pointing to such an interpretation has been discussed recently by Lewis and Crowe (1958), Lewis (1960), and Pandey (1961).

IV. Chiasma Frequencies and Cytological Evidence

A. Natural Selection and the Chiasma Frequency

So far, we have considered evidence which has been mainly genetic. In this section we shall discuss work on chiasma frequencies and related cytological variables in natural and artificial populations, with special reference to differences in breeding systems. Evidence points to a reasonable correlation between chiasma and recombination frequency (see Darlington, 1931a; Mather, 1938; Sturtevant, 1951; Brown and Zohary, 1955), although some evidence is rather conflicting. Furthermore, different theories exist on the mechanism of genetic recombination, such as classical breakage and reunion with chiasmata (Darlington, 1931a), and the copy-choice mechanism (Lederberg, 1955). Since, however, chiasmata occur in all higher organisms and lead to recombination, factors affecting chiasma frequencies must be considered.

Darlington (1932) reviewed early evidence showing that the behavior of chromosomes in the resting nucleus, in mitosis, and in meiosis, is controlled by the genotype, and emphasized that such variation must be examined in the light of evolutionary theory. He pointed out that high chiasma frequencies lead to the rapid production of variants by recombination, and thus that the rate of release of new variability by recombination could be controlled by controlling chiasma frequency. He found clones of *Fritillaria imperialis* (Darlington, 1930) which differed in chiasma frequency at meiosis. Each clone had a characteristic chiasma frequency which could only be attributed to genetic control. Other early evidence indicating genetic control of chromosome behavior is presented by Peto (1933) in *Lolium-Festuca* hybrids, Maeda (1937) and Emsweller and Jones (1945) in *Allium* species, Prakken and Müntzing (1942) in rye, and others.

Lamm (1936) studied the cytological effects of inbreeding rye and showed that the chiasma frequency of pollen mother cells of inbred rye was usually lower than that of the relatively outbred population from which the inbred parents were derived. Furthermore, the variability of chiasma frequency was probably greater in inbred rye, which had frequent disturbances of meiosis.

Outbred rye with a high chiasma frequency showed a negative correlation between chiasma frequencies for different chromosomes, whereas inbred rye with a lower chiasma frequency did not. Mather (1936) postulated that there is an upper limit to the number of chiasmata in a cell. In the outbred populations of rye, we are close to this limit, so competition between bivalents for a limited number of chiasmata occurs, thus leading to a negative correlation between numbers of chiasmata on different chromosomes. In inbred rye, however, we are further away from this limit, so that competition is less intense, and consequently the negative correlation would not be expected.

Alopecurus myosuroides, a typically outbred species, shows reduced fertility and also cytological abnormalities on inbreeding, whereas F_1 families between plants of different origin do not (Johnsson, 1944). Stickiness, first described by Beadle (1932) in *Zea mays*, and also asynapsis were common in some inbred families. Müntzing and Akdik (1948) studied the degree of pairing in the first three inbred generations of *Secale cereale* as compared with population (outbred) plants. As inbreeding proceeded, the degree of pairing was reduced, and meiotic abnormalities occurred. A positive correlation between chromosome pairing and male and female fertility was established. Wide outcrosses, for example between *Allium fistulosum* and *A. cepa*, may lead to great variability in chiasma frequency in the F_2, and of other meiotic characteristics (Levan, 1941; Maeda, 1937; Emsweller and Jones, 1945).

These studies show that inbreeding a naturally outbred species leads to a fall in chiasma frequency, and increased cytological abnormalities which are correlated with a decrease in general fitness. Wide outcrosses may have similar effects. Thus chiasma frequency and variability are of adaptive significance. Chiasma frequency is controlled by natural selection, and is an important facet of the breeding system of an organism, for it controls the recombination of factors on the same chromosome and so the nature of the balance discussed in Section II.

Some extreme mutants suppress chiasma formation and metaphase pairing. Examples occur in *Zea mays* (Beadle, 1932) and in *Crepis capillaris* (see Darlington, 1958). In *Drosophila melanogaster* there is a mutant in the third chromosome suppressing crossing-over in females (Gowen, 1933). In males, of course, crossing-over is normally entirely inhibited. Furthermore, crossing-over in different sections of chromosomes may be affected by the degree of localization of pairing and of chiasmata (Darlington, 1931a, 1958).

More recently Rees (1955a,b, 1957), Rees and Thompson (1956, 1958), and Thompson (1956) have carried out extensive studies on cytological variables, including chiasma frequencies, of the pollen mother cell nuclei in inbred lines of rye and their hybrids. Variations occurred between inbred lines for chiasma frequency, terminalization and other nuclear characters. These characters were shown to be controlled polygenically. In a diallel cross of four inbred lines and their hybrids, the average chiasma frequency in the hybrids was greater than in the inbred lines. This was probably in part due to non-allelic interaction. Furthermore, chiasma frequencies in the heterozygous hybrids were less variable than in the homozygous inbred lines (Fig. 6). Thus the developmental stability of the heterozygotes with respect to chiasma frequency was greater than the homozygotes. Equally homozygous inbred lines differed in their stability of chromosome behavior, so that stability cannot be directly related to the degree of heterozygosity in the genotype.

It appears, therefore, that the high degree of developmental stability associated with heterozygotes has been achieved in the populations by natural selection for particular balanced genotypes, and that, on inbreeding, this balance is broken down, exposing genotypes not previously subject to natural selection. In species which normally inbreed, the chromosomes should behave perfectly well in homozygous genotypes which have been exposed to natural selection.

Natural selection acts on chiasma frequencies and hence recombination at three levels, the plant, the cell, and the bivalent. At all three levels chiasma frequency is capable of adjustment, since stability at all three levels was shown to be genotypically controlled, but the control differed between levels.

In a further experiment (Rees, 1957), chiasma frequency and the frequency of pre-meiotic errors were scored in families of the F_3, F_4, and F_5 selfed generations from a cross between two inbred lines. It was found that there were highly significant differences between families within these generations, and parent-offspring regressions showed these differences to be genotypically controlled. In particular, one family had significantly fewer chiasmata than either parent. This family must presumably have arisen by recombination. The wide range of chiasma frequencies and pre-meiotic errors in the F_3 and subsequent generations show the effect of

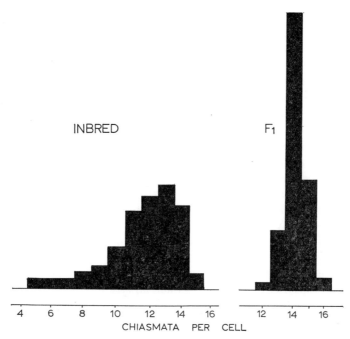

Fig. 6. The distributions of chiasma frequencies in 100 pollen mother cells of an inbred plant and an F_1 plant of rye. In the F_1 the mean is higher and the spread is smaller, and in the inbred plant the distribution is significantly skew (after Rees and Thompson, 1956).

recombination and segregation. The two characters, pre-meiotic errors and chiasma frequency, were inherited separately, since recombination was demonstrated for the characters considered jointly.

A study by Rees and Naylor (1960) shows how delicate is the control of chiasma frequency. It was found that within the rye anther, significant variation in chiasma frequency and of the timing sequence of meiotic divisions occurred. Variation in both characters is probably influenced by the system of distribution of nutrient or other substances to the anther during its development. Also, pollen mother cells reaching metaphase early had higher chiasma frequencies than those reaching metaphase later. As might be expected, the pattern of developmental variation in chiasma frequency was more consistent in anthers of heterozygous than homozygous plants. This illustrates the type of mechanism that may be involved in the control of chiasma frequency.

B. Supernumerary Chromosomes

In natural populations of grasshoppers (Acrididae), Barker (1960) showed that in *Myrmeleotettix maculatus* supernumerary or B-chromosomes were associated with a high chiasma frequency in one locality, whereas the absence of supernumeraries was associated with a lower chiasma frequency in a different locality. The high chiasma frequency only occurred in individuals with supernumeraries. It was suggested that the supernumeraries were present because they affected chiasma frequency. Mather (1939a) found that the addition of an extra chromosome in maize increased chiasma frequency. In *Chorthippus parallelus*, Barker showed that variation in chiasma frequency similar to that in *Myrmeleotettix* occurred but in this case without supernumeraries. Thus similar regional differences were probably produced by different mechanisms. It is not known exactly what the function of supernumerary chromosomes is, but it is supposed by Darlington (1956) that they boost the variability and hence the flexibility of a species. This, according to Darlington, explains why they are rare in polyploids, which have more hidden variability than diploids.

On inbreeding rye, the number of supernumerary chromosomes falls (Müntzing, 1954). Generally, in commercial strains supernumerary chromosomes are rare or absent, but are more frequent in the Asian primitive strains (Müntzing, 1958). In rye, as we have seen from the experiments of Rees and his collaborators, chiasma frequency falls and becomes more variable on inbreeding. It seems plausible, therefore, that supernumerary chromosomes are concerned with the flexibility of a species under natural selection, and that under cultivation, when natural selection is relaxed and inbreeding occurs, they will fall in number since flexibility to adapt to new conditions is of less importance. It is therefore not surprising that there is some evidence for correlations between supernumerary chromosomes and chiasma frequencies.

C. The Recombination Index

In this section, we have reviewed evidence suggesting that chiasma frequency, and the various cytological phenomena associated with meiosis, are under genotypic control, and that the frequency and variability of chiasma formation are an integral part of the breeding system. Chiasma frequency and related cytological phenomena are obviously closely connected with the fitness of a species in the broadest sense. There is a direct relation between the behavior of these cytological phenomena and both the immediate viability and potential fertility of a species. Cytological phenomena must therefore be considered when discussing the over-all fitness of a species (Darlington, 1939; Mather, 1943a; Stebbins, 1950; Thoday, 1953).

Chiasma frequency, which is one of the components of Darlington's recombination index, is subject to modification. The other component is chromosome number, and is, as Darlington (1958) points out, much more constant. In some invertebrate groups centric fusions may occur, reducing the chromosome number as in some grasshoppers (family Acrididae), and so increasing the size of linkage groups (White, 1954).

In the common shrew, *Sorex araneus*, which is a rodent, chromosome numbers vary from 22 to 27, but the total number of autosomal arms is always 36 (Sharman, 1956; Ford et al., 1957). Some chromosomal elements are present in either of two forms: as single metacentric chromosomes or as pairs of acrocentric chromosomes, so that the total number of autosomal arms is constant. Thus we have a chromosomal polymorphism, one function of which must be to exert some control over recombination. A similar system occurs in the rodent *Gerbillus* (Wahrman and Zahavi, 1955).

Most evolutionary change in recombination is, however, likely to occur as a result of variation in chiasma frequency and localization, or because of various cytological aberrations which reduce crossing-over such as the inversion heterozygote in *D. pseudoobscura*, or the complex translocation heterozygote in *Oenothera*. Complex interchange heterozygotes, which are at a selective advantage in inbred populations, have been recently found in *Periplaneta americana*, the American cockroach (John and Lewis, 1957, 1958) and in *Blaberus discoidalis* (John and Lewis, 1959). Generally, in the best understood cases, the establishment of interchange heterozygosity is favored by a change from outbreeding to inbreeding (Darlington and La Cour, 1950). Interchange heterozygosity preserves relational balance, characteristic of an outbred species. Furthermore, the linkage group is artificially extended beyond one chromosome.

Stebbins (1958b) has suggested that in the tribe Cichorieae of the Compositae, the primitive condition is an outcrossing population with a haploid number $x = 9$ giving free genetic recombination. With the occupation of more unstable habitats and the consequent greater fluctuations of population size, greater reproductive efficiency was evolved by reducing recombination, so leading to specialization. Recombination was postulated to be reduced by (1) a trend toward self-fertilization, (2) the development of apomixis, and (3) the reduction of the recombination index. Stebbins showed that the reduction of the recombination index included a reduction in chiasma frequency, increasing asymmetry of the karyotype, and a reduction of the chromosome number. He pointed out that the last two tendencies are intimately connected, and involve sequences of pericentric inversions, unequal reciprocal translocations, and the loss of centromeres. He found that karyotype asymmetry was greatest for $x \leqslant 5$ and least for $x = 9$, and that there was a tendency from cross- to self-fertilization as x decreased from 9. Correlated with this was an increasing instability of habitat. Similar systems occur in other groups of plants. Generally animals appear to have more asymmetrical karyotypes than plants, except perhaps in primitive invertebrate phyla.

A review by Grant (1958) discusses more generally how recombination in plants is affected by various ecological and genetic factors. For example, the long generation of woody plants is often associated with a high chromosome number, outcrossing, and relative freedom from inversions and translocations which restrict crossing-over. Shorter lived herbaceous plants more frequently have lower chromosome numbers, are more inbred, and have inversions and translocations. An extreme situation is represented by the complex translocation heterozygotes of *Oenothera* (Darlington, 1931b; Catcheside, 1940) and a few other plants (see Stebbins, 1950). Restriction of recombination cytologically or otherwise is desirable in herbaceous plants, as the adapted types are then more likely to be preserved and can exploit their habitats more efficiently. If recombination were not restricted, a variety of recombinant types would be generated, many of which would be disadvantageous. This latter system is, however, advantageous in longer lived plants, where many zygotes are produced, but most are selectively eliminated. The constancy of the woody species can therefore be maintained by intense selection from a spectrum of constantly produced recombinant types, whereas in herbaceous plants restricted recombination, preserving and allowing the desirable types to multiply rapidly, is at a premium. In the herbaceous species, therefore, immediate fitness is gained by restricting recombination at the expense of the flexibility that aids adaptation to new environments.

We must, however, contrast herbaceous ephemeral populations, which consist of dispersed groups with almost identical habitats, where selection for constancy is predominant, so leading to apomixis and other methods of reducing or eliminating recombination, and marginal populations of various *Drosophila* species studied by Carson and others (see Section II,C,1) in which inversion heterozygosity is low, so allowing a high level of recombination for the exploitation of *new* and *varied* environments. In the latter case free recombination is at a premium, since always the emphasis is on the exploitation of new and different environments, whereas in the ephemerals, constancy of environment and the creation of large numbers of identical genotypes in a short space of time is at a premium (see Thoday, 1953). In marginal populations, it is possible that once a certain recombinant type becomes established and predominant, then selection for restricted recombination may commence as is often found at the center of distribution of such species (Section II,C,1).

These rather general correlations are examples of the way in which recombination is a function of the breeding system. To a certain extent, the recombination index is a useful measure of total recombination. Within groups of plants, as in the Cichorieae, changes in the index may be correlated with changes in the breeding system, and changes in the requirements for new variability produced by recombination. Wider comparisons become more difficult because of factors such as cytological inhibition of crossovers and variable karyotype asymmetry.

V. Genotypic and Environmental Variables Affecting Recombination

A. Factors Affecting Recombination

Classical linkage experiments done under diverse environmental conditions and differing genetic backgrounds provide abundant evidence for the variability of recombination.

The effect of temperature on recombination is perhaps the best studied variable. Plough (1917) found that the recombination fractions in chromosomes II and III of *Drosophila melanogaster* were increased when the temperature at which the female fly developed departed in either direction from the normal culture temperature of 25°C. The greatest variation occurred in centromeric regions. The effect of temperature on recombination fractions in the acrocentric X-chromosome is also localized close to the centromere (Stern, 1926; Mather, 1939b). The observed pattern of variation may be interpreted as being due to natural selection, favoring an increased range of recombinants under abnormal conditions in order that the organism may better adapt itself to the new conditions. Similar variation was obtained by Hayman and Parsons (1961) for a segment near the centromere of the X-chromosome when flies were developed at 30°C, using 20°C as a control. As well as the expected increase in recombination near the centromere, heterogeneity of the data was very much increased at 30°C compared with 20°C. This was attributed to a deterioration in developmental stability at the higher temperature. That developmental stability breaks down in general at high temperatures has been shown by Thoday (1955), who demonstrated an increase in sterno-pleural chaeta asymmetry at 30°C. In Rees and Thompson's (1956) experiments, inbreeding led to greater variability of chiasma frequency. There is a close parallel here, for inbreeding causes an alteration in genetic environment, whereas temperature variation changes the physical environment.

Temperature has been shown to affect chiasma frequency in meiosis in some plants (Elliott, 1955; Dowrick, 1957; Wilson, 1959) and in the Orthoptera (White, 1934; Moffett, 1936). In *Neurospora crassa*, Rifaat (1959) has shown that temperature affects crossing-over and that the effect depends on the genotype of the stock used.

The influence of parental age on recombination is well known in *Drosophila melanogaster* (Bridges, 1915, 1929; Plough, 1917). A significant decrease in recombination with parental age has been demonstrated in mice (Fisher, 1949; Bodmer, 1961). The decrease in recombination is probably associated with less efficient chromosome pairing, demonstrating in yet another way the influence of developmental differences on chromosome behavior.

Other variables known to affect recombination in *Drosophila* are X-rays (Muller, 1925), calcium ions (Levine, 1955), the chelating agent ethylenediaminetetraacetic acid, EDTA (Levine, 1955), and exposure of pupae to extreme low humidity (Levine, 1955). In *Chlamydomonas reinhardi*, variations in recombination have followed treatment with EDTA and manganese chloride (Eversole and Tatum, 1956). In bacteriophage and yeast, ultraviolet light has been shown to increase recombination (Jacob and Wollman, 1955; Roman and Jacob, 1958). Other agents affecting recombination have been reported in these and other organisms, but will not be discussed here.

Differences in recombination between sexes may have large effects on the release of variability. In *Drosophila* there is an extreme sex difference, as there is normally no crossing-over in the male, although it can be induced by X-rays (Patterson and Suche, 1934), temperature (Shull and Whittinghill, 1934), formaldehyde (Sobels, 1956; Whittinghill and Lewis, 1961), and various other mutagens. In mice only comparatively minor sex differences have been reported, the majority, though not all, showing greater recombination in the female (see Castle, 1925; Crew and Koller, 1932; Mallyon, 1951; Michie, 1955; Parsons, 1958a). Crew and Koller (1932) also claim from cytological observations that the chiasma frequency is higher in females. Small sex differences occasionally occur in maize (Stadler, 1926) and in many other organisms.

From the point of view of recombination in natural populations, perhaps the most important specific effect is that of heterozygous inversions in reducing crossing-over in the chromosome on which they occur. Sturtevant (1917, 1919) was obviously studying the effect of inversion heterozygosity on recombination in his work on linkage variations in chromosomes II and III of *D. melanogaster*. Inversion heterozygotes in populations lead to the evolution of stable complexes of balanced interacting genes (Section II,C), and also increased crossing-over in non-homologous chromosomes (see Schultz and Redfield, 1951). Increased crossing-over in the non-homologous chromosomes may well be a means of compensating for the inhibition of crossing-over in the chromosome with the heterozygous inversion. Sturtevant and Mather (1938) pointed out that an organism may be able to adapt itself best to the necessary compromise between fitness in one environment and flexibility to adapt to others, by restricting inversion heterozygosity to one chromosome. In fact, they argue that new inversions will be accumulated on chromosomes in which inversions have already occurred and will not spread at random throughout the genome. Other cytological mechanisms such as complex translocation heterozygotes also modify crossing-over (see Section IV).

There is necessarily less information on the variability of genetic interference. Hayman and Parsons (1961), for example, studied the simultaneous effects of temperature, age, and the presence of a non-homologous heterozygous inversion on recombination in the proximal half of the X-chromosome of *D. melanogaster*. The order of the loci was *v-sd-car-*centromere. Temperature affected interference throughout the whole segment *v-car*, but recombination only for the segment *sd-car*, which is close to the centromere. A similar effect became apparent on recalculating Graubard's (1932) data for chromosome II. Age and the presence of the inversion, however, altered recombination throughout the whole segment *v-car*, but interference was not affected significantly. Thus here we have clear evidence of different factors affecting the distribution of crossovers in different ways. In some earlier data, however, interference has been altered slightly by age and the presence of inversions (see, for example, Bridges, 1929; Steinberg and Fraser, 1944; Rendel, 1957, 1958), but the effect of temperature seems much greater.

The extreme variability of interference in different *Drosophila* species has been stressed by Parsons (1958b). In this context, it is interesting to note that interference in the mouse is probably more intense than in either *D. melanogaster* or maize (Wallace, 1957; Parsons, 1958a,c,d). The degree of interference is clearly a significant component of the over-all amount of recombination. High levels of interference will help in preventing the breakdown of linked balanced interacting complexes by recombination. Thus the high level of interference in the mouse reflects a history of selection against free recombination. So far naturally occurring inversion heterozygotes analogous to those found in *Drosophila* species have not been found in mice. Furthermore, the sex difference in crossing-over in mice is not as extreme as in *Drosophila*.

From this survey, it is clear that the response to the variables discussed must, in general, have been adjusted by natural selection, although the way in which some of the variables (e.g., age, temperature, and sex difference) affect recombination is still obscure.

B. Genotypic Variation

Levine and Levine (1955) studied crossing-over in the third chromosome of various strains of *D. pseudoobscura*. They found evidence for genotypic control of crossing-over. Furthermore they showed that the amount of crossing-over in the X-chromosome can be modified by different third chromosome inversion homozygotes and heterozygotes (Levine and Levine, 1954). Lawrence (1958) studied crossing-over in the X-chromosome of *D. melanogaster* Samarkand and Oregon inbred lines and their F_1's under four different temperature regimes. He found that Oregon showed consistently higher recombination than Samarkand. The F_1, while exceeding the mid-parent, nowhere significantly exceeded both parents. Furthermore, the regional sensitivity of crossing-over to a change in environment varied markedly with the temperature regime. His conclusion was that the genotype controls not only the gross amount of recombination, but also plays a part, with the cytoplasm, in determining the pattern of response of meiosis to change in environment. Green (1959) studied crossing-over as affected by wild type alleles at the white locus (w) of Canton and Oregon stocks and found that crossing-over was affected in the region y (yellow)-w^{ch}-spl (split) according to whether the right hand segment of the white locus was heterozygous Canton or Oregon. Similar results were obtained from a comparison between Crimea and Formosa stocks.

Thoday and Boam (1956) inbred lines heterozygous for the sex-linked loci w, m, and f (forked) by mating nonrecombinant females and recessive brothers in each generation. They found a decline in recombination frequency that seemed to be due in part to heritable cytoplasmic differences, picked out by the selection of eggs whose cytoplasm produces relatively low recombination. Hence, recombination frequencies are under the control of the nucleus and cytoplasm, as might be expected. Differences between recombination which can be specifically ascribed to the cytoplasm are, of course, more difficult to detect.

In *Neurospora*, too, recombination frequencies have been shown to depend in part on the genetic background (Fincham, 1951; Frost, 1961; Rifaat, 1959).

C. Changing Recombination by Selection

On the basis of the evidence in favor of some genetic control of recombination, it seems reasonable to expect that crossing-over can be altered by selection. Gowen (1919) was unable to alter recombination by selecting for low and high crossing-over using six chromosome III markers in *D. melanogaster*, which were spread along the whole chromosome. Detlefsen and Roberts (1921), however, selected for high and low crossing-over in *D. melanogaster* between the sex-linked mutants white (w) and miniature (m), and found selection to be effective in reducing crossing-over from just under 30% to 2% in one line and to 8% in another. In the former case, crossing-over dropped so rapidly that it seems that an inversion may have occurred. A further likely explanation of at least part of the responses is that non-disjunction of the chromosomes became comparatively frequent. A correlation between decrease of crossing-over and increase on non-disjunction was reported by Anderson (1929) and Dobzhansky (1933), and its significance as a measure of the association between efficient chromosome pairing and crossing-over was stressed by Mather (1938).

In the line in which recombination dropped to 8%, the major part of the decrease occurred in six consecutive generations of the twenty-nine generations of selection. This suggests the occurrence of an "accelerated response" to selection and by analogy with the chaeta selection experiments already discussed, the occurrence and selection of rare recombinant genotypes which themselves modify recombination. Detlefsen (1920) crossed this selected line back to the base stock and analyzed recombination in the F_1 and F_2. He suggested that the extreme variability found in the F_2 compared with the parents and F_1 was due to modifiers which had been selected during the course of the selection experiment.

Selection for increased crossing-over between the genes was, however, ineffective. The intensity of selection between two widely separated markers is not very great, and there is no strong cytological mechanism by which crossing-over can be increased except for inversion heterozygosity in a non-homologous chromosome. As mentioned above, it is well known that environmental and genetic variables affect crossing-over mainly in the centromeric region in *Drosophila* (see, e.g., Schultz and Redfield, 1951; Mather, 1939b; Bridges, 1915; Plough, 1917). For these reasons Parsons (1958e) chose to select for recombinants in the segment black (b)-purple (pr), which is close to the centromere of linkage group II in *D. melanogaster*. After nine generations the b-pr recombination fraction was increased from 4.93 to 7.38 ($P < 0.05$). In the linked segment pr-vg (vestigial), in which nonrecombinants were necessarily selected, the recombination fraction was reduced from 8.77 to 6.94, and over the whole segment b-vg there was little alteration. Thus the distribution of crossovers was probably altered.

Recently Mukherjee (1961) reduced crossing-over in the males of *D.*

ananassae between two second chromosome markers perhaps by picking up an inversion. Selection for high crossing-over, however, gave little response. In *D. melanogaster* Acton (1961) was unsuccessful in reducing recombination by selection between the markers *cn* (cinnabar)-*vg*, but presumably he did not pick up an inversion. In conclusion, crossing-over can be increased or decreased by selection, but it does not appear, *a priori*, possible to predict the type of response, if any.

D. Conclusions

In conclusion, recombination frequencies and hence chiasma frequencies are affected by differences of temperature, age, genetic background, and other variables. The responses to such variations are most probably adjusted by natural selection balancing the need to preserve linked complexes which are of immediate advantage to a species, and allowing recombination to produce the variability which is essential for the long-term flexibility to adapt to new environments (Mather, 1943a). All the variables discussed affect the recombination index, since they alter the frequency and distribution of chiasmata. Localized differences in recombination, caused for example by inversion heterozygosity, translocations, and other crossover suppressors or enhancers, will be more especially important with respect to the balanced chromosome, for these variables can alter recombination between specific interacting polygenes in the direction favored by natural selection. The occurrence of an inversion, and the subsequent tightening of linkage, is for example a common mechanism for building up balanced genotypes and polymorphic linked complexes.

VI. Linkage and Recombination in Microorganisms

A. Genetic Systems

In higher organisms the recombination of genetic material occurs almost exclusively by sexual reproduction. One of the most striking contributions of modern microbial genetics is the discovery of diverse alternatives to the sexual system for the promotion of genetic recombination. Genetic analysis in bacteria can now be done by the three processes of transduction, transformation, and the quasi-sexual mating system of *E. coli*. The work of Pontecorvo and his colleagues has uncovered a process in fungi leading to recombination via mitosis and this has been called the parasexual cycle.

The relation between these various systems has been discussed by Pontecorvo (1959). He pointed out that there are two major features of any system promoting recombination. The first is the necessity for transfer of genetic material and the second is the nature of the recombination which follows the transfer. Thus in transduction, transformation, and the *E. coli* mating system, only a part of the total genetic material is transferred, whereas in the parasexual cycle, as in the normal sexual cycle, the whole of the genetic material may take part in the recombination process. However, recombination and subsequent reduction are uncoordinated in the parasexual cycle and may occur independently and in different nuclei. It is also possible for sexual (meiotic) and parasexual (mitotic) recombination to occur side by side in the same organism. It must be emphasized that the linkage relationships obtained from these different recombination processes are by no means comparable. Where a comparison can be made, as in *Aspergillus*, between the sexual and asexual cycle (Pontecorvo, 1959), very different linkage maps can be given by two different processes. We cannot discuss these systems in any further detail here, but wish to emphasize that they must provide ample means for the occurrence in nature of recombination in microorganisms. A major difference in this respect between microorganisms and most higher organisms is that recombination in the former, whether by a sexual process or otherwise, probably occurs at infrequent intervals which are interspersed by periods of purely vegetative propagation. During these periods, the only means of evolutionary progress may be by selection between clones of mutations. Clearly, as in the higher organisms, recombination in microorganisms by whatever process must be the major factor providing new genetic combinations for the action of natural selection.

Most microorganisms are haploid under normal conditions. Without special mechanisms for association between haploid cells of different genetic constitution, this precludes the storage of potential genetic variability in heterozygotes, which has been demonstrated to be so widespread, at least in *Drosophila* species (Tschetwerikoff, 1926, and others). Syntrophism, the simplest example of an association between haploid cells, is a situation such that two types of cells with complementary nutritional requirements can supplement each other when growing close together. In filamentous fungi the phenomenon of heterokaryosis, by which nuclei of unlike genotype may be maintained in the mycelium within the same cell, provides a precise mechanism for the more or less permanent association of nuclei of different genotypes.

When the diverse genotypes so maintained are complementary in their nutritional requirements, heterokaryosis provides an efficient mechanism for the storage of variability which is comparable to dominance in a diploid organism. Following the pioneer work of Dodge (1942), Beadle and Coonradt (1944), and Pontecorvo (1946), the "balanced" heterokaryon has been a valuable tool for the analysis of biochemical mutants produced in the laboratory. Clear evidence for the natural occurrence of heterokaryosis was given by Hansen and Snyder (1943) and later by Jinks (1952). The latter was able to show that isolates of *Penicillium* obtained from the wild contained mixtures of two karyotypes, neither of which had as high a growth rate as the heterokaryotic combination. Three out of six isolates investigated turned out to be heterokaryotic. It should be mentioned that heterokaryosis is an essential part of the parasexual cycle, which therefore provides an almost complete substitute for sexual reproduction in a diploid organism.

Heterokaryosis in filamentous fungi is undoubtedly a most important mechanism both for the storage of potential variability and for the segregation and re-assortment of nuclear ratios in accord with the immediate environmental needs of the fungus. Microorganisms are likely to encounter rapidly changing environmental conditions in which the ability to adapt by the alteration of nuclear ratios in a heterokaryon may be at a considerable premium (see Pontecorvo, 1946). In a fungus the population unit for evolutionary purposes is undoubtedly the whole mycelial colony rather than a single cell or nucleus within the mycelium. This is, in a rather loose sense, the "organism" whose evolution must be considered. Its genotype is the sum total of nuclear genotypes it contains which may, in a heterokaryon, show some considerable diversity. Segregation of nuclear ratios to produce heterokaryotic colonies with differing nuclear constitutions is a genuine mechanism for genetic segregation when considered in terms of the colony. A similar situation may hold for the case of syntrophism between cells in a bacterial colony.

The advantages of heterokaryosis in a mycelial colony may well lie at the basis of the evolution of the fungi. The limitations of this process of segregation are that it cannot combine mutations of independent origin within the same nucleus, and so on its own lead to the building up of any sort of polygenic complexes, other than by a series of appropriate successive mutations. Thus in order to build up a balanced genotype it is probably essential to have also a means for conventional genetic recombination.

Pontecorvo (1959) has estimated that in *Aspergillus nidulans*, which has both a sexual and a parasexual cycle, the amount of recombination occurring through the sexual cycle is about 500 times that occurring through the parasexual cycle. However, from evidence from studies of the parasexual cycle in asexual species he suggests that it might be more active in the production of recombinants in those species where there is no sexual cycle and may then be able to provide a level of recombination comparable with that provided by the sexual cycle. In fact Pontecorvo

(1959) even suggests that this higher activity of the parasexual in the absence of the sexual cycle may reflect an evolutionary balance between the two systems.

The arguments for the need to store potential genetic variability are as cogent for microorganisms as for any other organisms. The only mechanism capable of concealing such variability in the absence of widespread heterozygosis or heterokaryosis is the balanced polygenic complex. For such a haploid organism, these complexes must show internal balance, as do inbred species. Where a recombination process occurs in conjunction with some form of heterokaryosis, one might also expect the development of relational balance between the different nuclei contained within the same colony. There is no doubt, therefore, that both types of balanced complexes may be expected to occur in all species of microorganisms.

B. Polygenic Variation

Most of the variation that has been studied in microorganisms is not polygenic, being derived from clear-cut biochemical or "resistance" mutants. There is, however, a fair amount of work which indicates the existence of genes of minor effect in microorganisms. The phenomenon of "leaky" mutants is well known (see Bonner et al., 1952) and some work on modifiers affecting such a system has been done (Haskins and Mitchell, 1952). In only very few cases has modifying action (as opposed to complete suppression) been analyzed specifically from a genetic or a biochemical point of view. Eberhart and Tatum (1959) showed that a modifier of the thiamine-1 locus in Neurospora probably occurred at the thiamine-4 locus.

Markert (1950) and Markert and Owen (1954) have shown that melanin production in Glomerella is associated with a number of independent loci, most of which seem to have some effect on the activity of the enzyme tyrosinase. A similar situation is shown to occur in Neurospora by recent work of Horowitz et al. (1960) here one locus is thought to control the specificity of the enzyme and two other loci have been found which affect the inducibility of the enzyme. Differences in tyrosinase activity have also been claimed in association with wild type strains differing only at the mating type loci (Fox and Gray, 1950), but these are probably connected with the function of tyrosinase in perithecium development (Hirsch, 1954). Certainly variability in melanin pigments is notoriously ubiquitous. In Drosophila, for example, pigment formation is probably controlled by many loci, most of which have some effect on tyrosinase activity (Parsons, unpublished).

Probably the first specific suggestion of a polygenic system in microorganisms is Demerec's (1945) analysis of multi-step resistance to penicillin in *Staphylococcus*. He was able to show that the gradual acquisition of resistance to higher concentrations of penicillin was almost certainly due to the selection of resistant mutants. The final observed resistance was caused by a number of genetic factors which had been accumulated by selection.

A more detailed genetic analysis of a similar pattern of resistance to the antibiotic Chloromycetin in *E. coli* has been done by Cavalli and Maccacaro (1952). They produced strains highly resistant to Chloromycetin by the process of "training," which involves serial subcultures on successively higher concentrations of the antibiotic. Crosses of resistant strains carrying marker genes showed segregation of resistance to different concentrations of Chloromycetin, and linkage of the genes conferring resistance with the marker genes. Crosses between resistant strains gave recombinants with much lower resistance than the parents, which indicated some non-additive interactions between genes in the determination of resistance. Sparse growth was generally associated with high resistance. Sensitives from crosses between resistant strains still grew sparsely, providing further indication of gene interaction and also demonstrating the occurrence of a correlated response as discussed by Mather and Wigan (1942). "Single step" resistants can be obtained by plating a large number of bacteria on agar containing an appropriate inhibitory amount of Chloromycetin. The resistance of such strains is presumably due to a single mutation. Crosses between fifteen such mutants demonstrated the existence of at least four or five different loci. Some of these crosses produced progeny which were twice as resistant as either parent, demonstrating that single step mutants can be combined to give a higher level of resistance. Cavalli and Maccacaro (1952) suggest that the build-up of resistance involves the accumulation of successively interacting factors. Recombination will break down the interacting system and cause a severe reduction in resistance to the antibiotic. These studies, which have been largely overlooked, are a remarkable testimony to the possibilities for polygenic analysis in microorganisms. They suffered, however, from the lack of knowledge at that time concerning the mating system in *E. coli*.

Studies with host range mutants in phage by Baylor *et al.* (1957) and by Jinks (1961) indicate that these may in certain circumstances behave as polygenic systems. Combinations of mutants are shown to have cumulative effects and the mutants may map over a considerable region of the phage chromosome. However, as pointed out above, it is difficult to know how to correlate genetic distance in phage with genetic distance in higher organisms. Certainly host range is a character likely to be affected by many biochemical and physiological variables which may account for the polygenic nature of its inheritance. So far these genetic studies have not been combined with a biochemical analysis of the functions of the various mutants.

Pateman (1955, 1959), Lee and Pateman (1959, 1961), and Pateman and Lee (1960) have studied the inheritance of ascospore size in *Neurospora crassa*. They have shown that ascospore size is under polygenic control, demonstrated linkage of polygenes and marker genes, and also interaction between polygenes. Tetrad analysis in *Neurospora* provides a most powerful extra tool for the study of polygenic variation. Pateman and Lee (1960) have provided the first conclusive demonstration of regular segregation of polygenes in ordered tetrads. However, the character they studied suffers from the major disadvantage that it can only be measured in crosses, which involves extensive progeny testing to standard reference stocks.

For a comprehensive polygenic analysis aimed at studying the mode of action of individual polygenes it is necessary to have a simple character which can be measured accurately with little trouble, and whose physiological basis is amenable to biochemical analysis. The variation associated with a mutation causing an alteration in an enzyme which leads to a requirement for a particular supplement is an obvious candidate for such a study. The definition of a polygene (or modifier) in this context simply depends on the resolution of the available phenotypic analysis. Ultimately, with a combination of suitably refined biochemical and biometrical techniques, any polygenic segregation should, in principle, be resolvable into a series of single factor segregations with apparently major effects. There is clearly more hope of achieving this sort of resolution with a comparatively simple and well-defined biochemical system.

Preliminary results have been obtained from an analysis of modifiers affecting a proline requiring mutant in *Neurospora crassa* (Bodmer, unpublished). This mutant is known to cause an alteration in the enzyme responsible for the last step in proline biosynthesis. Variation between strains in the ability to grow on selective media is measured with the aid of a simple device for measuring colony size. The proline mutant was introduced into a wild type by successive backcrossing, and the variation that has been studied is that produced by backcrosses of the wild type carrying the proline mutant to derivatives of the original mutant strain. Three different conditions of growth were used, namely, ability to grow at 25°C in the absence of proline and in the presence of small amounts of proline, and at 35°C with limiting amounts of proline.

Specific modifiers affecting all three conditions of growth have been found and shown to segregate in tetrads. Of particular interest is one modifier which *reduces* the amount of growth with small amounts of proline. This modifier may well be concerned with control of enzyme formation through a repressor system such as has been studied in *E. coli* by Pardee *et al.* (1959). The enzyme concerned in the proline system is temperature sensitive and two of the modifiers studied affect this temperature sensitivity, inhibiting growth at 35°C. One of these modifiers has been shown to occur in a wild type appearing in the ancestry of the derived proline strains used in the original backcrosses. Three of the modifiers are probably linked to the mating type locus and two closely linked modifiers occur on another chromosome. Using appropriate biochemical markers it is possible to force recombination in the region of the mating type locus and demonstrate the production of unusual combinations of characters by recombination. The two closely linked modifiers produce recombinant "high" types in combination with a third factor with a frequency of about 1 to 5%. They are able to grow appreciably in the absence of proline. Evidently we have here a linked balanced complex of modifiers affecting this one simple biochemical character. The mode of action of the modifiers and their particular interactions are amenable to biochemical analysis. It seems likely that systematic polygenic analyses along these lines will reveal a type of genetic variation qualitatively different from most of the classical biochemical mutants, which are specifically selected for their clear-cut effects. Moreover, such studies should provide a model on which to base our ideas on the mode of action of polygenes and the relationship between their genetic and biochemical interactions.

C. Linkage between Related Biochemical Loci

The occurrence of cases of close linkage between loci with related phenotypic effect in *Drosophila* was pointed out some time ago by Grüneberg (1937) and his observations were extended to other organisms by Komai (1950). Their examples are now somewhat confused by modern knowledge of genetic fine structure, a topic which is not in the scope of this review and which has been recently reviewed by Pontecorvo (1959). Pontecorvo (1950) and Catcheside (1951) first suggested that there might be linkage between genes controlling successive steps in the biosynthesis of certain types of metabolites. This view is now untenable as a general hypothesis, but has gained remarkable support in certain cases, from the work of Demerec and his colleagues on *Salmonella typhimurium*. They have found a number of examples of very closely linked loci which occur sequentially and adjacent to each other in the order of the biochemical reactions they control (see Demerec and Hartman, 1959). For example, four loci, A, B, C, D, affecting tryptophan synthesis are known. The sequence of the biochemical steps they control is:

$$\begin{array}{cccc} \twoheadrightarrow \text{anthranilic} & \twoheadrightarrow \text{indole-glycerol} & \rightarrow \text{indole} & \rightarrow \text{tryptophan} \\ A \quad \text{acid} & B \quad \text{phosphate} & \underbrace{\phantom{\text{indole}}}_{C \text{ and } D} & \end{array}$$

where C and D determine the specifity of the two components of the enzyme tryptophan synthetase (see Yanofsky, 1960). These four loci are very closely linked, probably adjacent, in the order of their biochemical reactions. Such a cluster of four loci in biochemical sequence has also been shown to exist for tryptophan biosynthesis in *E. coli* (see Yanofsky, 1960), and more recently in *B. subtilis* (Anagnostopoulos and Crawford, 1961). However, the same steps in tryptophan biosynthesis are controlled by four *unlinked* loci in *Neurospora crassa*. There is clearly a fundamental difference here between *Neurospora* and the bacteria. Presumably this difference must have some sort of evolutionary basis.

Many cases of clusters are now known in *Salmonella* and some in *E. coli*. However, in both these organisms cases are also known of sequential biochemical steps which are controlled by loosely linked loci. There are also several cases suggestive of clustering in the fungi. The best established of these is the work of Gross and Fein (1960) on mutants in *Neurospora* affecting the synthesis of aromatic compounds. They found two closely linked (probably adjacent) mutants on chromosome II affecting enzymes involved in successive biochemical reactions. A very closely linked aberration possibly overlapping these mutants was found to affect simultaneously these same two reactions, and also a further two reactions immediately preceding them. It thus seems likely that at least four successive enzymatic steps are controlled by closely linked loci in this genetic region. A further locus controlling a later biosynthetic step occurs on the same chromosome at a distance of 10–15 units from this cluster.

In Table 3 a summary is given, from published sources, of data on the linkage relationship between loci controlling related biochemical steps in various microorganisms. In cases where the biochemistry has not been worked out in sufficient detail, a common biochemical requirement is taken as the criterion for relationship. The linkage relations are classified broadly as clustered, linked, loosely linked, and unlinked with the reservation set out at the foot of the table. Out of thirteen systems studied in *Salmonella*, ten or possibly eleven show evidence of clustering, whereas only five out of sixteen in *Neurospora* show evidence of clustering. *E. coli* also shows a high frequency of clustering: three well-established cases out of six investigated. There are some interesting similarities and differences. The case of the tryptophan loci has already been discussed.

TABLE 3

Linkage between Related Biochemical Loci in Micro-organisms:
Details of Number of Loci Clustered, Linked, Loosely Linked, and Unlinked*†

Biochemical requirement	Salmonella typhimurium	Escherichia coli	Neurospora crassa	Aspergillus nidulans	Other organisms
Tryptophan	Cluster of 4 in biochemical sequence	Cluster of 4, as in Salmonella (some evidence for linked "constitutive" locus)	4 unlinked	—	Bacillus subtilis cluster of 4 as in Salmonella
Histidine	Cluster of 6 in biochemical sequence (some evidence for "constitutive" locus and linked "operator" locus)	Only 1 mapped	2 linked, 3 unlinked	—	B. Subtilis Preliminary evidence suggests cluster of at least 3. Streptomyces coelicolor 2 unlinked. Schizosaccharomyces pombe 3 linked, 3 or more unlinked
Methionine	2 linked (?cluster) (out of biochemical sequence), 3 unlinked	Cluster (?) of 2 (same as in Salmonella) 1 loosely linked	3 linked, 2 linked, 3 unlinked	2 unlinked	Streptomyces coelicolor 2 loosely linked(?), 2 unlinked
Adenine	2 linked (one also requires thiamine) with further linked locus for guanine, 2 linked (require thiamine), 4 unlinked (1 requires thiamine)	3 loosely linked	2 loosely linked, 2 linked, 4 unlinked	Cluster(?) of 2, 3 loosely linked, 1 unlinked	Saccharomyces cluster of 2, 2 loosely linked, 2 unlinked

Proline	*Cluster* of 3 (or 4) in biochemical sequence	Only 1 mapped	Only 1 mapped; 3(?) *linked* modifiers, 2 *linked* modifiers	—
Pantothenic acid	*Cluster* of 3	Only 1 mapped	Only 1 mapped	—
Isoleucine/valine	*Cluster* of 5 (1 isoleucine and valine), probably in biochemical sequence	2 *linked* (1 isoleucine, 1 isoleucine and valine)	*Cluster*(?) of 3 (2 isoleucine and valine, 1 valine)	—
Threonine	*Cluster* of 4 in biochemical sequence	Only 1 mapped	*Cluster*(?) of 2, 1 *unlinked*	—
Ornithine, citrulline, and arginine	3 *linked*, 4 *loosely linked*	Only 1 mapped	*Cluster*(?) of 2, *cluster*(?) of 2, 4 *unlinked*	3 *unlinked*
Cystine	*Cluster* of 2 (adjacent biochemical steps), 3 *loosely linked*	—	*Cluster* of 2	Only 1 mapped
Aromatic compounds	*Cluster*(?) of 2, 3 "*unlinked*"	—	*Cluster* of 2, also evidence from deletion (or "operator" gene?) for at least 2 more in same cluster. 1 locus *linked* to cluster. 2 *linked*, 2 *unlinked*	—
Leucine	*Cluster* of 1 and a suppressor	Only 1 mapped	2 *loosely linked*, 2 *loosely linked*, 1 *unlinked*	Only 1 mapped
Lysine	—	Only 1 mapped	2 *loosely linked*, 2 *loosely linked*, 1 *unlinked*	2 *unlinked*
Pyrimidine	—	Only 1 mapped(?)	3 *loosely linked*, 2 *loosely linked*, 1 *unlinked*	—
Nicotinic acid	Only 1 known (requires thiamine and nicotinic acid)	Only 1 mapped		3 *unlinked*

TABLE 3 (Continued)
Details of Number of Loci Clustered, Linked, Loosely Linked, and Unlinked*†

Biochemical requirement	Salmonella typhimurium	Escherichia coli	Neurospora crassa	Aspergillus nidulans	Other organisms
Thiamine	See adenine and nicotinic acid	—	2 loosely linked, 2 unlinked	—	—
Inability to utilize lactose	Only 1 mapped (probably homologous with E. coli)	Cluster of 4: 2 structural, 1 "constitutive", and 1 "operator" (Jacob et al., 1960—Operon)	—	2 unlinked	—
Inability to utilize galactose	Only 1 mapped	Cluster of 5 or more (1 may be either "constitutive" or "operator")	—	4 or 5 unlinked	—
Inability to utilize arabinose	Cluster(?) of 2	Cluster of 3, 1 loosely linked	—	—	—
p-Aminobenzoic acid	—	—	2 linked	Only 1 mapped	—
Riboflavin	—	—	—	3 unlinked	—

* "*Cluster*" refers to cases where loci are adjacent. When queried the evidence is not conclusive, but comes from very close linkage with no intervening loci but good evidence against "allelism."
  Cases of "*unlinked*" in *Salmonella* are those in which no linkage has been detected in transduction experiments, and loci have not been mapped by hybridization with *E. coli*. Otherwise "*unlinked*" means no linkage with any other loci having similar requirements.
  "*Loosely linked*" refers to sets of loci extending over more than about 15–20 map units, and "*linked*" to sets lying within this range. Separate mention of sets of loci means that they occur on different chromosomes.
  An attempt has been made to include all published data on systems with 2 loci or more in one organism and at least one locus in another organism.

† Sources of data—*Salmonella*: Demerec et al. (1956, 1957, 1958, 1960); Demerec and Hartman (1959); Hartman et al. (1960b); Hartman et al. (1960a); Smith (1961); Ames et al. (1960); Dawson and Smith-Keary (1960). *E. coli*: Gross and Englesberg (1959); Pardee et al. (1959); Jacob et al. (1960); Lederberg (1960); Yanofsky (1960). *Neurospora*: Barratt et al. (1954); Perkins (1959); Barratt et al. (1961); Bodmer (unpublished); Gross and Fein (1960); Murray (1960). *Aspergillus*: Pontecorvo (1959); Forbes

Histidine mutants, another well-studied category, show a similar disparity between *Salmonella* and *B. subtilis* as compared with *Neurospora*. However, the mycobacterium *Streptomyces coelicolor* shows no evidence of clustering. It is notable that all but one of the cases of clustering in *Neurospora* are also associated with clustering in *Salmonella*. In spite of the likely effect of transduction in biasing, to some extent, observed linkage relationships, these observations undoubtedly reflect a major feature of the organization of the genetic material.

More difficult to assess than the clustering is whether there is any tendency for "more than random" linkage between loci which are not clustered. The difficulties inherent in such an investigation have been emphasized by Pontecorvo (1959) in his stimulating discussion of this problem. The biochemical mutants obtained by the usual procedures are a highly selected group and may, in any case, occur more frequently in regions of chromosome with higher than average mutability. Nevertheless, there is a definite suggestion that such nonrandom linkages may occur, as for example the modifiers of a proline locus in *Neurospora* (Section VI,B—Bodmer, unpublished). Murray (1960) has shown that three out of the eight methionine loci which have been mapped in *Neurospora* occur within a region of 10–15 map units, though not in a cluster. It is possible to calculate the probability with which a given number of loci will be distributed among a given number of chromosomes, if one makes the simplifying assumption that loci are equally likely to occur independently and at random on any of the chromosomes. These assumptions though oversimplified provide at least a crude measure for comparison. Thus, for example, the probability that the eight methionine loci of *Neurospora* will be distributed in one linked group of three, one linked group of two, and three unlinked loci is 0.245. On this basis alone, there is no reason whatsoever to suspect this distribution of loci as showing more than a "random amount" of linkage. However, the fact that the three loci which are linked occur so close together certainly decreases the probability. For the proline modifiers, three linked and two linked loci give a significance level of about 3.7%. Thus even without the evidence for close linkage, the distribution of the proline modifiers is certainly suggestive of "more than average" linkage.

There is clearly a major qualitative difference between clustering of loci, as occurs in *Salmonella*, and the sort of close linkage discussed above. Although clustering is not exclusive in *Salmonella*, and clusters probably do occur in Neurospora, there is certainly a basic difference between these two species in this respect. It has been suggested (see Thoday in discussion on Demerec and Demerec, 1956) that this difference may be a reflection of the fact that bacteria have no nuclear membrane and show no clear distinction between nucleus and cytoplasm, or, in other words, a reflection of a comparatively low level of organization within the bacterial cell. The analogs to the clusters of *Salmonella* may therefore be found in cytoplasmic structures in higher organisms. The clustering must clearly have a direct physiological advantage, whether occurring in the chromosomes or in a cytoplasmic equivalent. Close linkage, however, such as occurs in *Neurospora* for the proline modifiers, is much more likely to reflect an evolutionary molding of the genetic material along the lines of Mather's (1943a) concepts of the evolution of balanced polygenic complexes. The intervention of a nuclear membrane and regular mitosis and meiosis has relaxed the physiological necessity of clustering of loci on the chromosome and, perhaps, allowed the delegation of this to some cytoplasmic equivalent structure (ribosomes or messenger RNA?). This has allowed the reorganization of the genetic material according to the more long-term evolutionary needs outlined by Darlington (1958) and Mather (1943a).

A probable explanation for the integration of clusters of loci such as occur in *Salmonella* and *E. coli* is given by the remarkable work of Pardee et al. (1959) and Jacob et al. (1960) on the β-galactosidase system in *E. coli* (for a recent review see Jacob and Monod, 1961). Two very closely linked loci z and y control the structure of the inducible enzymes β-galactosidase, and galactoside-permease which is responsible for allowing the entry of galactosides into the cell. A further closely linked locus i is such that i^- mutants convert both of the inducible enzymes into the constitutive state, in which their synthesis does not need to be initiated by the action of an inducer. Pardee et al. (1959) have shown that this locus i most probably functions by producing a repressor which generally inhibits enzyme synthesis and whose inhibition is counteracted by the inducer. Constitutive mutants i^- have lost the capacity to synthesize active repressor. (Genes such as the i gene have been called "regulator" genes.) Jacob et al. (1960) have discovered a further class of mutants which convert the enzymes into their constitutive state. The cardinal feature of these mutants is that they only affect the enzymes or proteins controlled by y and z mutants in the *coupling arrangement*, when carried in partial heterozygotes for this region of the *E. coli* chromosome. These mutants have been designated o for "operator" and are closely linked to y, z, and i, the order being y-z-o-i. Thus if o^+ represents the normal condition and o^c the mutant constitutive condition, heterozygotes $i^+o^+z^-y^-/i^+o^cz^+y^+$ will allow constitutive production of both enzymes, whereas heterozygotes $i^+o^+z^+y^+/i^+o^cz^-y^-$ will not. This is a remarkable type of position effect *between loci*.

It must be emphasized that the term locus here corresponds, more or less, to Benzer's (1957) cistron, and not to the mutable site (muton or recon in Benzer's terminology). Thus the position effect that has been demonstrated by Jacob et al. (1960) is not the same as a position effect between mutations which have occurred within the same cistron. If the operator loci were not so closely linked to the structural loci in higher organisms as they seem to be in microorganisms, then these might easily be associated with position effects at a distance.

Jacob et al. (1960) suggest that the repressor synthesized by the regulator locus acts directly on the operator locus. They call such an integrated group of genes an "operon." It seems likely that most cases of clusters are operons. Preliminary evidence for "operator" loci lies in the occurrence of single site mutants or restricted deletions which affect the activity of more than one enzyme of a cluster. Such evidence exists for the histidine system in *Salmonella* (Ames et al., 1960) and the galactose cluster in *E. coli* (Lederberg, 1960). The aberration described by Gross and Fein (1960) affecting four of the enzymes in the aromatic pathway of *Neurospora* may also be associated with an operator locus. Several cases are now known where regulator loci are not linked to their respective structural loci (see Jacob and Monod, 1961). The "operon" concept, however, as defined by Jacob et al. (1960), requires that the operator locus be closely linked to the structural loci. It remains to be seen whether the concept holds in this form for higher organisms.

The further analysis of such systems in microorganisms and higher organisms combined with a study of their comparative biochemical genetics not only opens up an exciting prospect for the integration of biochemical and developmental genetics, but also clearly brings us up against the possibility of integrating biochemical and population genetics since linkage is involved in both.

VII. Theoretical Genetics of the Linked Gene Complex

A. Introduction

The Hardy-Weinberg law shows that for two alleles A and a at a single locus, the expected frequencies of the three genotypes AA, Aa, and aa after one generation of random mating are given by p^2, $2pq$ and q^2, respectively, where p and q $(p + q = 1)$ are the frequencies of alleles A and a. When there are no differences in fitness between the three genotypes, equilibrium is reached after only one generation of random mating. For the case of two loci each with two alleles, say A, a and B, b, it is no longer true that, without selective differences between genotypes, equilibrium is reached after one generation of random mating. The general equations for this situation were given by Jennings (1917) and references to other early literature are given by Geiringer (1944). In this case the frequencies of the gametes AB, Ab, aB, and ab gradually approach equilibrium values given by the product of the frequencies of the two alleles contained in a gamete. The rate of approach to the equilibrium is given by r, the recombination fraction between the two loci. Thus if p_1, q_1 and p_2, q_2 represent the equilibrium frequencies of genes A, a and B, b, respectively (where $p_1 + q_1 = 1 = p_2 + q_2$), the equilibrium frequencies of gametes AB, Ab, aB, and ab, are p_1p_2, p_1q_2, q_1p_2, and q_1q_2, respectively. We shall call such an equilibrium a "gene frequency equilibrium." As in the situation for only one locus, such a gene frequency equilibrium no longer exists, in general, when there is any departure from random mating. The case of mixed selfing and random mating has been considered by Bennett and Binet (1956).

When there are selective differences between genotypes, it is also no longer true that random mating necessarily leads to a gene frequency equilibrium. The first consideration of the way in which selection may affect the equilibrium gametic frequencies is due to Fisher (1930a). He suggested a model of interacting selective values, as described in Section I, which would, in general, lead to an excess of either coupling (AB and ab) or repulsion (Ab and aB) gametes over the expected frequency given by a gene frequency equilibrium. When natural selection favors, say, repulsion gametes as opposed to coupling gametes, this will lead to selection for reduced recombination between the two loci. Mather's theory of balanced polygenic complexes provides a more comprehensive and elaborate verbal consideration of the relationship between interacting selective values and recombination.

The first work on a mathematical model for the changes in frequency of the four gametes formed by two alleles at each of two loci, which takes account of selective differences between genotypes, is due to Wright (1952). He assumed a simplified model of viabilities based on Mather's concept of balance. Under his assumptions, the recombination fraction must be of the order of the selective differences between the genotypes for linkage to cause an appreciable departure from a gene frequency equilibrium. It should be noted that the characteristic of a gene frequency equilibrium is that the frequencies of coupling and repulsion heterozygotes are equal. Departure from a gene frequency equilibrium leads to an excess, at equilibrium, of either coupling or repulsion heterozygotes.

Kimura (1956) considered a model, suggested by Sheppard (1955—see also Section III, C) and based on Fisher's (1930a) discussion, in which there is selection for a locus linked to one that is maintained in a polymorphic equilibrium. He showed that with interacting viabilities of the sort proposed by Fisher, the second locus will remain polymorphic only if it is sufficiently closely linked to the already polymorphic locus. He also pointed out that under these conditions, any inversion or crossover reducing mechanism bringing the two loci closer together will be favored by natural selection. He derived the general equations for changes in gametic frequencies with any selective values, using a continuous time model. Exactly analogous equations for a discrete time model are given by Lewontin and Kojima (1961). They consider the equilibrium conditions for some simple symmetric sets of viabilities and come to the conclusion that linkage is only important when there are viability interactions. The magnitude of linkage required to cause departure from gene frequency equilibrium (linkage equilibrium in their terminology) is of the order of what they call the "epistatic deviations" in selective value. Their viability models, however, are not intended to bear any relation to the type of viability interactions considered by Fisher and Mather.

There are three well-defined questions for which it seems reasonable to expect theory to provide some sort of an answer. The first is the one which has been considered with certain simplifying assumptions as described above, by Wright (1952), Kimura (1956), and Lewontin and Kojima (1961); namely, under what conditions will linkage cause an equilibrium which departs appreciably from gene frequency equilibrium or, equivalently, cause appreciable difference in frequency, at equilibrium, between coupling and repulsion heterozygotes. The second question considered to some extent by Kimura (1956) is that of the conditions under which selection will favor closer linkage between two loci. The third and by no means least important question is that of the initial conditions for the establishment of the various gametic genotypes.

In the rest of this section we shall outline some theoretical work done by one of us (Bodmer) in an attempt to answer these questions under somewhat more general conditions than have been considered by other workers.

B. Formulation of the Two Locus Theory

There are four gametic genotypes which can be formed with two alleles at each of two loci, A, a and B, b, namely, coupling gametes AB, ab and repulsion gametes Ab and aB. If the loci affect a quantitative character giving optimum fitness at intermediate values and alleles A and B act to increase the quantitative character whereas a and b act to decrease it, then in Mather's terminology the repulsion gametes will be balanced and coupling gametes will be unbalanced. Ten zygotic genotypes can be formed from these four gametic genotypes. They fall into three categories:

(a) Four genotypes homozygous at both loci of which two are homozygous for the balanced gametes and two for the unbalanced gametes.

(b) Four genotypes heterozygous at a single locus of which two carry only one of the alleles A and B, whereas the other two carry three of these alleles.

(c) The two double heterozygotes, distinguishing coupling and repulsion arrangements. We shall designate the four gametes AB, ab, Ab, aB by suffices 1, 2, 3, and 4, respectively, so that x_2, for example, is the frequency of gamete $2 \equiv ab$ and a_{23} is the relative viability of the zygotic genotype carrying gametes 2 and 3, i.e., ab/Ab. The notation for designating the frequencies and viabilities of the various genotypes is illustrated in Table 4.

TABLE 4
General Scheme of Genotypes, Their Frequencies, and Viabilities for the Two Locus Theory

	Gametes			
	Unbalanced		Balanced	
	AB	ab	Ab	aB
Frequencies:	x_1	x_2	x_3	x_4 ($x_1 + x_2 + x_3 + x_4 = 1$)

Zygotic genotypes

									Double heterozygotes:	
	Homozygotes				Single heterozygotes				Coupling	Repulsion
	$\frac{AB}{AB}$	$\frac{ab}{ab}$	$\frac{Ab}{Ab}$	$\frac{aB}{aB}$	$\frac{AB}{Ab}$	$\frac{AB}{aB}$	$\frac{ab}{Ab}$	$\frac{ab}{aB}$	$\frac{AB}{ab}$	$\frac{Ab}{aB}$
Viabilities:	a_{11}	a_{22}	a_{33}	a_{44}	a_{13}	a_{14}	a_{23}	a_{24}	a_{12}	a_{34}

It is possible to show under quite general conditions that random mating is equivalent to random union of gametes, so that, for example, the frequency of genotype $12 \equiv AB/ab_2$ before selection is $2x_1x_2$. It is this simple almost obvious fact which makes it possible to consider changes in zygotic genotype frequency entirely in terms of the gametic frequencies. As soon as there is any departure from random mating, this is, in general, no longer true. The expected gametic frequencies in one generation can be obtained in terms of those in the previous generation by considering the frequency with which each genotype produces any particular gamete and combining the results from all the genotypes. Thus, for example, the genotype Ab/aB occurs with frequency $2x_3x_4a_{34}$ after selection and has a gametic output

$$\tfrac{1}{2}rAB + \tfrac{1}{2}rab + \tfrac{1}{2}(1-r)Ab + \tfrac{1}{2}(1-r)aB$$

where r is the recombination fraction between the two loci. It therefore contributes amounts $ra_{34}x_3x_4$ to the frequency of gametes AB and ab in the next generation and $(1-r)a_{34}x_3x_4$ to the frequency of gametes Ab and aB. We thus obtain the following equations for the frequencies x_1', etc., in one generation in terms of the frequencies x_1, etc., in the previous generation:

$$\begin{aligned} Tx_1' &= a_{11}x_1{}^2 + a_{13}x_1x_3 + a_{14}x_1x_4 + (1-r)a_{12}x_1x_2 + ra_{34}x_3x_4 \\ Tx_2' &= a_{22}x_2{}^2 + a_{23}x_2x_3 + a_{24}x_2x_4 + (1-r)a_{12}x_1x_2 + ra_{34}x_3x_4 \\ Tx_3' &= a_{33}x_3{}^2 + a_{13}x_1x_3 + a_{23}x_2x_3 + ra_{12}x_1x_2 + (1-r)a_{34}x_3x_4 \\ Tx_4' &= a_{44}x_4{}^2 + a_{14}x_1x_4 + a_{24}x_2x_4 + ra_{12}x_1x_2 + (1-r)a_{34}x_3x_4 \end{aligned} \quad (1)$$

Here $x_1 + x_2 + x_3 + x_4 = 1$ and T is such that $x_1' + x_2' + x_3' + x_4' = 1$ and is the mean population fitness. If we write:

$$a_1 = a_{11}x_1 + a_{12}x_2 + a_{13}x_3 + a_{14}x_4, \text{ etc.}$$

so that a_1, a_2, a_3, and a_4 are, respectively, the average selective values of the gametes AB, ab, Ab, and aB for given gametic frequencies, then Equations (1) can be rewritten in the form:

$$\begin{aligned} Tx_1' &= a_1x_1 + r(a_{34}x_3x_4 - a_{12}x_1x_2) \\ Tx_2' &= a_2x_2 + r(a_{34}x_3x_4 - a_{12}x_1x_2) \\ Tx_3' &= a_3x_3 - r(a_{34}x_3x_4 - a_{12}x_1x_2) \\ Tx_4' &= a_4x_4 - r(a_{34}x_3x_4 - a_{12}x_1x_2) \end{aligned} \quad (2)$$

These are the equations which form the basis of most of our further calculations. They are analogous to Equations (1) given by Kimura (1956), and equivalent to Equations 11.a–11.d given by Lewontin and Kojima (1961).

When $r = 0$ and there is complete linkage between the two loci, the equations are equivalent to those obtained for the changes in frequency of four alleles at a single locus, as might be expected. In general, equilibrium frequencies are given by putting $x_1' = x_1$, etc. When these are such that

$$a_{34}x_3x_4 - a_{12}x_1x_2 = 0,$$

they are identical with the equilibrium frequencies for four alleles at a single locus ($r = 0$) under similar conditions.

As pointed out above, a gene frequency equilibrium, which is obtained for all values of r when all the viabilities a_{12}, etc., are unity so that there are no selective differences between the genotypes, implies that the equilibrium frequencies of coupling and repulsion heterozygotes are equal, i.e., that $x_1x_2 = x_3x_4$. This condition, together with the fact that $x_1 + x_2 + x_3 + x_4 = 1$, is sufficient to enable the gametic genotype frequencies to be expressed in terms of products of the appropriate gene frequencies, where for example $x_1 + x_3$ is the frequency of A.

When $a_{12} = a_{34}$, i.e., coupling and repulsion heterozygotes have equal viabilities, the gene frequency equilibrium is given by the equations:

$$T = a_1 = a_2 = a_3 = a_4,$$
$$x_1 x_2 = x_3 x_4 \quad \text{and} \quad x_1 + x_2 + x_3 + x_4 = 1.$$

Since there are five equations for four unknowns, the existence of a gene frequency equilibrium implies a restriction on the viabilities which can easily be derived but will not be given here. This imposes a limitation on the description of zygotic genotype frequency changes only in terms of gene frequencies. It is probable that many configurations of viabilities will allow an equilibrium which departs only little from a gene frequency equilibrium (a quasi-gene frequency equilibrium). The stability of such equilibriums will depend on the value of the recombination fraction in relation to the viabilities. When there are considerable viability interactions between the loci, as noted by Wright (1952) and Lewontin and Kojima (1961), there may not be a stable quasi-gene frequency equilibrium. In such a case the description of the genetic constitution of a population must be in terms of *gametic* frequencies. The specification only of gene frequencies is no longer adequate. It is such considerations which lie at the basis of Fisher's criticism (see Fisher, 1941) of Wright's concept of adaptive peaks and valleys (see Dobzhansky, 1951; Wright, 1940).

A general analytic solution of Equations (2) does not seem to be practicable. We shall therefore consider approximate solutions under various simplifying assumptions and then exact solutions for appropriate simple sets of viabilities.

C. Initial Progress of New Gametic Combinations

1. *Introduction*

The classical approach to the analysis of theoretical population genetic models is to determine their non-trivial equilibrium points and ascertain under what conditions these are stable. However, the conditions for the establishment of a new gene (or gamete) may often be of more importance in the evolutionary history of a population than the conditions for a stable equilibrium. By assuming that the initial frequency of the new gamete is small, it is possible to obtain linear approximations for a general set of equations such as (2) above. From these it is easy to obtain general conditions in terms of the viabilities and the recombination fraction under which the new gamete will increase. This approach to the study of population genetic models was developed by Bodmer (1960b) in the study of the population genetics of homostyle primroses and has been extended to a number of other problems (Bodmer, 1960c; Bodmer and Parsons, 1960; Parsons and Bodmer, 1961; Parsons, 1961b).

2. *Increase of a Gene Linked to a Polymorphic Locus*

Following Kimura (1956), we consider first the case in which a locus A-a is kept polymorphic by overdominance of the heterozygote Aa while the allele b is predominant at a second locus. We wish to determine the conditions under which a mutation B at the second locus will increase in frequency. We assume that, before the introduction of B, genotypes ab/ab, Ab/Ab, and Ab/ab with viabilities $a_{22} = 1 - \alpha$, $a_{33} = 1 - \beta$, and $a_{23} = 1$, were at an equilibrium with frequency of $ab = x_2 = p = \beta/\alpha + \beta$ and the frequency of $Ab = x_3 = q = 1 - p = \alpha/\alpha + \beta$ (see, e.g., Fisher, 1922). After the introduction of B, the gametic frequencies for AB, ab, Ab, and aB are x_1, $x_2 = p - X_2$, $x_3 = q - X_3$, and x_4, where x_1, X_2, X_3, and x_4 are assumed to be small so that squares and higher powers can be neglected and are such that $x_1 + x_4 = X_2 + X_3$. If we write:

$$W_1 = pa_{12} + qa_{13}$$
$$W_4 = pa_{24} + qa_{34},$$

these being the initial gamete viabilities for the new gametes AB and aB, respectively, and $U = W_1 - W_4$, then it turns out that linkage is only important in the establishment of B if

$$r < |U| \qquad (3)$$

where $|U|$ represents the positive value of U.

The initial rate of increase of B can be expressed in two different forms according to whether the ratio $r/|U|$ is big or small. If we assume, without loss of generality, that $W_1 > W_4$, then for large r/U the initial rate of increase is given approximately by λ_1^*, where

$$2W\lambda_1^* \doteq W_1 + W_4 + \frac{U(qa_{34} - pa_{12})}{pa_{12} + qa_{34}}$$

and we neglect U^2/r^2, etc. For small r/U the initial rate of increase is given by λ_1 where

$$W\lambda_1 = W_1 - rpa_{12}$$

neglecting r^2/U^2. In both cases $W = 1 - \alpha p^2 - \beta q^2$ is the mean fitness of the equilibrium population before the introduction of B.

When $r = 0$, λ_1 reduces to the expression for the rate of increase of a third allele at a locus already polymorphic for two alleles (Bodmer and

Parsons, 1960; Haldane, 1957). In either case B will only increase if λ_1 or $\lambda_1^* > 1$. It can be shown that in general if $r < U$ then $\lambda_1 > \lambda_1^*$, and vice versa. Thus if the viabilities are such that $\lambda_1^* \leq 1$ and $U \neq 0$, it is nevertheless possible in certain cases that $\lambda_1 > 1$ for sufficiently small r. In such a situation B will only increase in frequency if it is sufficiently closely linked to A.

If $a_{12} = a_{34}$ (coupling and repulsion viabilities equal) and also $a_{13} = a_{24}$, then $U = 0$ when $p = q$. The model considered by Lewontin and Kojima (1961) incorporates these assumptions and is therefore an unfavorable model from the point of view of the initial importance of linkage. Although their model exhibits a certain type of departure from additivity of viability effects of the two loci, it does not seem to incorporate conditions really appropriate for the importance of linkage. They assume that all four homozygous genotypes have the same viability so that their model is, in fact, incompatible with Mather's concept of balance. There is no doubt that in many cases it is misleading to consider genotype viabilities in terms of main effects and interactions between loci. Emphasis must, in general, be placed on the particular *gametic* combinations involved.

The set of viabilities assumed by Kimura (1956) is given in our notation, for a discrete generation model, in Table 5. For this model, $\lambda_1^* < 1$ and the new gene B will only increase in frequency if it is sufficiently closely linked to the original polymorphic locus. The condition on r for the increase of B implies the condition given by Kimura (1956) that all four gametes should be maintained in stable equilibrium, but the reverse is not true.

TABLE 5
Viability Model Following Kimura (1956—Table 2)

	Genotypes		
	AA	Aa	aa
BB	$1 + s$	$1 + t$	$1 - s$
Bb	1	$1 + t$	1
bb	$1 - s$	$1 + t$	$1 + s$

For balanced polymorphism of A-a, $t > s$. B will only increase in frequency in a population polymorphic for gametes Ab and ab if

$$r < s(t - s)/(1 + t)(t + s)$$

where r is the recombination fraction between the two loci.

It is possible, under certain circumstances, that a "position effect," i.e., a difference between the coupling and repulsion viabilities, will allow B to increase provided linkage is sufficiently close, but not otherwise.

For simplicity assume that

(i) $\alpha = \beta$, so that $p = q = \frac{1}{2}$.
(ii) $a_{13} = a_{24} = 1 - \alpha_{13}$
(iii) $a_{12} = 1$ (coupling viability)
and (iv) $a_{34} = 1 - \alpha_{34}$ where $\alpha_{34} > 0$ (repulsion viability).

Then we have $\lambda_1 > 1$ if $\alpha > r + \alpha_{13}$
and $\lambda_1^* < 1$ if $\alpha < \alpha_{34}/2 + \alpha_{13}$.

Thus if $\alpha > \alpha_{13}$ but $\alpha < \alpha_{13} + \alpha_{34}/2$, B will only increase if $r < \alpha_{34}/2$. *In other words the second locus will only increase in frequency if the recombination fraction between the two loci is less than half the difference in viability between coupling and repulsion heterozygotes.* The remarkable pair of chaeta genes analyzed by Gibson and Thoday (1959, 1961) may fit into this category. Further examples involving different viability models will be considered later in this section.

3. *Simultaneous Increase of New Alleles at Each of Two Loci*

We now assume that ab/ab is the predominant genotype and that $x_2 = 1 - X_2$ is the frequency of gamete ab, where X_2, x_1, x_3, x_4 are all small. This may represent the situation in which neither A nor B can increase on their own but the gamete AB is advantageous, so that AB will increase in frequency provided linkage between the loci is sufficiently tight. Alternatively it may represent the situation where either one or both of A and B can increase alone, but where their frequencies are still small. On neglecting squares of x_1, etc., we obtain linear equations for the frequencies of gametes AB, Ab, aB in one generation in terms of their frequencies in the previous generation. Gametes Ab and aB will increase in frequency if $a_{22} < a_{23}$ or a_{24}, respectively, where as before, a_{22}, a_{23}, a_{24} are the viabilities of genotypes ab/ab, Ab/ab and aB/ab. These are simply the conditions that either genotype Aa or Bb shows initial heterozygote advantage with respect to fitness. Gamete AB will increase in frequency so long as

$$a_{12} > a_{22}/1 - r$$

where a_{12} is the viability of the coupling heterozygote AB/ab. This condition may be rewritten:

$$r < \frac{a_{12} - a_{22}}{a_{12}} \quad (4)$$

and holds irrespective of the values of a_{23} and a_{24}. *In other words, if the coupling double heterozygote is at an advantage, but single heterozygotes are no fitter than homozygote ab/ab, A and B will simultaneously increase in* *frequency only if linkage between the two loci is closer than the expression given by (4)*. This represents an automatic selection for close linkage between interacting loci in outbreeding populations, which is comparable in importance to the automatic selection for initial heterozygote advantage which has been discussed by Bodmer (1960b), Bodmer and Parsons (1960), and Parsons and Bodmer (1961). For small $a_{12} - a_{22}$ inequality (4) gives the same result as does (3) when the equilibrium frequency of Ab is small.

This type of model was suggested by Sheppard (1959) as a possible mechanism for the evolution of the supergene controlling mimicry patterns in the moth *Papilio dardanus*. As a specific example he gave the situation discussed in Section III, C where a moth carrying both a mutation to extend the costal bar and one to change the pale area of the wing from yellow to white would be at an advantage. A moth carrying only one of these mutations would not be at an advantage as compared with the prevailing genotype. Thus, as analyzed above, these two mutants would only be selected through the coupling gamete if they were sufficiently closely linked.

It is possible to obtain complete solutions to the linear equations in x_1, x_3, and x_4. If $x_1^{(n)}$, $x_3^{(n)}$ represent the frequencies of gametes AB and Ab n generations after these had values $x_1^{(0)}$ and $x_3^{(0)}$ then we have

$$x_1^{(n)} = \left\{\frac{a_{12}(1-r)}{a_{22}}\right\}^n x_1^{(0)}$$

and

$$[(a_{23} - a_{12}(1-r)]x_3^{(n)} = \left\{\frac{a_{23}}{a_{22}}\right\}^n [a_{23} - a_{12}(1-r)]x_3^{(0)} + ra_{12}x_1^{(0)}\left[\left\{\frac{a_{23}}{a_{22}}\right\}^n - \left\{\frac{a_{12}(1-r)}{a_{22}}\right\}^n\right] \quad (5)$$

Thus even if $a_{23} > a_{22}$, so that x_3 can increase on its own, the increase in x_1 will nevertheless predominate if $a_{23} < a_{12}(1-r)$ or if

$$r < \frac{a_{12} - a_{23}}{a_{22}} \quad (6)$$

This inequality is closely analogous to (4), and when $a_{23} > a_{22}$, if r satisfies (6) it also satisfies (4), namely the condition that x_1 should increase in frequency. When $a_{23} > a_{12}(1-r)$, the frequency of Ab is determined largely by its increase due to the heterozygote advantage of Ab/ab. When on the other hand, $a_{23} < a_{12}(1-r)$, the frequency of Ab is determined largely by its rate of production by recombination from the heterozygote AB/ab. Thus, if also $a_{24} < a_{22}$, this shows that *a modifier B whose only effect is to enhance the heterozygote advantage of Aa will only be selected if it is sufficiently closely linked to locus A-a.* If r does not satisfy inequality (6), the gamete AB will not be able to increase in frequency appreciably during the more rapid increase of gamete Ab. This is exactly the situation predicted by Parsons and Bodmer (1961) in their discussion of the evolution of overdominance. They stated that modifiers of the fitness of a heterozygote at a locus A-a would be selected far more rapidly if they were linked to this locus. In fact, unless the linkage is sufficiently close, the increase in frequency of the modifier will be of the second order as compared with the increase in frequency of A at the primary locus. We have here an appropriate model for the evolution of linked polymorphic systems as discussed first by Fisher (1930b) and later amplified by Sheppard (1953).

Suppose, for example, that $a_{12} = 1.15$, $a_{23} = 1.05$, and $r = 5\%$ where $a_{22} = 1$ is the viability of ab/ab. Then the excess advantage of the double heterozygote AB/ab over the single heterozygote Ab/ab is 10%, and the rate of increase of the coupling gamete AB is given by $a_{12}(1 - r) = 1.0925$, whereas the rate of increase of Ab is given by $a_{23} = 1.05$. The number of generations needed for AB to increase by factors of 100, 1000, and 10,000 together with the factor increase in Ab for the same numbers of generations are given in Table 6. If, for example, $x_1{}^0 = 10^{-5}$ and $x_2{}^0 = 10^{-3}$, then after 104 generations the frequency of AB is about 10% and the frequency of Ab is still only about 16%.

TABLE 6
Selection for a Linked Modifier of Heterozygote Advantage

Factor increase in AB	Factor increase in Ab	Number of generations
100	12.66	52.08
1,000	45.04	78.12
10,000	160.3	104.16

ab/ab is the prevailing genotype and gametes AB, Ab are rare.

Genotypes:	ab/ab	AB/ab	Ab/ab
Have viabilities:	1	1.15	1.05
Recombination fraction $r = 5\%$			

In the above discussion we have, of course, assumed that the initial frequency of AB is much lower than that of Ab. When the gamete AB is advantageous as compared with ab, perhaps even when neither Ab nor aB are, it is clearly desirable that the rate of production of gametes AB is as high as possible. These can be produced either by mutation to A or

B in gametes aB or Ab, or by recombination in the repulsion heterozygote Ab/aB. If the mutation rates $a \to A$ and $b \to B$ are μ_1, μ_2, respectively, and the viabilities of heterozygotes Ab/ab and aB/ab are $1 - \alpha_{23}$ and $1 - \alpha_{24}$ relative to a viability of 1 for ab/ab, then the frequencies of these heterozygotes as maintained by the balance of mutation against selection are μ_1/α_{23} and μ_2/α_{24}, respectively (see, for example, Fisher, 1930a). Thus the rate of production of AB by mutation is $\mu_1\mu_2/\alpha_{23}$ or $\mu_1\mu_2/\alpha_{24}$, and by recombination is $r\mu_1\mu_2/\alpha_{23}\alpha_{24}$, since the frequency of repulsion heterozygotes Ab/aB is $2\mu_1\mu_2/\alpha_{23}\alpha_{24}$. The rate of production by recombination is higher than that by mutation if $r > \alpha_{23}$ or α_{24}. Thus, when a higher rate of production of gamete AB is at a premium, this places a lower limit on the optimum value of the recombination fraction. When the viability of Ab/aB is $a_{12} = 1 - \alpha_{12}$, then the condition for AB to increase is that $r < \alpha_{12}$ so that optimum recombination fractions are restricted by the inequality

$$\alpha_{12} > r > \alpha_{23}, \alpha_{24}.$$

This is perhaps the simplest example of the way in which natural selection will adjust recombination so as to achieve a balance between flexibility in producing new variation, and the necessity for the sake of immediate fitness, to maintain intact advantageous gametes. It is this simple balance which Fisher (1930a) used to illustrate the way in which recombination will be adjusted by natural selection, as discussed in earlier sections. The advantage of recombination will be particularly favorable where genes of small effect combine to produce a large effect, i.e., where α_{12} is considerably larger than either α_{23} or α_{24}. If, for example, α_{23}, $\alpha_{24} = 1\%$ and $r = 10\%$, then the rate of production of AB by recombination is ten times the rate of production by mutation, and AB/ab must have an excess viability of at least 10% for gamete AB to increase in frequency. When Ab and aB show some heterozygote advantage but AB is better than either of them, as discussed above, the rate of production of AB by recombination in general far exceeds the rate of production by mutation.

As a further example of a model leading to the evolution of linked polymorphic complexes, we shall consider the set of viabilities given by Kimura (1956) to explain the polymorphism of the linked loci pink versus yellow and banded versus unbanded in the land snail *Cepaea nemoralis*. This model, which was suggested by Sheppard, is illustrated in our notation in Table 7. The conditions yellow and banded are determined by recessive genes a and b. The viabilities are such that pink unbanded and yellow banded phenotypes are both fitter than pink banded or yellow unbanded. There is also some overdominance with respect to fitness for heterozygotes Aa. Thus Aa is overdominant in the presence of B for all $t > 0$, but is only overdominant in the presence of bb if $t > s_3 + s_4$.

TABLE 7
Viability Model for Banding and Shell Color in *Cepaea*
(Based on Kimura, 1956)

	Pink		Yellow
	AA	Aa	aa
Unbanded			
BB	$1 + s_1$	$1 + s_1 + t$	$1 - s_2$
Bb	$1 + s_1$	$1 + s_1 + t$	$1 - s_2$
Banded			
bb	$1 - s_3$	$1 - s_3 + t$	$1 + s_4$

Consider first a population containing only pink and yellow unbanded snails in overdominance polymorphism for gametes AB and aB. The banded gene b will only increase when introduced into such a population if $s_1 > s_2$. Whatever the values of s_1 and s_2, the initial viabilities of the new gametes Ab and ab are equal, so that linkage cannot affect the increase of b [cf. inequality (3)]. On the other hand, when $t > s_3 + s_4$ and only banded pink and yellow snails occur in the population, linkage may influence the increase of the unbanded gene B. In the simple case when $s_1 = s_2 = s_3 = s_4 = s$, if $\frac{1}{2} > s/t > 1 - (1/\sqrt{2})$, B will only increase in frequency if r is sufficiently small. If $s/t < 1 - (1/\sqrt{2})$, B will increase in frequency for any value of r, but when

$$r < \frac{2s(t - 2s)}{t(1 + s + t)}$$

linkage will enhance the rate of increase of B. If $t < s_3 + s_4$, and the population consists almost entirely of yellow banded snails, neither the pink nor the unbanded genes can increase on their own. These genes will increase in frequency because of the advantage of the coupling gamete AB only if linkage between the two loci is sufficiently tight. More specifically, the gamete AB will only increase in frequency if

$$r < \frac{t + s_1 - s_4}{1 + s_1 + t}$$

[cf. inequality (4)]

or, approximately, if the recombination fraction is less than the excess fitness of pink unbanded heterozygotes (AB/ab) over yellow banded homozygotes ab/ab. In this case the population may eventually become homozygous AB/AB pink unbanded, unless $s_1 > s_2$ as discussed above, and, also r is sufficiently small. As pointed out by Kimura (1956), any gamete carrying a crossover reducing mechanism is likely to be selected for. From this analysis it is clear that linkage is only likely to have been of importance with such a viability model if the original "wild type" was yellow and banded. If we assume a fairly direct relationship between the effect of genes A and B on banding and color and their effects on fitness, then this can be reconciled with the fact that the original wild type is the multiple recessive. We have shown that one of the conditions for these genes to increase in a population which is predominantly yellow and banded is that the coupling gamete AB should be at least partially dominant with respect to fitness. This implies at least partial dominance of these factors with respect to their effects on shell color and banding if these are directly associated with their effects on fitness. No doubt further possibly linked modifiers enhancing the dominance of these factors may have been selected during the course of their establishment in the population.

D. Equilibrium Conditions for Viabilities Which Correspond to Mather's Concept of Balance

We consider first the situation when r is small and balanced gametes Ab and aB are predominant. If the viability of Ab/aB (which is a_{34}) is bigger than either a_{33} (Ab/Ab) or a_{44} (aB/aB), then when $r = 0$, Ab and aB can be maintained in polymorphic equilibrium with frequencies p and q, say. We can then solve approximately for the equilibrium frequencies x_1, x_2 of the unbalanced gametes AB, ab, if we assume that x_1, x_2 are proportional to r and neglect r^2 and higher powers. We obtain:

$$x_1 = \frac{ra_{34}pq}{W - W_1}, \quad \text{and} \quad x_2 = \frac{ra_{34}pq}{W - W_2}$$

where, as before, a_{34} is the viability of the repulsion heterozygote Ab/aB, W is the mean fitness of a population containing only gametes Ab and aB at their equilibrium frequencies p and q, and W_1 and W_2 are the mean viabilities of unbalanced gametes AB and ab, respectively. It can be shown that such an equilibrium exists and is stable provided r is sufficiently small, and provided W_1 and W_2 are less than W. It is, however, difficult to obtain the upper limit for r which permits this equilibrium. The conditions W_1, $W_2 < W$ imply, in general, that the viabilities of the single heterozygotes AB/Ab, etc., are lower than those of the balanced homozygotes Ab/Ab and aB/aB, which is exactly the situation envisaged by Mather in his discussion of balance. We shall refer to an equilibrium which is such that the frequencies of either balanced or unbalanced gametes are proportional to the recombination fraction, as a "linkage balance equilibrium." *Thus, whenever r is sufficiently small and viabilities correspond to Mather's balance, a linkage balance equilibrium exists and is stable.*

Even for small r, it is difficult to investigate the equilibriums for general sets of viabilities. We shall therefore assume, now, a simple symmetric viability model which illustrates Mather's concept of balance. This model is essentially the same as one considered by Wright (1952) and is given in Table 8. When α is positive, double heterozygotes have a higher viability

TABLE 8
Symmetric Viability Model According to Mather's Concept of Balance

	Balanced homozygotes		Double heterozygotes		Unbalanced homozygotes	
	$\dfrac{Ab}{Ab}$	$\dfrac{aB}{aB}$	$\dfrac{Ab}{aB}$	$\dfrac{AB}{ab}$	$\dfrac{AB}{AB}$	$\dfrac{ab}{ab}$
Viabilities:	$a_{33}=1$	$a_{44}=1$	$a_{12}=1+\alpha$	$a_{34}=1+\alpha$	$a_{11}=1-\gamma$	$a_{22}=1-\gamma$

	Single heterozygotes			
	$\dfrac{aB}{ab}$	$\dfrac{Ab}{ab}$	$\dfrac{AB}{aB}$	$\dfrac{AB}{Ab}$
Viabilities:	$a_{24}=1-\beta$	$a_{23}=1-\beta$	$a_{14}=1-\beta$	$a_{13}=1-\beta$

than balanced homozygotes. If also $\gamma > \beta > 0$, unbalanced homozygotes are less fit than single heterozygotes which are themselves less fit than the balanced homozygotes. If $\beta < 0$ but is less in magnitude than α, the model represents cumulative heterozygote advantage, which does not correspond to balance in the strict sense. The particular cases considered by Wright (1952) were when $\alpha = 0$, $\gamma = 4\beta$ and when $\alpha = 2d$, $\gamma = 4s$ and $\beta = s - d$. When $\gamma = 0$ these viabilities correspond to a symmetrical version of the model considered by Lewontin and Kojima (1961). On substituting the viabilities given in Table 8 into the general equations (1) or (2), it is easily shown that, at equilibrium, $x_1 = x_2$ and $x_3 = x_4$, so that

$$x_1 = \tfrac{1}{2} - x_3$$

The equilibrium frequencies of the unbalanced gametes are given by the cubic equation:

$$8x^3(4\beta + 2\alpha - \gamma) + 4x^2(\gamma - 6\beta - 3\alpha) \\ + 2x[2\beta + \alpha + 2r(1+\alpha)] - r(1+\alpha) = 0 \quad (7)$$

The equilibrium frequencies of the balanced gametes are then given by $\frac{1}{2} - x$ where x is a solution of Equation (7). It should be noted that for the special cases considered by Wright (1952), $4\beta + 2\alpha - \gamma = 0$ so that Equation (7) reduces to a quadratic. If $\alpha = \beta = \gamma = 0$, (7) only has the root $x_1 = \frac{1}{4}$ giving, as expected, a gene frequency equilibrium, for the equilibrium gene frequencies are both $\frac{1}{2}$. When $r = 0$, $x = 0$ or $\frac{1}{2}$, giving either only balanced or only unbalanced gametes; or

$$x = \frac{2\beta + \alpha}{4(2\beta + \alpha - \gamma/2)} \doteq \frac{1}{4}\left(1 + \frac{\gamma}{2(2\beta + \alpha)}\right) \text{ for small } \frac{\gamma}{2\beta + \alpha},$$

giving a quasi-gene frequency equilibrium. In general for fairly small $\gamma/(2\beta + \alpha)$ and $r(1 + \alpha)/(2\beta + \alpha)$, the three solutions of Equation (7) are approximately:

$\frac{r(1 + \alpha)}{2(2\beta + \alpha)}$ corresponding to a linkage balance equilibrium with unbalanced gametes in the minority,

$\frac{1}{2} - \frac{r(1 + \alpha)}{2(2\beta + \alpha)}$ corresponding to a linkage balance equilibrium with balanced gametes in the minority, and

$\frac{1}{4} + \frac{\gamma}{8(2\beta + \alpha)}$ corresponding to a quasi-gene frequency equilibrium.

It should be noted that unless $\gamma = 0$ there is no actual gene frequency equilibrium. If γ is not small compared with $2\beta + \alpha$ and $r(1 + \alpha)$, then there may be no quasi-gene frequency equilibrium. This is a situation not covered by Wright's (1952) assumptions, but which is exemplified by recombinational lethals and sub-vitals as discussed in Section II.

By considering small departures from an equilibrium position, it can be shown that the equilibrium will be stable if:

$$r(1 + \alpha) > -\frac{\alpha}{2} - \beta + 2x[3(2\beta + \alpha) - \gamma] + 6x^2[\gamma - 2(2\beta + \alpha)] \quad (8)$$

where x is the equilibrium frequency of unbalanced gametes. When γ is small both the linkage balance equilibriums, with the frequency of unbalanced gametes in the majority or in the minority, exist and are stable if

$$r(1 + \alpha) < \frac{\alpha + 2\beta}{4} - \frac{\gamma}{8} \quad (9)$$

and then the quasi-gene frequency equilibrium is unstable. If, however, the recombination fraction is bigger than the limit given by (9), then only the quasi-gene frequency equilibrium is stable. For larger values of γ, the quasi-gene frequency equilibrium does not generally exist, and the linkage balance equilibrium is stable only if r is sufficiently small. The linkage balance equilibrium with unbalanced gametes in the majority is only stable for sufficiently small values of γ and r. If the recombination fraction is too large, then in some cases it may not be possible to have any equilibriums in which all four gametes are maintained in the population.

It is of some interest to consider under what conditions a gamete carrying a crossover reducer will be selected for. Suppose that $\overline{Ab}$ is an Ab gamete which is such that the recombination fraction in $\overline{Ab}/aB$ heteroxygotes is now r' instead of r, but whose viabilities are just the same in combination with the other gametes as for gamete Ab. Then assuming $\overline{Ab}$ is introduced with small frequency into a population in equilibrium for gametes AB, Ab, aB, and ab, it can be shown that the new gamete will increase in frequency if:

$$r'(1 + \alpha)(\tfrac{1}{2} - x) < x(2\beta + \alpha) + 2x^2[\gamma - 2(2\beta + \alpha)] \qquad (9)$$

where, as before, x is the equilibrium frequency of unbalanced gametes. When x is near $\tfrac{1}{4}$ giving a quasi-gene frequency equilibrium we must have

$$r'(1 + \alpha) < \frac{\gamma^2}{16(2\beta + \alpha)},$$

where $\gamma/(2\beta + \alpha)$ is small. This implies that for the new gamete to increase, the reduced recombination fraction must be very small. If x is proportional to the recombination fraction, giving a linkage balance equilibrium, then the new gamete will increase if $r' < r$. *In other words, when there is already a linkage balance equilibrium, natural selection will in general favor any balanced gamete giving a reduced recombination fraction.*

E. The Evolution of the Balanced Complex

In order to consider how a population in linkage balance equilibrium may have evolved, it is necessary to relate the viabilities corresponding to a balance model to the conditions for the increase of newly occurring gametes, as discussed in Section VII,C. We assume the symmetric model given in Table 8 of the previous section.

If the unbalanced gamete ab predominates, and A and B are rare, then they will increase in frequency alone in the balanced gametes aB, Ab, provided $\gamma > \beta$. When, however,

$$r(1 + \alpha) < \alpha + \beta \text{ [cf. inequality (6)]}$$

the increase of A and B is dominated by the increase in frequency of the unbalanced gamete AB. Thus from the previous section, if r is sufficiently small, a linkage balance equilibrium will, in general, be obtained with balanced gametes in the majority. In other words, if linkage is sufficiently tight, equilibrium for a balanced complex may evolve through the initial selection of an unbalanced gamete, which later gives rise to appreciable frequencies of balanced gametes by recombination.

The situation with the balanced gamete Ab predominating and both a and B rare is probably of more importance. So long as $\beta > 0$, neither a nor B can increase on their own. Only if

$$r < \alpha/1 + \alpha$$

will a and B increase through the increase of the balanced gamete aB. Thus a new balanced gamete will only increase in a population predominantly homozygous for an existing balanced gamete if the balanced heterozygote shows some overdominance for fitness and if the recombination fraction is less than the excess fitness of the balanced heterozygote. As pointed out before, this will automatically lead to an observed association between heterozygote advantage and linkage. The evolution of overdominance in its most general form as discussed by Parsons and Bodmer (1961) depends on the selection of sets of linked genes which form overdominant complexes with respect to fitness. It seems that the extension to the case of more than one locus, of the principle that a new gene will only increase in an outbreeding population if it is advantageous in the heterozygote, involves in many instances the need for sufficiently close linkage between the loci. When $\beta < 0$, and we have the case of cumulative overdominance, then a and B can increase on their own, but linkage will enhance the selection of the balanced gamete aB if $r < (\alpha - |\beta|)/(1 + \alpha)$, where $|\beta|$ is the positive value of β.

In Table 9 are given some numerical examples of the equilibrium conditions given by the viability model we have considered above. For all the examples we have taken $\alpha = 10\%$, $\beta = 5\%$ so that $\alpha + 2\beta = \frac{1}{5}$. The examples cover a range of values of γ, the viability deficiency of unbalanced homozygotes. We have only considered cases where $r \leq \alpha$, which are the cases when a new balanced gamete will be selected for in a population predominantly homozygous for an existing balanced gamete. In general there is a linkage balance equilibrium with the frequency of unbalanced gametes in the minority. In none of the cases considered does a quasi-gene frequency equilibrium exist. All but the last case allow for selection of gametes exhibiting closer linkage. This last example is particularly interesting as with small values for γ and r, it allows for stability of both types of linkage balance equilibrium. However, the equilibrium with unbalanced gametes in the majority does not allow for selection of gametes giving low recombination fractions. This equilibrium is in fact only just stable and depends on a delicate balance between the values of the recombination fraction and the viability deficiency of unbalanced homozygotes.

It is clear that even considering only the case of two loci, there are many situations in which linkage can make a profound difference to the selection and ultimate equilibrium of new gametic combinations. This is particularly so when the viabilities are in line with Mather's concept of balance and there is also some excess advantage of balanced heterozygotes as compared with balanced homozygotes (i.e., relational balance).

TABLE 9
Numerical Examples of Equilibrium Conditions with Symmetrical Viabilities Exhibiting Balance

α	β	γ	r	Equilibrium frequencies of the unbalanced gametes	Upper limit on r' for increase of new gametes with reduced recombination fraction r'
		(%)			
10	5	80	9	11.70% stable	$r' < 8.15\%$
10	5	80	5	8.28% stable	$r' < 4.80\%$
10	5	40	5	10.22% stable	$r' < 4.67\%$
10	5	40	9	14.57% stable	$r' < 7.48\%$
10	5	20	9	17.53% stable	$r' < 6.37\%$
10	5	20	1	2.74% stable	$r' < 1.00\%$
10	5	5	10	22.67% stable	$r' < 3.12\%$
10	5	5	1	29.83% unstable	—
				45.87% stable	No r' selected
				2.87% stable	$r' < 1.00\%$

$1 + \alpha$ = viability of double heterozygotes.
$1 - \beta$ = viability of single heterozygotes.
1 = viability of balanced homozygotes.
$1 - \gamma$ = viability of unbalanced homozygotes.
r = recombination fraction.
Where only one equilibrium is given, this is the only one which exists.

It seems unlikely that consideration of models involving more than two loci will alter the broad qualitative conclusions reached above although they might materially affect the quantitative results obtained. There is certainly a need for more work along the lines we have indicated. An investigation of, for example, the effects of interaction between the sexes with respect to fitness together with a sex difference in recombination should provide a more realistic model for the situation in *Drosophila*. It is also important to know the effect of departure from random mating and the way in which internal as opposed to relational balance may evolve. However, especially if we consider the comparatively high selective values which are now becoming common in experimental population genetics, there seems to be no justification for claiming, as has so often been done in the past, that theoretical population genetic studies do not fit in with the concept of the importance of linkage as developed by Fisher, Mather, and others.

VIII. Discussion

In spite of a great deal of experimental evidence pointing to the importance of linkage in evolution, there are few experimental studies showing how linked systems are composed. However, earlier theoretical discussions indicated the type of organization that might be expected. Fisher (1930a) showed how interaction between two genetic factors, with respect to fitness, may be associated with selection for closer linkage, and he also considered under what circumstances looser linkage may be advantageous. Darlington (1932), in discussing cytological evidence, argued that linkage could control the antagonistic evolutionary requirements of immediate fitness and future flexibility, and Mather's (1943a) specific model of the balanced genotype showed more precisely how linked interacting systems could accommodate these antagonistic requirements. We have reviewed evidence suggesting that recombination under diverse conditions is sufficiently variable to provide ample opportunity for such control to take place.

However, it is only recently that studies have become detailed enough to locate specific modifiers affecting a quantitative trait and to find out how these modifiers interact. Artificial selection has been used for this purpose. If continued long enough, selection may establish extreme probably homozygous genotypes produced from balanced genotypes by recombination. The genes responsible may then be amenable to analysis, by statistical or biochemical techniques, if they cannot be analyzed directly. It is really a question of finding the correct analytical technique to amplify the effect of these genes so that they behave, to all intents and purposes, as major genes. From the results of selection experiments and their analysis, we can infer the genetic architecture of the original population. Such experiments provide abundant evidence for the existence of the linked balanced polygenic complex as predicted by Mather.

Evidence for balanced complexes in nature comes from the analysis of inversion complexes, the breakdown of fitness by recombination, and from the discovery of the factors responsible for an observed response to selection in the wild stock which formed the base population of a selection line (Gibson and Thoday, 1962). All these studies serve to emphasize the large store of potential variability in natural populations. Much of this can only be uncovered by selection experiments which break down linked polygenic complexes by recombination. There is a lesson to be learned by the practical breeder who may give up selection before he has been able to change the genetic constitution of his line so as to favor the occurrence of an appropriate recombinational product.

Heritability, which is often used as a measure of the genetic variance available for the action of selection, is only meaningful in the stages of selection before the production of recombinants. It cannot predict the nature of the variability produced by recombination. The breeder needs the recombinants not only to achieve the selection aim he has in mind, for example, increased yield, but also to overcome the problem of correlated responses to selection. As yield is increased, fertility may fall and only a recombinational event may break up the association between fertility and yield genes permitting simultaneous selection for high fertility and yield.

Most of the specific evidence for the existence of linked polygenic complexes comes from work with *Drosophila*. However, we have presented some detailed evidence for their existence both in the house mouse and in *Neurospora*. There is less direct evidence, for a number of other species from quantitative genetic studies and the results of selection experiments. Balanced polymorphism for overdominant gene complexes is, however, a widespread phenomenon. Overdominant polymorphic gene complexes probably represent an extreme form of the balanced complex with very tight linkage between the factors involved.

Polymorphic overdominant complexes have almost certainly evolved by the accumulation of linked interacting modifiers (see Parsons and Bodmer, 1961 for discussion). Overdominance is probably not associated with a genetic difference at a single site, but is associated with interaction between linked genes, the degree of linkage being adjusted by natural selection. The evolution of the balanced genotype is qualitatively little different. We have shown, for example, how two interacting genes, not individually showing heterozygote advantage, may increase in frequency and so help to build up the balanced genotype we observe. Of special note is that even quite loose linkage may permit this to occur, although the degree of interaction needed varies with the linkage value. With tight linkage, interactions can be slight, but with loose linkage, interactions must be more extreme, as has been found with the two balanced factors of Gibson and Thoday (1959) which are separated by a map distance of about 20 cM. One of the striking features of modern population genetics is the variability of selective values of genotypes, and the common occurrence of very extreme values. It is therefore reasonable to expect that recombination fractions and nonrandomness of linkage phase of all orders of magnitude between interacting genes may occur.

Our theoretical studies are exclusively concerned with outbreeding populations, and so the evolution of relational as opposed to internal balance. Mather pointed out that the type of balanced complex evolved would depend on the breeding system. Inbreeding must lead to complexes which are fit when homozygous whereas outbreeding produces complexes which are at an advantage in the heterozygote, thus leading, effectively, to a generalized evolution of overdominance for fitness.

Complex characters in higher organisms such as viability or size must be affected by a vast constellation of basic biochemical differences, so that polygenic activity controlling such characters will be ubiquitous. Specific modifiers affecting such characters may be difficult to detect. As already pointed out, chaeta number in *Drosophila* is rather less complex and is more amenable to specific analysis. Even less complex in organization will be the control of a specific biochemical step by an enzyme.

It is in the microorganisms that we may expect to find information on the basic organization of genes, and their interactions within and between linkage groups. It is possible to study in detail the modifiers of a simple biochemical character, perhaps associated with a single enzyme difference. The work of Jacob and his colleagues on the genetic mechanisms concerned in the control of enzyme formation may prove to be a type of genetic difference often associated with the control of quantitative differences. Such studies will be a necessary prerequisite for the extension of work to developmentally more complex organisms.

A striking discovery by Demerec and his colleagues, which must be of evolutionary significance, is clustering between loci having related biochemical effects in *Salmonella*, a phenomenon which seems to be more widespread in bacteria than in higher organisms. Comparisons between microorganisms of the genetic organization of specific biosynthetic pathways will be of extreme evolutionary interest, especially when they are related to the nature of the microorganisms, their breeding structure, and their mode of recombination.

However, we must distinguish between sets of closely linked genes whose effects interact in respect to fitness, but which are not necessarily related physiologically, and the clustering of loci having related biochemical effect. An example of the former is provided by heterostyly in *Primula*, where several closely linked loci controlling style length, anther height, pollen grain size, and incompatibility interact to promote outbreeding.

Linkage plays a major part in maintaining the organization of simple biochemical systems in microorganisms, as it does in the organization of the development of higher organisms. The evidence reviewed shows abundantly that the genetic organization of complex characters is very much dependent on linkage between interacting factors. With the refinement of genetic analysis in microorganisms, where the relation between gene structure and function can be determined very accurately, there is an exciting prospect in the integration of careful detailed work on specific loci, with equally careful work on the modifiers affecting these loci. Such work should be of extreme evolutionary interest and should deepen our understanding both of the basic mechanisms involved in the evolution of individual cistrons, and of the genetic architecture of whole linkage groups.

Acknowledgments

We would like to thank Professor J. M. Thoday for many stimulating discussions and for helpful criticism of the manuscript. We would also like to thank Dr. J. A. Pateman for helpful discussion and criticism in relation to Section VI, Dr. J. L. Jinks for his criticism of the manuscript, and Mrs. V. C. Fyfe for help with the figures and references.

We wish to thank the following authors and publishers for permission to reproduce various figures and tables: Figures 1 and 3 which were modified from *Heredity*—Oliver & Boyd, Ltd. and Professor K. Mather; Figure 5 from *Genetics*—the Editors of *Genetics* and Professor Th. Dobzhansky; Figure 6 from *Heredity*—Oliver & Boyd, Ltd. and Dr. H. Rees; Figures 2 and 4—Professor J. M. Thoday; and Table 2—Blackwell Scientific Publications, Ltd. and Dr. R. R. Race.

References

Acton, A. B. 1957. Chromosome inversions in natural populations of *Chironomus tentans*. *J. Genet.* **55**, 61–94.

Acton, A. B. 1961. An unsuccessful attempt to reduce recombination by selection. *Am. Naturalist* **95**, 119–120.

Aird, I., Bentall, H. H., and Roberts, J. A. F. 1953. A relationship between cancer of the stomach and the *ABO* blood groups. *Brit. Med. J.* I, 799–801.

Ames, B. N., Garry, B., and Herzenberg, L. A. 1960. The genetic control of the enzymes of histidine biosynthesis in *Salmonella typhimurium*. *J. Gen. Microbiol.* **22**, 369–378.

Anagnostopoulos, C., and Crawford, I. P. 1961. Transformation studies on the linkage of markers in the tryptophan pathway in *Bacillus subtilis*. *Proc. Natl. Acad. Sci. U.S.* **47**, 378–390.

Anderson, E. G. 1929. Studies on a case of high non-disjunction in *Drosophila melanogaster*. *Z. Induktive Abstammungs-u. Vererbungslehre* **51**, 397–441.

Barker, J. F. 1960. Variation of chiasma frequency in and between natural populations of Acrididae. *Heredity* **14**, 211–214.

Barratt, R. W., Newmeyer, D., Perkins, D. D., and Garnjobst, L. 1954. Map construction in *Neurospora crassa*. *Advances in Genet.* **6**, 1–93.

Barratt, R. W., Strickland, W. N., and Ogata, W. N. 1961. Stock lists and linkage maps of *Neurospora*. Distributed by Fungal Genetics Stock Center, Dartmouth College, Hanover, New Hampshire.

Bateman, N. 1960. Selective fertilization at the *T*-locus of the mouse. *Genet. Research Cambridge* **1**, 226–238.

Baylor, M. B., Hurst, D. D., Allen, S. L. and Bertani, E. T. 1957. The frequency and distribution of loci affecting host range in the Coliphage T2H. *Genetics* **42**, 104–120.

Beadle, G. W. 1932. A gene for sticky chromosomes in *Zea mays. Z. Induktive Abstammungs-u. Vererbungslehre.* **63**, 195–217.

Beadle, G. W., and Coonradt, V. L. 1944. Heterocaryosis in *Neurospora crassa. Genetics* **29**, 291–308.

Bennett, J. H., and Binet, F. E. 1956. Association between mendelian factors with mixed selfing and random mating. *Heredity* **10**, 51–55.

Benzer, S. 1957. The elementary units of heredity. In "The Chemical Basis of Heredity" (W. D. McElroy and B. Glass, eds.), pp. 70–93. Johns Hopkins Press, Baltimore, Maryland.

Bodmer, W. F. 1958. Natural crossing between homostyle plants of *Primula vulgaris. Heredity* **12**, 363–370.

Bodmer, W. F. 1960a. Interaction of modifiers: The effect of *pallid* and *fidget* on *polydactyly* in the mouse. *Heredity* **14**, 445–448.

Bodmer, W. F. 1960b. The genetics of homostyly in populations of *Primula vulgaris. Phil. Trans. Roy. Soc.* **B242**, 517–549.

Bodmer, W. F. 1960c. Discrete stochastic processes in population genetics. *J. Roy. Stat. Soc.* **B22**, 218–244.

Bodmer, W. F. 1961. Viability effects and recombination differences in a linkage test with pallid and fidget in the house mouse. *Heredity* **16**, 485–495.

Bodmer, W. F., and Parsons, P. A. 1960. The initial progress of new genes with various genetic systems. *Heredity* **15**, 283–299.

Bonner, D. M., Yanofsky, C., and Partridge, C. W. H. 1952. Incomplete genetic blocks in biochemical mutants of *Neurospora. Proc. Natl. Acad. Sci. U.S.* **38**, 25–34.

Breese, E. L., and Mather, K. 1957. The organisation of polygenic activity within a chromosome in *Drosophila.* I. Hair characters. *Heredity* **11**, 373–395.

Breese, E. L., and Mather, K. 1960. The organisation of polygenic activity within a chromosome in *Drosophila.* II. Viability. *Heredity* **14**, 375–399.

Bridges, C. B. 1915. A linkage variation in *Drosophila. J. Exptl. Zool.* **19**, 1–21.

Bridges, C. B. 1922. The origin of variations in sexual and sex-limited characters. *Am. Naturalist* **56**, 51–63.

Bridges, C. B. 1929. Variation in crossing over in relation to age of female in *Drosophila melanogaster. Carnegie Inst. Wash. Publ.* **399**, 63–89.

Bridges, C. B. 1939. Cytological and genetic basis of sex. In "Sex and Internal Secretions," (E. Allen, ed.), 2nd ed., pp. 15–63. Williams & Wilkins, Baltimore, Maryland.

Brncic, D. 1961. Non random association of inversions in *Drosophila pavani. Genetics* **46**, 401–406.

Brown, S. W., and Zohary, D. 1955. The relationship of chiasmata and crossing over in *Lilium formosanum. Genetics* **40**, 850–873.

Cain, A. J., and Sheppard, P. M. 1950. Selection in the polymorphic land snail *Cepaea nemoralis. Heredity* **4**, 275–294.

Cain, A. J., and Sheppard, P. M. 1954. Natural selection in *Cepaea. Genetics* **39**, 89–116.

Carlson, E. A. 1958. The bearing of a complex-locus in *Drosophila* on the interpretation of the *Rh* series. *Am. J. Human Genet.* **10**, 465–473.

Carson, H. L. 1955. Variation in genetic recombination in natural populations. *J. Cellular Comp. Physiol.* **45** (Suppl. 2), 221–236.

Carson, H. L. 1958. Response to selection under different conditions of recombination in *Drosophila. Cold Spring Harbor Symposia Quant. Biol.* **23**, 291–306.

Castle, W. E. 1925. A sex difference in linkage in rats and mice. *Genetics* **10**, 580–582.

Catcheside, D. G. 1940. Structural analysis of *Oenothera* complexes. *Proc. Roy. Soc.* **B128**, 509–535.
Catcheside, D. G. 1951. "The Genetics of Micro-organisms," 223 pp. Pitman, London.
Cavalli, L. L., and Maccacaro, G. A. 1952. Polygenic inheritance of drug-resistance in the bacterium *Escherichia coli*. *Heredity* **6**, 311–331.
Chung, C. S., and Morton, N. E. 1961. Selection at the *ABO* locus. *Am. J. Human Genet.* **13**, 9-27.
Clarke, C. A., and Sheppard, P. M. 1960a. The evolution of dominance under disruptive selection. *Heredity* **14**, 73–87.
Clarke, C. A., and Sheppard, P. M. 1960b. The evolution of mimicry in the butterfly *Papilio dardanus*. *Heredity* **14**, 163–173.
Clarke, C. A., and Sheppard, P. M. 1960c. Super-genes and mimicry. *Heredity* **14**, 175–185.
Cooper, J. P. 1959. Selection and population structure in *Lolium*. III. Selection for date of ear emergence. *Heredity* **13**, 461–479.
Cooper, J. P. 1960. Selection and population structure in *Lolium*. IV. Correlated response to selection. *Heredity* **14**, 229–246.
Crew, F. A. E., and Koller, P. C. 1932. The sex incidence of chiasma frequency and genetical crossing-over in the mouse. *J. Genet.* **26**, 359–383.
Crosby, J. L. 1940. High proportions of homostyle plants in populations of *Primula vulgaris*. *Nature* **145**, 672.
Crosby, J. L. 1949. Selection of an unfavourable gene-complex. *Evolution* **3**, 212–230.
Crowe, L. K. 1960. The exchange of genes between nuclei of a dikaryon. *Heredity* **15**, 397–405.
da Cunha, A. B. 1955. Chromosomal polymorphism in the Diptera. *Advances in Genet.* **7**, 93–138.
da Cunha, A. B., and Dobzhansky, T. 1954. A further study of chromosomal polymorphism in *Drosophila willistoni* in its relation to environment. *Evolution* **8**, 119–134.
Darlington, C. D. 1930. Chromosome studies in *Fritillaria*. III. Chiasma formation and chromosome pairing in *Fritillaria imperialis*. *Cytologia* **2**, 37–55.
Darlington, C. D. 1931a. Meiosis. *Biol. Revs. Cambridge Phil. Soc.* **6**, 221–264.
Darlington, C. D. 1931b. The cytological theory of inheritance in *Oenothera*. *J. Genet.* **24**, 405–474.
Darlington, C. D. 1932. The control of the chromosomes by the genotype and its bearing on some evolutionary problems. *Am. Naturalist* **66**, 25–51.
Darlington, C. D. 1939. "The Evolution of Genetic Systems," 1st ed., 149 pp. Cambridge Univ. Press, London and New York.
Darlington, C. D. 1941. The causal sequence of meiosis. II. Contact points and crossing-over potential in a triploid *Fritillaria*. *J. Genet.* **41**, 35–48.
Darlington, C. D. 1956. "Chromosome Botany," 186 pp. Allen & Unwin, London.
Darlington, C. D. 1958. "Evolution of Genetic Systems," 2nd ed., 265 pp. Oliver & Boyd, Edinburgh.
Darlington, C. D., and La Cour, L. F. 1950. Hybridity selection in *Campanula*. *Heredity* **4**, 217–248.
Dawson, G. W. P., and Smith-Keary, P. F. 1960. Analysis of the su-$leuA$ locus in *Salmonella typhimurium*. *Heredity* **15**, 339–350.
Day, P. R. 1961. Quoted by Lewis, D. Growth and genetics of higher fungi. *Nature Lond.* **190**, 399–400.

Demerec, M. 1945. Production of staphylococcus strains resistant to various concentrations of penicillin. *Proc. Natl. Acad. Sci. U.S.* **31**, 16–24.

Demerec, M., and Demerec, Z. E. 1956. Analysis of linkage relationships in *Salmonella* by transduction techniques. *Brookhaven Symposia in Biol.* **8**, 75–84.

Demerec, M., and Hartman, P. E. 1959. Complex loci in microorganisms. *Ann. Rev. Microbiol.* **13**, 377–406.

Demerec, M., Moser, H., Clowes, R. C., Lahr, E. L., Ozeki, H., and Vielmetter, W. 1956. Bacterial genetics. *Ann. Rept. Director Dept. Genet. Carnegie Inst. Wash. Yearbook* **55**, 301–315.

Demerec, M., Lahr, E. L., Ozeki, H., Goldman, I., Howarth, S., and Djordjevic, B. 1957. Bacterial genetics. *Ann. Rept. Director Dept. Genet., Carnegie Inst. Wash. Yearbook* **56**, 368–376.

Demerec, M., Lahr, E. L., Miyake, T., Goldman, I., Balbinder, E., Banič, S., Hashimoto, K., Glanville, E. V., and Gross, J. D. 1958. Bacterial genetics. *Ann. Rept. Director Dept. Genet., Carnegie Inst. Wash. Yearbook* **57**, 390–406.

Demerec, M., Lahr, E. L., Balbinder, E., Miyake, T., Ishidsu, J., Mizobuchi, K., and Mahler, B. 1960. Bacterial genetics. *Ann. Rept. Director Dept. Genet. Carnegie Inst. Wash. Yearbook* **59**, 426–641.

Detlefsen, J. A. 1920. Is crossing over a function of distance? *Proc. Natl. Acad. Sci. U.S.* **6**, 663–670.

Detlefsen, J. A., and Roberts, E. 1921. Studies on crossing over. I. The effect of selection on crossover values. *J. Exptl. Zool.* **32**, 333–354.

Diver, C. 1929. Fossil records of mendelian mutants. *Nature* **124**, 183.

Dobzhansky, T. 1933. Studies on chromosome conjugation. II. The relation between crossing-over and disjunction of chromosomes. *Z. Induktive Abstammungs-u. Vererbungslehre* **64**, 269–309.

Dobzhansky, T. 1946. Genetics of natural populations. XIII. Recombination and variability in populations of *Drosophila pseudoobscura*. *Genetics* **31**, 269–290.

Dobzhansky, T. 1950. Genetics of natural populations. XIX. Origin of heterosis through natural selection in populations of *Drosophila pseudoobscura*. *Genetics* **35**, 288–302.

Dobzhansky, T. 1951. "Genetics and the Origin of Species," 3rd ed., 364 pp. Columbia Univ. Press, New York.

Dobzhansky, T. 1957. Mendelian populations as genetic systems. *Cold Spring Harbor Symposia Quant. Biol.* **22**, 385–393.

Dobzhansky, T., and Epling, C. 1948. The suppression of crossing over in inversion heterozygotes of *Drosophila pseudoobscura*. *Proc. Natl. Acad. Sci. U.S.* **34**, 137–141.

Dobzhansky, T., and Levene, H. 1951. Development of heterosis through natural selection in experimental populations of *Drosophila pseudoobscura*. *Am. Naturalist* **85**, 247–264.

Dobzhansky, T., and Levene, H. 1955. Genetics of natural populations. XXIV. Developmental homeostasis in natural populations of *Drosophila pseudoobscura*. *Genetics* **40**, 797–808.

Dobzhansky, T., and Pavlovsky, O. 1958. Interracial hybridization and breakdown of coadapted gene complexes in *Drosophila paulistorum* and *Drosophila willistoni*. *Proc. Natl. Acad. Sci. U.S.* **44**, 622–629.

Dobzhansky, T., and Sturtevant, A. H. 1938. Inversions in the chromosomes of *Drosophila pseudoobscura*. *Genetics* **23**, 28–64.

Dobzhansky, T., and Wallace, B. 1953. The genetics of homeostasis in *Drosophila*. *Proc. Natl. Acad. Sci. U.S.* **39**, 162–171.

Dobzhansky, T., Burla, H., and da Cunha, A. B. 1950. A comparative study of chromosomal polymorphism in sibling species of the *willistoni* group of *Drosophila*. *Am. Naturalist* **84**, 229–236.

Dobzhansky, T., Levene, H., and Spassky, B., and Spassky, N. 1959. Release of genetic variability through recombination. III. *Drosophila prosaltans*. *Genetics* **44**, 75–92.

Dodge, B. O. 1942. Heterocaryotic vigour in *Neurospora*. *Bull. Torrey Bot. Club* **69**, 75–91.

Dowrick, G. J. 1957. The influence of temperature on meiosis. *Heredity* **11**, 37–49.

Dowrick, V. P. J. 1956. Heterostyly and homostyly in *Primula obconica*. *Heredity* **10**, 219–236.

Dubinin, N. P., and Tiniakov, G. G. 1946. Inversion gradients and natural selection in ecological races of *Drosophila funebris*. *Genetics* **31**, 537–545.

Dunn, L. C. 1956. Analysis of a complex gene in the house mouse. *Cold Spring Harbor Symposia Quant. Biol.* **21**, 187–195.

Dunn, L. C. 1957. Evidence of evolutionary forces leading to the spread of lethal genes in wild populations of house mice. *Proc. Natl. Acad. Sci. U.S.* **43**, 158–163.

Dunn, L. C., Beasley, A. B., and Tinker, H. 1958. Relative fitness of wild house mice heterozygous for a lethal allele. *Am. Naturalist* **92**, 215–220.

Eberhart, B. M., and Tatum, E. L. 1959. A gene modifying the thiamine requirement of strains of *Neurospora crassa*. *J. Gen. Microbiol.* **20**, 43–53.

Elliott, C. G. 1955. The effect of temperature on chiasma frequency. *Heredity* **9**, 385–398.

Emsweller, S. L., and Jones, H. A. 1945. Further studies on the chiasmata of the *Allium cepa* × *Allium fistulosum* hybrid and its derivatives. *Am. J. Botany* **32**, 370–379.

Ephrati-Elizur, E., Srinivasan, P. R., and Zamenhof, S. 1961. Genetic analysis, by means of transformation, of histidine linkage groups in *Bacillus subtilis*. *Proc. Natl. Acad. Sci. U.S.* **47**, 56–63.

Ernst, A. 1933. Weitere Untersuchungen zur Phänanalyse, zum Fertilitätsproblem und zur Genetik heterostyler Primeln. I. *Primula viscosa* All. *Arch. Julius Klaus-Stift. Vererbungsforsch. Sozialanthropol u. Rassenhyg.* **8**, 1–215.

Ernst, A. 1936. Weitere Untersuchungen zur Phänanalyse, zum Fertilitätsproblem und zur Genetik heterostyler Primeln. II. *Primula hortensis* Wettstein. *Arch. Julius Klaus-Stift. Vererbungsforsch. Sozialanthropol. u. Rassenhyg.* **11**, 1–280.

Eversole, R. A., and Tatum, E. L. 1956. Chemical alteration of crossing-over frequency in *Chlamydomonas*. *Proc. Natl. Acad. Sci. U.S.* **42**, 68–73.

Falconer, D. S. 1955. Patterns of response in selection experiments with mice. *Cold Spring Harbor Symposia Quant. Biol.* **20**, 178–196.

Fincham, J. R. S. 1951. A comparative genetic study of the mating-type chromosomes of two species of *Neurospora*. *J. Genet.* **50**, 221–229.

Fisher, R. A. 1922. On the dominance ratio. *Proc. Roy. Soc. Edinburgh* **42**, 321–341.

Fisher, R. A. 1930a. "The Genetical Theory of Natural Selection." 272 pp. Clarendon Press. Oxford.

Fisher, R. A. 1930b. The evolution of dominance in certain polymorphic species. *Am. Naturalist* **64**, 385–406.

Fisher, R. A. 1933. Selection in the production of eversporting stocks. *Ann. Botany (London)* **188**, 727–733.

Fisher, R. A. 1941. Average excess and average effect of a gene substitution. *Ann. Eugen. (London)* **11**, 53–63.

Fisher, R. A. 1947. The Rhesus factor; a study in scientific method. *Am. Scientist* **35**, 95–103.

Fisher, R. A. 1949. A preliminary linkage test with *agouti* and *undulated* mice. *Heredity* **3**, 229–241.
Fisher, R. A. 1950. Polydactyly in mice. *Nature* **165**, 407.
Fisher, R. A., and Diver, C. 1934. Crossing-over in the land snail *Cepaea nemoralis* L. *Nature* **133**, 834.
Forbes, E. 1959. Use of mitotic segregation for assigning genes to linkage groups in *Aspergillus nidulans*. *Heredity* **13**, 67–80.
Ford, C. E., Hamerton, J. L., and Sharman, G. B. 1957. Chromosome polymorphism in the common shrew. *Nature* **180**, 392–393.
Ford, E. B. 1945. Polymorphism. *Biol. Revs. Cambridge Phil. Soc.* **20**, 73–88.
Ford, E. B. 1955. Rapid evolution and the conditions which make it possible. *Cold Spring Harbor Symposia Quant. Biol.* **20**, 230–238.
Fox, A. S., and Gray, W. D. 1950. Immunogenetic and biochemical studies of *Neurospora crassa*: differences in tyrosinase activity between mating types of strain 15300 (albino-2). *Proc. Natl. Acad. Sci. U.S.* **36**, 538–546.
Frost, L. C. 1961. Heterogeneity in recombination frequencies in *Neurospora crassa*. *Genet. Research Cambridge* **2**, 43-62.
Geiringer, H. 1944. On the probability theory of linkage in Mendelian heredity. *Ann. Math. Stat.* **15**, 25–57.
Gibson, J. B., and Thoday, J. M. 1959. Recombinational lethals in a polymorphic population. *Nature* **184**, 1593–1594.
Gibson, J. B., and Thoday, J. M. 1962. Effects of disruptive selection VI. A second chromosome polymorphism. *Heredity* **17**, 1–26.
Gibson, J. B., Parsons, P. A., and Spickett, S. G. 1961. Correlations between chaeta number and fly size in *Drosophila melanogaster*. *Heredity* **16**, 349–354.
Gluecksohn-Waelsch, S. (1954). Some genetic aspects of development. *Cold Spring Harbor Symposia Quant. Biol.* **19**, 41–49.
Goodale, H. D. 1937. Can artificial selection produce unlimited change? *Am. Naturalist* **71**, 433–459.
Goodale, H. D. 1938. A study of the inheritance of body weight in the albino mouse by selection. *J. Heredity* **29**, 101–112.
Gowen, J. W. 1919. A biometrical study of crossing over. On the mechanism of crossing over in the third chromosome of *Drosophila melanogaster*. *Genetics* **4**, 205–250.
Gowen, J. W. 1933. Meiosis as a genetic character in *Drosophila melanogaster*. *J. Exptl. zool.* **65**, 83–106.
Grant, V. 1958. The regulation of recombination in plants. *Cold Spring Harbor Symposia Quant. Biol.* **23**, 337–363.
Graubard, M. A. 1932. Inversion in *Drosophila melanogaster*. *Genetics* **17**, 81–105.
Green, M. M. 1959. Effect of different wild-type isoalleles on crossing-over in *Drosophila melanogaster*. *Nature* **184**, 294.
Gross, J., and Englesberg, E. 1959. Determination of the order of mutational sites governing L-arabinose utilization in *Escherichia coli* B/r by transduction with phage P1bt. *Virology* **9**, 314–331.
Gross, S. R., and Fein, A. 1960. Linkage and function in *Neurospora*. *Genetics* **45**, 885–904.
Grüneberg, H. 1937. Gene doublets as evidence for adjacent small duplications in *Drosophila*. *Nature* **140**, 932.
Haldane, J. B. S. 1957. The conditions for coadaptation in polymorphism for inversions. *J. Genet.* **55**, 218–225.
Hansen, H. N., and Snyder, W. C. 1943. The dual phenomenon and sex in *Hypomyces solanix cucurbitae*. *Am. J. Botany* **30**, 419–422.

Harrison, B. J., and Mather, K. 1950. Polygenic variability in chromosomes of *Drosophila melanogaster* obtained from the wild. *Heredity* **4**, 295–312.

Hartman, P. E., Hartman Z., and Serman, D. 1960a. Complementation mapping by abortive transduction of histidine-requiring *Salmonella* mutants. *J. Gen. Microbiol.* **22**, 354–368.

Hartman, P. E., Loper, J. C., and Serman, D. 1960b. Fine structure mapping by complete transduction between histidine-requiring *Salmonella* mutants. *J. Gen. Microbiol.* **22**, 323–353.

Haskell, G. 1954. Correlated responses to polygenic selection in animals and plants. *Am. Naturalist* **88**, 5–20.

Haskell, G. 1959. Further evidence against pleiotropic gene action in correlated responses to selection. *Genetica* **30**, 140–151.

Haskins, F. A., and Mitchell, H. K. 1952. An example of the influence of modifying genes in *Neurospora*. *Am. Naturalist* **86**, 231–238.

Hayman, D. L., and Parsons, P. A. 1961. The effect of temperature, age and an inversion on recombination values and interference in the X-chromosome of *Drosophila melanogaster*. *Genetica* **31**, 1–15.

Hildreth, P. E. 1955. A test for recombinational lethals in the X-chromosome of *Drosophila melanogaster*. *Proc. Natl. Acad. Sci. U.S.* **41**, 20–24.

Hildreth, P. E. 1956. The problem of synthetic lethals in *Drosophila melanogaster*. *Genetics* **41**, 729–742.

Hirsch, H. M. 1954. Environmental factors influencing the differentiation of protoperithecia and their relation to tyrosinase and melanin formation in *Neurospora crassa*. *Physiol. Plantarum* **7**, 72–97.

Holt, S. B. 1945. A polydactyl gene in mice capable of nearly regular manifestation. *Ann. Eugen. (London)* **12**, 220–249.

Hopwood, D. A. 1959. Linkage and the mechanism of recombination in *Streptomyces coelicolor*. *Ann. N.Y. Acad. Sci.* **81**, 887–898.

Horowitz, N. H., Fling, M., Macleod, H. L., and Sueoka, N. 1960. Genetic determination and enzymatic induction of tyrosinase in *Neurospora*. *J. Mol. Biol.* **2**, 96–104.

Jacob, F., and Monod, J. 1961. Genetic regulatory mechanisms in the synthesis of proteins *J. Mol. Biol.* **3**, 318—356.

Jacob, F., and Wollman, E. L. 1955. Etude génétique d'un bactériophage tempéré d'*Escherichia coli* III. Effect du rayonnement ultraviolet sur la recombinaison génétique. *Ann. inst. Pasteur* **88**, 724–749.

Jacob, F., Perrin, D., Sanchez, C., and Monod, J. 1960. L'opéron: groupe de gènes à expression coordonnée par un opérateur. *Compt. rend. acad. sci.* **250**, 1727–1729.

Jennings, H. S. 1917. The numerical results of diverse systems of breeding, with respect to two pairs of characters, linked or independent, with special relation to the effects of linkage. *Genetics* **2**, 97–154.

Jinks, J. L. 1952. Heterocaryosis in wild *Penicillium*. *Heredity* **6**, 77–87.

Jinks, J. L. 1961. Gene structure and function in the *h*III region of bacteriophage T4. *Heredity* **16**, 153–168.

John, B., and Lewis, K. R. 1957. Studies on *Periplaneta americana*. I. Experimental analysis of male meiosis. II. Interchange heterozygosity in isolated populations. *Heredity* **11**, 1–22.

John, B., and Lewis, K. R. 1958. Studies on *Periplaneta americana*. III. Selection for heterozygosity. *Heredity* **12**, 185–197.

John, B., and Lewis, K. R. 1959. Selection for interchange heterozygosity in an inbred culture of *Blaberus discoidalis* (Serville). *Genetics* **44**, 251–267.

Johnsson, H. 1944. Meiotic aberrations and sterility in *Alopecurus myosuroides*. *Hereditas* **30**, 469–566.
Kimura, M. 1956. A model of a genetic system which leads to closer linkage by natural selection. *Evolution* **10**, 278–287.
Komai, T. 1950. Semi-allelic genes. *Am. Naturalist* **84**, 381–392.
Lamm, R. 1936. Cytological studies on inbred rye. *Hereditas* **22**, 217–240.
Lawrence, M. J., 1958. Genotypic control of crossing-over on the first chromosome of *Drosophila melanogaster*. *Nature* **182**, 889–890.
Lederberg, E. M. 1960. Genetic and functional aspects of galactose metabolism in *Escherichia coli* K-12. *Symposium Soc. Gen. Microbiol.* 10th pp. 115–131.
Lederberg, J. 1955. Recombination mechanisms in bacteria. *J. Cellular Comp. Physiol.* **45** (Suppl. 2), 75–107.
Lee, B. T. O., and Pateman, J. A. 1959. Linkage of polygenes controlling size of ascospores in *Neurospora crassa*. *Nature* **183**, 698–699.
Lee, B. T. O., and Pateman, J. A. 1961. Studies concerning the inheritance of ascospore length in *Neurospora crassa*. *Australian J. Biol. Sci.* **14**, 223–230.
Leupold, U. 1958. Studies on recombination in *Schizosaccharomyces pombe*. *Cold Spring Harbor Symposia Quant. Biol.* **23**, 161–170.
Levan, A. 1941. The cytology of the species hybrid *Allium cepa* × *fistulosum* and its polyploid derivatives. *Hereditas* **27**, 253–272.
Levene, H. 1959. Release of genetic variability through recombination. IV. Statistical theory. *Genetics* **44**, 93–104.
Levene, H., Pavlovsky, O., and Dobzhansky, T. 1954. Interaction of the adaptive values in polymorphic experimental populations of *Drosophila pseudoobscura*. *Evolution* **8**, 335–349.
Levine, R. P. 1955. Chromosome structure and the mechanism of crossing over. *Proc. Natl. Acad. Sci. U.S.* **41**, 727–730.
Levine, R. P., and Levine, E. E. 1954. The genotypic control of crossing over in *Drosophila pseudoobscura*. *Genetics* **39**, 677–691.
Levine, R. P. and Levine, E. E. 1955. Variable crossing over arising in different strains of *Drosophila pseudoobscura*. *Genetics* **40**, 399–405.
Levitan, M. 1955. Studies of linkage in populations I. Associations of second chromosome inversions in *Drosophila robusta*. *Evolution* **9**, 62–74.
Levitan, M. 1958a. Non-random associations of inversions. *Cold Spring Harbor Symposia Quant. Biol.* **23**, 251–268.
Levitan, M. 1958b. Studies of linkage in populations II. Recombination between linked inversions of *D. robusta*. *Genetics* **43**, 620–633.
Levitan, M., and Salzano, F. M., 1959. Studies of linkage in populations III. An association of linked inversions in *Drosophila guaramunu*. *Heredity* **13**, 243–248.
Lewis, D. 1952. Serological reactions of pollen incompatibility substances. *Proc. Roy. Soc.* **B140**, 127–135.
Lewis, D. 1954. Comparative incompatibility in Angiosperms and Fungi. *Advances in Genet.* **6**, 235–285.
Lewis, D. 1960. Genetic control of specificity and activity of the S antigen in plants. *Proc. Roy. Soc.* **B151**, 468–477.
Lewis, D., and Crowe, L. K. 1958. Unilateral interspecific incompatibility in flowering plants. *Heredity* **12**, 233–256.
Lewontin, R. C., and Kojima, K. 1961. The evolutionary dynamics of complex polymorphisms. *Evolution* **14**, 458–472.
Lewontin, R. C., and White, M. J. D. 1960. Interaction between inversion polymor-

phisms of two chromosome pairs in the grass-hopper, *Moraba scurra*. *Evolution* **14**, 116–129.

Maeda, T. 1937. Chiasma studies in *Allium fistulosum*, *Allium cepa* and their F_1, F_2 and backcross hybrids. *Japan. J. Genet.* **13**, 146–159.

Mallyon, S. A. 1951. A pronounced sex difference in recombination values in the sixth chromosome of the house mouse. *Nature* **168**, 118.

Markert, C. L. 1950. The effects of genetic changes on tyrosinase activity in *Glomerella*. *Genetics* **35**, 60–75.

Markert, C. L., and Owen, R. D. 1954. Immunogenetic studies of tyrosinase specificity. *Genetics* **39**, 818–835.

Mather, K. 1936. Competition between bivalents during chiasma formation. *Proc. Roy. Soc.* **B120**, 208–227.

Mather, K. 1938. Crossing-over. *Biol. Revs. Cambridge Phil. Soc.* **13**, 252–292.

Mather, K. 1939a. Competition for chiasmata in diploid and trisomic maize. *Chromosoma* **1**, 119–129.

Mather, K. 1939b. Crossing over and heterochromatin in the X chromosome of *Drosophila melanogaster*. *Genetics* **24**, 413–435.

Mather, K. 1941. Variation and selection of polygenic characters. *J. Genet.* **41**, 159–193.

Mather, K. 1942. The balance of polygenic combinations. *J. Genet.* **43**, 309–336.

Mather, K. 1943a. Polygenic inheritance and natural selection. *Biol. Revs. Cambridge Phil. Soc.* **18**, 32–64.

Mather, K. 1943b. Species differences in *Petunia*. I. Incompatibility. *J. Genet.* **45**, 215–235.

Mather, K. 1953. The genetical structure of populations. *Symposia Soc. Exptl. Biol.* **7**, 66–95.

Mather, K. 1955a. The genetical basis of heterosis. *Proc. Roy. Soc.* **B144**, 143–150.

Mather, K. 1955b. Polymorphism as an outcome of disruptive selection. *Evolution* **9**, 52–61.

Mather, K., and De Winton, D. 1941. Adaptation and counteradaptation of the breeding system in *Primula*. *Ann. Botany (London)* **5**, 297–311.

Mather, K., and Harrison, B. J. 1949. The manifold effect of selection. *Heredity* **3**, 1–52, 131–162.

Mather, K., and Vines, A. 1951. Species crosses in *Antirrhinum* II. Cleistogamy in the derivatives of *A. majus* × *A. glutinosum*. *Heredity* **5**, 195–214.

Mather, K., and Wigan, L. G. 1942. The selection of invisible mutations. *Proc. Roy. Soc.* **131**, 50–64.

Michie, D. 1955. Genetical studies with "vestigial tail" mice. I. The sex difference in crossing-over between vestigial and Rex. *J. Genet.* **53**, 270–279.

Millicent, E., and Thoday, J. M. 1960. Gene flow and divergence under disruptive selection. *Science* **131**, 1311–1312.

Millicent, E., and Thoday, J. M. 1961. Effects of disruptive selection IV. Gene-flow and divergence. *Heredity* **16**, 199–218.

Misro, B. 1949. Crossing-over as a source of new variation. *Proc. Intern. Congr. Genet. 8th Congr. Stockholm* 1949 pp. 629–630.

Moffett, A. A. 1936. The origin and behaviour of chiasmata XIII. Diploid and tetraploid *Culex pipiens*. *Cytologia (Tokyo)* **7**, 184–197.

Müntzing, A. 1954. Cyto-genetics of accessory chromosomes (B-chromosomes). *Caryologia* **6**, (Suppl.) 282–301.

Müntzing, A. 1958. Accessory chromosomes. *Trans. Bose Research Inst.* **22**, 1–15.

Müntzing, A., and Akdik, S. 1948. Cytological disturbances in the first inbred generations of rye. *Hereditas* **34**, 485–509.

Mukherjee, A. S. 1961. Effect of selection on crossing over in the males of *Drosophila ananassae*. *Am. Naturalist* **95**, 57–59.

Muller, H. J. 1925. The regionally differential effect of X rays on crossing over in autosomes of *Drosophila*. *Genetics* **10**, 470–507.

Murray, N. E. 1960. The distribution of methionine loci in *Neurospora crassa*. *Heredity* **15**, 199–206.

Pandey, K. K. 1961. S allele mutations and components of the S gene. *Heredity* **16**, 239.

Parag, Y., and Raper, J. R. 1960. Genetic recombination in a common B cross of *Schizophyllum commune*. *Nature* **188**, 765–766.

Pardee, A. B., Jacob, F., and Monod, J. 1959. The genetic control and cytoplasmic expression of "inducibility" in the synthesis of β-galactosidase by *E. coli*. *J. Mol. Biol.* **1**, 165–178.

Parsons, P. A. 1958a. A balanced four-point linkage experiment for linkage group XIII of the house mouse. *Heredity* **12**, 77-95.

Parsons, P. A. 1958b. Genetical interference in *Drosophila* spp. *Nature* **182**, 1815–1816.

Parsons, P. A. 1958c. Additional three-point data for linkage group V of the mouse. *Heredity* **12**, 357–362.

Parsons, P. A. 1958d. A survey of genetical interference in maize. *Genetica* **29**, 222–237.

Parsons, P. A. 1958e. Selection for increased recombination in *Drosophila melanogaster*. *Am. Naturalist* **92**, 255–256.

Parsons, P. A. 1961a. Fly size, emergence time and sterno-pleural chaeta number in *Drosophila*. *Heredity* **16**, 455–473.

Parsons, P. A. 1961b. The initial progress of new genes with viability differences between sexes and with sex linkage. *Heredity* **16**, 103–107.

Parsons, P. A., and Bodmer, W. F. 1961. The evolution of overdominance: natural selection and heterozygote advantage. *Nature* **190**, 7–12.

Pateman, J. A. 1955. Polygenic inheritance in *Neurospora*. *Nature* **176**, 1274–1275.

Pateman, J. A. 1959. The effect of selection on ascospore size in *Neurospora crassa*. *Heredity* **13**, 1–21.

Pateman, J. A., and Lee, B. T. O. 1960. Segregation of polygenes in ordered tetrads. *Heredity* **15**, 351–361.

Patterson, J. T., and Suche, M. L. 1934. Crossing over induced by X rays in *Drosophila* males. *Genetics* **19**, 223–236.

Payne, F. 1918. An experiment to test the nature of the variations on which selection acts. *Indiana Univ. Studies* **5** (36), 1–45.

Payne, F. 1920. Selection for high and low bristle number in the mutant strain "reduced." *Genetics* **5**, 501–542.

Perkins, D. D. 1959. New markers and multiple point linkage data in *Neurospora*. *Genetics* **44**, 1185–1226.

Peto, F. H. 1933. The cytology of certain intergeneric hybrids between *Festuca* and *Lolium*. *J. Genet.* **28**, 113–156.

Plough, H. H. 1917. The effect of temperature on crossing over in *Drosophila*. *J. Exptl. Zool.* **24**, 147–209.

Pontecorvo, G. 1946. Genetic systems based on heterocaryosis. *Cold Spring Harbor Symposia Quant. Biol.* **11**, 193–201.

Pontecorvo, G. 1950. New fields in the biochemical genetics of micro-organisms. *Biochem. Soc. Symposia (Cambridge, Engl.)* **4**, 40–50.

Pontecorvo, G. 1959. "Trends in Genetic Analysis," 145 pp. Columbia Univ. Press, New York.
Prakken, R., and Müntzing, A. 1942. A meiotic peculiarity in rye, simulating a terminal centromere. *Hereditas* **28**, 441–482.
Race, R. R., and Sanger, R. 1958. "Blood Groups in Man," 3rd ed., 377 pp. Blackwell, Oxford, England.
Race, R. R., and Sanger, R. 1959. The inheritance of blood groups. *Brit. Med. Bull.* **15**, 99–109.
Rees, H. 1955a. Heterosis in chromosome behaviour. *Proc. Roy. Soc.* **B144**, 150–159.
Rees, H. 1955b. Genotypic control of chromosome behaviour in rye. I. Inbred lines. *Heredity* **9**, 93–116.
Rees, H. 1957. Genotypic control of chromosome behaviour in rye. IV. The origin of new variation. *Heredity* **11**, 185–193.
Rees, H., and Naylor, B. 1960. Developmental variation in chromosome behaviour. *Heredity* **15**, 17–27.
Rees, H., and Thompson, J. B. 1956. Genotypic control of chromosome behaviour in rye. III. Chiasma frequency in homozygotes and heterozygotes. *Heredity* **10**, 409–424.
Rees, H., and Thompson, J. B. 1958. Genotypic control of chromosome behaviour in rye. V. The distribution pattern of chiasmata between pollen mother cells. *Heredity* **12**, 101–111.
Rendel, J. M. 1957. Relationship between coincidence and crossing-over in *Drosophila*. *J. Genet.* **55**, 95–99.
Rendel, J. M. 1958. The effect of age on the relationship between coincidence and crossing over in *Drosophila melanogaster*. *Genetics* **43**, 207–214.
Rifaat, O. M. 1959. Effect of temperature on crossing-over in *Neurospora crassa*. *Genetica* **30**, 312–323.
Roberts, C. F. 1961. Genetic analysis of sugar mutants. *Aspergillus News Letter*, Spring 1961.
Roberts, J. A. F. 1959. Some associations between blood groups and disease. *Brit. Med. Bull.* **15**, 129–133.
Roman, H. 1956. Studies of gene mutation in *Saccharomyces*. *Cold Spring Harbor Symposia Quant. Biol.* **21**, 175–185.
Roman, H., and Jacob, F. 1958. A comparison of spontaneous and ultraviolet-induced allelic recombination with reference to the recombination of outside markers. *Cold Spring Harbor Symposia Quant. Biol.* **23**, 155–160.
Saunders, E. R. 1911. Further experiments on the inheritance of "doubleness" and other characters in stocks. *J. Genet.* **1**, 303–376.
Schultz, J., and Redfield, H. 1951. Interchromosomal effect on crossing over in *Drosophila*. *Cold Spring Harbor Symposia Quant. Biol.* **16**, 175–197.
Sharman, G. B. 1956. Chromosomes of the common shrew. *Nature* **177**, 941–942.
Sheppard, P. M. 1953. Polymorphism, linkage and the blood groups. *Am. Naturalist* **87**, 283–294.
Sheppard, P. M. 1955. Genetic variability and polymorphism: synthesis. *Cold Spring Harbor Symposia Quant. Biol.* **20**, 271–275.
Sheppard, P. M. 1959. The evolution of mimicry; a problem in ecology and genetics. *Cold Spring Harbor Symposia Quant. Biol.* **24**, 131–140.
Shull, A. F., and Whittinghill, M. 1934. Crossovers in male *Drosophila melanogaster* induced by heat. *Science* **80**, 103–104.

Sismanidis, A. 1942. Selection for an almost invariable character in *Drosophila*. *J. Genet.* **44**, 204–215.
Smith D. A. 1961. Some aspects of the genetics of methioneless mutants of *Salmonella typhimurium*. *J. Gen. Microbiol* **24**, 335–353.
Sobels, F. H. 1956. Studies on the mutagenic action of formaldehyde in *Drosophila*. II. The production of mutations in females and the induction of crossing-over. *Z. Induktive Abstammungs-u. Vererbungslehre* **87**, 743–752.
Spassky, B., Spassky, N. Levene, H., and Dobzhansky, T. 1958. Release of genetic variability through recombination. I. *Drosophila pseudoobscura*. *Genetics* **43**, 844–867.
Spiess, E. B., 1958. Effects of recombination on viability in *Drosophila*. *Cold Spring Harbor Symposia Quant. Biol.* **23**, 239–250.
Spiess, E. B. 1959. Release of genetic variability through recombination II. *Drosophila persimilis*. *Genetics* **44**, 43–58.
Stadler, L. J. 1926. The variability of crossing over in maize. *Genetics* **11**, 1–37.
Stebbins, G. L. 1950. "Variation and Evolution in Plants," 643 pp. Columbia Univ. Press, New York.
Stebbins, G. L. 1958a. The inviability, weakness, and sterility of interspecific hybrids. *Advances in Genet.* **9**, 147–215.
Stebbins, G. L. 1958b. Longevity, habitat, and release of genetic variability in the higher plants. *Cold Spring Harbor Symposia Quant. Biol.* **23**, 365–378.
Steinberg, A. G., and Fraser, F. C. 1944. Studies on the effect of X chromosome inversions on crossing over in the third chromosome of *Drosophila melanogaster*. *Genetics* **29**, 83–103.
Stern, C. 1926. An effect of temperature and age on crossing over in the first chromosome of *Drosophila melanogaster*. *Proc. Natl. Acad. Sci. U.S.* **12**, 530–532.
Sturtevant, A. H. 1917. Genetic factors affecting the strength of linkage in *Drosophila*. *Proc. Natl. Acad. Sci. U.S.* **3**, 555–558.
Sturtevant, A. H. 1918. An analysis of the effects of selection. *Carnegie Inst. Wash. Publ.* **264**, 1–68.
Sturtevant, A. H. 1919. Contributions to the genetics of *Drosophila melanogaster*. III. Inherited linkage variations in the second chromosome. *Carnegie Inst. Wash. Publ.* **278**, 305–341.
Sturtevant, A. H. 1951. The relation of genes and chromosomes. *In* "Genetics in the 20th Century" (L. C. Dunn, ed.), pp. 101–110. Macmillan, New York.
Sturtevant, A. H., and Mather, K. 1938. The interrelations of inversions, heterosis and recombination. *Am. Naturalist* **72**, 447–452.
Tebb, G., and Thoday, J. M. 1954. Stability in development and relational balance of X-chromosomes in *Drosophila melanogaster*. *Nature* **174**, 1109.
Thoday, J. M. 1953. Components of fitness. *Symposia Soc. Exptl. Biol.* **7**, 96–113.
Thoday, J. M. 1955. Balance, heterozygosity and developmental stability. *Cold Spring Harbor Symposia Quant. Biol.* **20**, 318–326.
Thoday, J. M. 1958. Effects of disruptive selection: the experimental production of a polymorphic population. *Nature* **181**, 1124–1125.
Thoday, J. M. 1959. Effects of disruptive selection. I. Genetic flexibility. *Heredity* **13**, 187–203.
Thoday, J. M. 1960. Effects of disruptive selection. III. Coupling and repulsion. *Heredity* **14**, 35–49.
Thoday, J. M. 1961. The location of polygenes. *Nature* **191**, 368–370.
Thoday, J. M., and Boam, T. B. 1956. A possible effect of the cytoplasm on recombination in *Drosophila melanogaster*. *J. Genet.* **54**, 456–461.

Thoday, J. M., and Boam, T. B. 1959. Effects of disruptive selection. II. Polymorphism and divergence without isolation. *Heredity* **13**, 205–218.
Thoday, J. M., and Boam, T. B. 1961. Regular responses to selection I. Description of responses. *Genet. Research Cambridge* **2**, 161–176.
Thompson, J. B., 1956. Genotypic control of chromosome behaviour in rye. II. Disjunction at meiosis in interchange heterozygotes. *Heredity* **10**, 99–108.
Tschetwerikoff, S. S. 1926. On certain features of the evolutionary process from the viewpoint of modern genetics. *J. Exptl. Biol. Med. U.S.S.R.* **2**, 3–54.
Vetukhiv, M. 1954. Integration of the genotype in local populations of three species of *Drosophila*. *Evolution* **8**, 241–251.
Vetukhiv, M. 1956. Fecundity of hybrids between geographic populations of *Drosophila pseudoobscura*. *Evolution* **10**, 139–146.
Vetukhiv, M. 1957. Longevity of hybrids between geographic populations of *Drosophila pseudoobscura*. *Evolution* **11**, 348–360.
Wahrman, J., and Zahavi, A. 1955. Cytological contributions to the phylogeny and classification of the rodent genus *Gerbillus*. *Nature* **175**, 600–602.
Wallace, B. 1953. On coadaptation in *Drosophila*. *Am. Naturalist* **87**, 343–358.
Wallace, B. 1955. Inter-population hybrids in *Drosophila melanogaster*. *Evolution* **9**, 302–316.
Wallace, B., and Vetukhiv, M. 1955. Adaptive organisation of the gene pools of *Drosophila* populations. *Cold Spring Harbor Symposia Quant. Biol.* **20**, 303–310.
Wallace, B., King, J. C., Madden, C. V., Kaufmann, B., and McGunnigle, E. C. 1953. An analysis of variability through recombination. *Genetics* **38**, 272–307.
Wallace, M. E. 1957. A balanced three-point experiment for linkage group V of the house mouse. *Heredity* **11**, 223–258.
White, M. J. D. 1934. The influence of temperature on chiasma frequency. *J. Genet.* **29**, 203–215.
White, M. J. D. 1954 "Animal Cytology and Evolution," 2nd ed., 454 pp. Cambridge Univ. Press, London and New York.
White, M. J. D. 1957. Cytogenetics of the grasshopper *Moraba scurra*. II. Heterotic systems and their interaction. *Australian J. Zool.* **5**, 305–337.
White, M. J. D. 1958. Restrictions on recombination in grasshopper populations and species. *Cold Spring Harbor Symposia Quant. Biol.* **23**, 307–317.
Whittinghill, M., and Lewis, B. M. 1961. Clustered cross-overs from male *Drosophila* raised on formaldehyde media. *Genetics* **46**, 459–462.
Wiener, A. S., and Wexler, I. B. 1958. "Heredity of the Blood Groups," 150 pp. Grune & Stratton, New York.
Wigan, L. G. 1944. Balance and potence in natural populations. *J. Genet.* **46**, 150–160.
Wigan, L. G. 1949a. The distribution of polygenic activity on the X chromosome of *Drosophila melanogaster*. *Heredity* **3**, 53–66.
Wigan, L. G. 1949b. Chromosome regions which give new variation by crossing over. *Proc. Intern. Congr. Genet. 8th Congr. Stockholm* 1949 pp. 686–687.
Wigan, L. G., and Mather, K. 1942. Correlated response to the selection of polygenic characters. *Ann. Eugen. (London)* **11**, 354–364.
Wilson, J. Y. 1959. Changes in the distribution of chiasmata in response to experimental factors in the bluebell, *Endymion nonscriptus* (L.) Garcke. *Genetica* **30**, 417–434.
Winge, Ö. 1931. The inheritance of double flowers and other characters in *Matthiola*. *Z. Zücht. Reihe A z. Pflanzenzucht.* **17**, 118–135.
Wright, S. 1940. The statistical consequences of Mendelian heredity in relation to speciation. "The New Systematics," pp. 161–183. Clarendon Press, Oxford.

Wright, S. 1952. The genetics of quantitative variability. "Quantitative Inheritance," pp. 1–41. H. M. Stationery Office, London.

Wright, S., and Dobzhansky, T. 1946. Genetics of natural populations. XII. Experimental reproduction of some of the changes caused by natural selection in certain populations of *Drosophila pseudoobscura*. *Genetics* **31**, 125–156.

Yanofsky, C. 1960. The tryptophan synthetase system. *Bacteriol. Revs.* **24**, 221–245.

EPISOMES

Allan M. Campbell

Department of Biology, University of Rochester, Rochester, New York

	Page
I. Introduction	101
A. General Properties of Episomes	101
B. Brief Description of Episomes Other than Phage	102
C. General Remarks	107
II. The Autonomous State	107
A. Vegetative Replication of Virulent and Temperate Bacteriophages	107
B. The Carrier State of Phage	108
C. Episomes Other than Phages	109
III. The Integrated State	109
A. Chromosomal Localization of Phage and Other Episomes	109
B. Mode of Attachment of the Prophage to the Bacterial Chromosome	111
IV. Immunity	114
A. Superinfection Immunity and Repression	114
B. Cytoplasmic Nature	115
C. Genetic Determination of Immunity	116
D. Conclusions	120
E. Episomes Other than Phage	120
V. Lysogenization. Transition from the Vegetative to the Integrated State	121
A. Infection of Sensitive Cells by Temperate Phage	121
B. The Decision Not to Lyse	122
C. Genetic Incorporation	123
D. Episomes Other than Phage	124
VI. Transition from the Integrated to the Autonomous State	124
VII. Curing	125
VIII. "Gene Pick-Up" by Episomes	126
A. Phage-Mediated Transduction	126
B. F-Mediated Transduction	129
C. Existing Episomes as Products of Gene Pick-Up	131
IX. Partial Diploidy in *Escherichia coli*	132
X. Transfer of Episomes by Other Episomes	134
XI. Relationship to Cellular Regulatory Mechanisms	135
XII. General Discussion	135
Acknowledgments	137
References	137

I. Introduction

A. General Properties of Episomes

The concept of the episome was introduced by Jacob and Wollman (1958b) as a generic term for those genetic elements able to exist in two alternative states: (1) an *integrated* state, in which the element is associated with some point on, or small region of, a chromosome, and apparently multiplies synchronously with it, and (2) an *autonomous* state, in which the element multiplies independently of, and frequently faster than, the cell which carries it. They further stipulate that the integrated state must represent an addition to the chromosome rather than a replacement of homologous genetic material.

Most of the episomes of bacteria fall into one of two classes: (1) temperate bacteriophages (or, strictly, the genetic material thereof), and (2) transfer factors, which can pass from cell to cell during conjugation independently of the bulk of the bacterial genome. Some transfer factors play a causative role in the conjugation process itself, such as the fertility factor (F) of *Escherichia coli* K12. We shall consider such independent transfer as indicating that a given factor is an episome even where an alternation between integrated and autonomous states has not been shown.

The episomic nature of temperate bacteriophage was established earlier than that of the transfer factors, and we have a great deal more information about them, which has been the subject of several excellent reviews (Lwoff, 1953; Bertani, 1958; Jacob and Wollman, 1959a). The purpose of the present review will be to ask some of the same questions about episomes in general which have been asked about phage in particular. Before proceeding, we shall give a brief historical account of the known episomes other than phage as well as some instances of possibly similar elements in higher organisms.

B. BRIEF DESCRIPTION OF EPISOMES OTHER THAN PHAGE

1. *The F Agent*

Genetic recombination resulting from cellular conjugation was discovered in *Escherichia coli* K12 in 1946 (Lederberg and Tatum, 1946), but it was several years before some aspects of the mechanism became clear. The initial observation leading to this clarification was that some strains had spontaneously lost their ability to mate with each other, although retaining the ability to mate with the parent strain. The parent strain was thus designated as fertile (F+) and the derived strain as infertile (F−). Subsequent studies showed that mating is an asymmetric process, wherein one partner functions as a donor and the other as a recipient of genetic material. Only F+ strains can function as donors in crosses (Lederberg *et al.*, 1952; Hayes, 1953a,b).

It was also found that the F+ property could be transferred from cell to cell at a rate in excess of the transfer of bacterial genes in general, and that F+ cells could be "cured" of the F property by exposure to acridine dyes (Hirota, 1960), a treatment known to remove nonchromosomal genetic determinants from yeast (Ephrussi et al., 1949) and other organisms (see Lederberg, 1952). From these facts, the existence of a discrete F factor could be inferred, which was carried by F+ cells but not by F− cells.

F+ strains give rise to variants (Hfr), in which the frequency of transfer of bacterial markers to F− cultures is greatly elevated, and that of the F property itself greatly depressed. It can be verified by microscopy and micromanipulation that when an Hfr and F− cell come together, they form a cytoplasmic connection, and that recombinants appear only among the descendants of the F− exconjugant (Lederberg, 1956, 1957; Anderson and Mazé, 1957). The mechanism of mating has been studied in some detail (Jacob and Wollman, 1956a,b, 1957a, 1958a; Skaar and Garen, 1956; Wollman and Jacob, 1955, 1958a,b). Starting from a given point of origin on a formally circular linkage structure, genetic markers enter the recipient in a temporal order which corresponds to their linkage relationships. The entire transfer requires about 2 hours, and the final character transferred is the fertility property itself. The point of origin and also the direction around the circle are characteristic of the particular Hfr isolate used. Hfr strains can revert with various degrees of ease to F+, demonstrating the presence of F as a genetic element in Hfr strains. Hfr strains are not cured by acridine dyes (Hirota, 1960). The F agent in the Hfr strains is thus in the integrated rather than the autonomous state.

The F agent seems to contain DNA, because incorporation of radioactive phosphorus into F (as judged by subsequent "suicidal" inactivation on storage due to disintegration of phosphorus atoms within the F particle) is inhibited by mitomycin (Driskell and Adelberg, 1961). Naturally occurring variants of F have been reported (Lederberg and Lederberg, 1956; Bernstein, 1958) and laboratory variants mediating high frequency gene transfer (Jacob and Adelberg, 1959) are now common. From phosphorus suicide experiments, the nucleic acid content of F can be estimated as 90–120 million molecular weight (Lavallé and Jacob, 1961). This calculation is made from the relative sensitivities of F and of bacteriophage lambda, assuming a molecular weight of 60×10^6 for lambda.

Another type of Hfr strain (very high frequency, Vhf) has recently been described (Taylor and Adelberg, 1960). These differ from ordinary Hfr in that the whole genome is transferred with high frequency. With these strains it has been possible to confirm by linkage studies the circular chromosome which Jacob and Wollman had proposed. It is not known whether the Vhf character is a genetic property of F, of the bacterium, or of the position or mode of attachment of F.

Many strains of *Escherichia coli* (Cavalli and Heslot, 1949; Lederberg *et al.*, 1952; de Haan, 1954, 1955; Lieb *et al.*, 1955; Bertani and Six, 1958; Ørskov and Ørskov, 1961) and the related genera *Salmonella* (Baron *et al.*, 1959b; Demerec *et al.*, 1960; Miyake and Demerec, 1959; Zinder, 1960a,b; Ørskov, *et al.*, 1961) and *Shigella* (Luria and Burrous, 1957) can act as recipients in crosses with F+ or Hfr strains. In most of these cases, it has been shown that the F agent is actually transferred to and subsequently carried by these strains, although in some instances it no longer functions as a fertility factor in the new environment and is detectable only by the ability of the strains which carry it to transfer the F property back into K12. Mating by conjugation has been described also in the genera *Pseudomonas* (Holloway, 1955, 1956) and *Serratia* (Belser and Bunting, 1956). In *Pseudomonas*, fertility is controlled by an infectious agent similar to F (Holloway and Jennings, 1958).

2. *The Colicin Factor(s)*

Certain strains of *Escherichia coli* produce proteins (colicins) which are highly toxic for other strains. The large literature on colicins and other bacteriocins will not be reviewed here (see Fredericq, 1957, 1958). We are concerned only with the genetic control of colicin biosynthesis.

In order that the symbols used should be clear, we explain first that colicins attach themselves to specific receptors on the surface of their victim, and that colicins are classified according to their attachment specificity. Thus, colicin V will attack a certain group of bacterial strains and colicin E another. If two different colicins are found with the same attachment specificity, they are called, for example, E1 and E2.

Much of the genetic work on colicinogeny concerns the colicin formed by strain K30 of *E. coli*, which belongs to the colicin "E" family and has been called ER, E(1) or E1 (Fredericq, 1957). It has also on occasions been improperly referred to as "colicin K30." It was found (Fredericq, 1953; Fredericq and Betz-Bareau, 1953a,b,c) that, in crosses between colicinogenic and non-colicinogenic parents, one could not map this colicin factor in relation to other markers.

After our understanding of and control over the mating process had been improved by Jacob and Wollman, more definitive experiments were possible. Using the colicin factor of strain K30, which had been transmitted into strain K12, Alfoldi *et al.* (1957, 1958) showed that the transfer of a *col*− determinant from the donor into a *col*+ recipient resulted in the death of the zygote—a curious situation whose cause remains unexplained but which allows genetic localization of the colicinogeny factor by time of entrance experiments. By making different Hfr strains colicinogenic, they showed that the time of entrance of the $col+$ determinant into the F− recipient depended on the Hfr strain used. The reciprocal crosses showed that lethal zygosis showed a similar dependence on time of entry. One can thus conclude that the $col+$ determinant occupies a chromosomal site in these strains. Single cell pedigree analysis revealed that all of the progeny of a single zygote which had received $col+$ were themselves $col+$. This is similar to the behavior of F (Lederberg, 1959) and in marked contrast to the results obtained in comparable experiments with ordinary genetic markers (Lederberg, 1957; Anderson and Mazé, 1957). It explains the absence of linkage observed by Fredericq, and necessitates the assumption that, following conjugation, the $col+$ determinant can multiply faster than the rest of the genome.

Is colicin E1 unique, or are all colicinogeny determinants episomes? Crosses involving colicinogenic parents are complicated by various effects on the fertility of crosses (Fredericq and Betz-Bareau, 1956; Fredericq and Papavassiliou, 1957), which probably reflect differences in the pattern of lethal zygosis (Fredericq, 1956a). However, the rapid transfer of colicinogeny in mixed culture between distantly related strains which has been shown for many different colicins (Fredericq, 1954a, 1956b, 1957; Hamon, 1956) is certainly suggestive. In some cases, including that of colicin E1, this transfer requires that the donor be F+ (Fredericq, 1954a). However, the transfer of colicinogeny was distinct from that of other markers. With strains which produced more than one colicin, frequently only one determinant was transferable.

The transfer of the colicin I factor to non-colicinogenic *Salmonella* does not require the presence of F (Stocker, 1960; Ozeki et al., 1961). Furthermore, colicinogenic strains can transfer other markers of the bacterium at low frequencies. The colicin I determinant has therefore as much right to be considered a fertility factor as does F itself. It differs from F in having a much higher rate of integration; so that rapid transfer only occurs from cell lines very recently made colicinogenic.

From its sensitivity to phosphorus decay, the nucleic acid content of the colicinogeny factor must be less than a molecular weight of 2.4 million (Lavallé and Jacob, 1961).

3. *The Resistance Transfer Factor*

Strains of pathogenic *Shigella* resistant to streptomycin, tetracycline, chloramphenicol, and sulfonamide have been shown to carry on agent which imparts all these resistances simultaneously (Harada et al., 1960; Mitsuhashi et al., 1960a,b). This agent resembles a fertility factor in causing its own transfer from cell to cell by conjugation, but differs in that it does not render any other genes of the bacterium transferable. If the resistance transfer factor (RTF) is introduced into an F+ or Hfr strain of *Escherichia coli*, the transfer of bacterial genes by the fertility factor is eliminated or strongly reduced. An F+ cell is thus converted into an F− phenocopy, and spontaneous loss of RTF is accompanied by recovery of the F+ character (Watanabe and Fukasawa, 1960h; 1961a,b,c,i).

Our accounts of this work will be based on the available American literature. References to the Japanese literature (Akiba et al., 1960; Ochiai et al., 1959; Watanabe and Fukasawa, 1960a–g; 1961d–h) are included for the linguistically competent reader.

RTF is transferable from *Shigella* into members of the genera *Escherichia* and *Salmonella*. It is susceptible to curing by acridine dyes. Both the ease of transfer and the curing indicate that it is ordinarily in the autonomous state. The results of Hfr(RTF) × F− crosses suggest that, in a fraction of the population, RTF is associated with the host chromosome at or close to the B_1 locus (Watanabe and Fukasawa, 1962). A reversible equilibrium between integrated and autonomous states is thus indicated.

The individual determinants of resistance can become stably, irreversibly integrated, probably by replacing their homologs in the bacterial chromosome. This process has no known connection with the integration of RTF.

4. *The Sporulation Factor*

It has been suggested (Jacob et al., 1960b) that the factor responsible for bacterial sporulation might be an episome. This should provide an interesting opportunity to study the role of an episome in a simple developmental process.

5. *The Mycelial Factor in Aspergillus*

Roper (1958) studied three strains of *Aspergillus* in which distinguishable obvious morphological variations had been induced by acriflavine. The first (M1) was examined in greatest detail. Heterokaryons were made between genetically marked wild type and M1 individuals. Conidial isolates from these were mostly of the two parental types. Some individuals had the genotype of the originally wild type nucleus, but had become M1. The reciprocal class was never found.

This indicates that M1 is a genetic element which is generally associated with the nucleus but which can, rather infrequently, infect another nucleus. It can be considered a transfer factor which operates between nuclei in a synkaryon rather than between isolated bacterial cells. One point which should be stressed is that M1 arose as an induced mutant of a normal cell type, and we are therefore not free to assume that the wild type lacks any homolog to it.

6. *Controlling Elements in Maize*

McClintock (1956) has investigated extensively a variety of controlling elements in maize, and has emphasized the difference between these elements and the normal genes of that organism. Genetically, controlling elements are unique in their ability to undergo transposition from one part of the genome to another. Functionally, controlling elements may (1) cause chromosome breakage in neighboring regions, (2) modify or suppress the activity of neighboring genes, or (3) activate other controlling elements far removed from them in the genetic complement. It is their transposability which justifies our inclusion of these elements as episomes. Their functional aspects make them more similar to bacterial regulators and operators, for which transposability has not been found (cf. McClintock 1961).

C. General Remarks

We see that not all of the episomes mentioned here have been shown to satisfy Jacob and Wollman's definition, nor is there good reason to suppose that they will. They all have in common the property of being associated with the normal genetic complement of a cell but also of being able on occasions to assort independently of it. The ability to *replicate* in alternative states of course implies more than this, and remains to be proven (or disproven) in several cases. At present it seems more important to look for similarities than for differences, and the definition may well be modified with increasing knowledge.

II. The Autonomous State

A. Vegetative Replication of Virulent and Temperate Bacteriophages

A large amount of chemical and genetic work has converged to form a picture of intracellular bacteriophage growth which is satisfying at a certain level (Levinthal, 1959). Most of the evidence comes from the virulent coliphages T2, T4, and T6. The application to other phage-bacterium systems seems a reasonable extrapolation. Following injection of the phage DNA into the bacterial cell, certain new enzymes not found in the non-infected cell begin to appear which are necessary for the synthesis of phage rather than bacterial DNA (Kozloff, 1960). Almost all such early enzymes characterized so far function in the diversion of the synthetic pathways toward the production of DNA containing hydroxymethylcytosine rather than cytosine. Since this base occurs only in these particular virulent phages, such steps seem irrelevant to the vegetative growth of temperate phages. However, some early protein synthesis is prerequisite to the vegetative multiplication of the temperate coliphage lambda (Thomas, 1959).

In T2-infected cells, this early protein synthesis is necessary for phage DNA synthesis, which can then proceed independently of additional protein synthesis (Hershey and Melechen, 1957). The phage DNA thus produced is genetic material and not merely a precursor thereof (Tomizawa, 1958). In the average infected cell the amount of free phage DNA rises to a maximum of about 40 phage equivalents per cell and remains roughly constant until lysis. This constancy does not result from a cessation of synthesis but rather from equality of the rates of synthesis of new phage DNA and its rate of removal from the vegetative state by maturation.

The existence of a vegetative pool can also be inferred from the occurrence of genetic recombinants among the earliest mature phage particles produced, and from the kinetics of accumulation of recombinants among mature phage. These constitute the principal *genetic* evidence for a pool. The consistency with the chemical evidence, together with the possibility of artificially divorcing vegetative multiplication from maturation by the addition of chloramphenicol, leave little doubt as to the correctness of the general picture.

In those systems where initial infection does not strongly exclude later superinfection, one can estimate the number of vegetative phage in an infected cell as a function of time by studying the contribution of genetically marked superinfecting phage to the final yield and assuming that the superinfecting phage participates on an equal basis with the vegetative phage present at the moment of superinfection. The method can give self-consistent results (Thomas, 1959).

B. The Carrier State of Phage

Following infection with a virulent phage of the T2 type, a cell does not undergo any further divisions but simply enlarges somewhat during virus growth and ultimately lyses. With temperate phages, a fraction of the infected cells also lyse without dividing. The remainder survive the infection and produce colonies of lysogenic cells in which the phage genome is integrated as prophage. There are some reports which suggest that both temperate phages and some types of weakly virulent mutants thereof can also continue indefinitely in a non-integrated "carrier" (or pseudolysogenic) state, in which the cells multiply and occasionally lyse and liberate phage (Zinder, 1958; Luria et al., 1958). In those cases most carefully studied (Li et al., 1961), the indefinite maintenance of the carrier state is destroyed by the presence of antiphage serum, which implies that it is an artifact of intraclonal reinfection. There seems little doubt, however, that, during a limited number of generations following infection, a transient state exists in which the cell continues to grow and phage genomes multiply within the cells in a manner not exactly comparable either to the vegetative or to the prophage state.

C. Episomes Other than Phages

The autonomous state of the fertility factor resembles superficially the carrier state of phage, and the highly contagious nature of the F+ property suggests the possibility that its maintenance likewise requires frequent reinfections between cells. There is little direct evidence on this point, but the high frequency of occurrence of F− individuals among motilized isolates from F+ cultures (Skaar et al., 1957) is suggestive. The autonomous state of the colicinogenic factor is only seen transiently (Stocker, 1960; Ozeki et al., 1961). Whether this reflects an incapacity for indefinite autonomous replication or merely a high but constant probability of integration is not certain. The number of F particles per cell is probably about three, as judged by the enzyme levels of cells made partially diploid by F-mediated transduction (Jacob and Monod, 1961; Jacob and Wollman, 1961).

III. The Integrated State

A. Chromosomal Localization of Phage and Other Episomes

A lysogenic culture is defined as one in which the capacity to form phage is perpetuated intracellularly and transmitted from one generation to the next. Since the phage has a specific set of genetic information, each cell of a lysogenic culture must contain one or more copies of this information, and to these copies the name *prophage* is given. It seemed of fundamental importance to know whether the prophage was present in a few copies, distributed regularly between the daughter cells at each division, or in many copies, distributed at random at division. Studies on superinfection of *Shigella* lysogenic for phage P2 (Bertani, 1953a) and of *E. coli* lysogenic for lambda (Jacob and Wollman, 1953) indicated that the former hypothesis was correct, and this was soon confirmed by the results of bacterial crosses of *Escherichia coli* (E. Lederberg and J. Lederberg, 1953; Appleyard, 1954a), which showed that, in a lysogenic cell, the genome of a temperate phage behaves as though it were located at a specific point on the bacterial chromosome. Crosses between lysogenic parents carrying genetically different lambda prophages revealed (1) that the attachment sites of the two prophages were allelic and (2) that there was linkage between the prophage and other genes of the bacterium. This observation has been generalized to other systems, although the facts do not completely justify such a generalization.

On the basis of those lysogenic systems for which crosses have been made, three types can be distinguished. (1) The lambda type, in which the phage has a unique point of attachment to the bacterial chromosome. Many different phages, some of them related to lambda, have been shown each to have its own specific attachment site (Jacob and Wollman, 1957b). (2) The P2 type, in which the phage preferentially occupies a particular site, but where it is possible to form, under appropriate conditions, quite stable lysogens in which the preferred site is vacant and one of the several possible secondary sites is instead occupied (Bertani and Six, 1958). (3) The P1 type, for which the segregation in crosses does not permit the assignment of any chromosomal location to the prophage (Jacob and Wollman, 1957b). A separate transfer mechanism for P1 is possible (Boice and Luria, 1961).

The difference between lambda and P2 may be technical rather than fundamental. These two phages differ in the behavior of homopolylysogens, which are bacterial lines perpetuating simultaneously more than one related prophage. In P2 double lysogens the two prophages are located at different places on the bacterial chromosome, whereas with lambda they occupy the same site (or extremely closely linked sites). Double lysogens of lambda are also less stable than those of P2. It seems possible that P2 can also form double lysogens of the lambda type, but that they are even more unstable and are never recovered, thereby rendering easier the isolation of double lysogens in which distinct locations are occupied.

In the case of prophage, one can infer a chromosomal location of the phage genome itself rather than of some other gene affecting prophage maintenance because each parent of a bacterial cross can be marked with a genetically different prophage. It has not been possible to do the analogous experiment with other episomes. One cannot therefore exclude totally the possibility that the $F+ \rightarrow$ Hfr change, for example, might involve, rather than the fixation of F itself at a particular site, a mutation at that site which affects the behavior of F during conjugation. Richter's (1957, 1961) finding of $F-$ recombinants which become Hfr upon infection with F fits this hypothesis. It can be explained alternatively by assuming that in these strains a *part* of F has adhered to the chromosome. Wollman and Jacob (1958a) have presented other evidence which can also be interpreted in this manner.

B. Mode of Attachment of the Prophage to the Bacterial Chromosome

The manner of attachment of the prophage to the bacterial chromosome has been much discussed in the past. At one time, the problem was how a small linear structure (the phage genome) could act as part of a large linear structure (the bacterial chromosome). The most appealing solution was that the small structure was inserted into the continuity of the large one so that one linear genome would result. In recent years, many data have appeared which seem to contradict such a co-linear

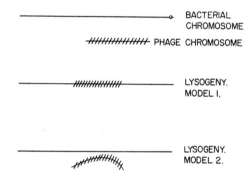

Fig. 1. Possible modes of attachment of prophage and bacterial chromosome. Model 1: Insertion. Model 2: Partial synapsis of central region of phage to bacterial chromosome.

structure, and popular taste has tended to favor some sort of branched model, in which the prophage is not really attached to the bacterial chromosome at all but rather portions of the two are permanently synapsed together. These two extreme cases are illustrated in Fig. 1 as Model 1 and Model 2, respectively. Bertani (1958) has categorized more completely the various possible modes of juxtaposition.

To evaluate much of the recent evidence on this subject, one must understand first that it is primarily evidence *against* the insertion hypothesis rather than *for* any particular kind of branched structure. One is thus really setting a model which makes quite specific predictions against *all* other possibilities, and whenever one prediction fails, the specific model is discarded. This seems a somewhat unfair procedure to the reviewer, and we will react to it by discussing primarily insertion hypotheses. To explain some of the facts in this way requires additional *ad hoc* assumptions, but insertion, with complications, is not inherently less desirable than an undefined model of branching or sticking together.

Besides such data which bear on the answer, there is also information which changes the question. Our large linear structure (the bacterial chromosome) behaves, on formal genetic analysis, not as a line but as a circle (Jacob and Wollman, 1957a). This really does not matter very much, but what if the small structure is also a circle? Detailed linkage studies lead to the conclusion that the genome of one phage (T4) is indeed circular Streisinger, Edgar, and Harrar, quoted by Stahl (1961). If circularity is a property of phages in general, the equivalent of the insertion hypothesis is to make one circle out of two, and we will discuss later the simplest model for accomplishing this.

The most obvious approach to distinguishing the various models is to examine the results of crosses between two lysogenic parents in which both the prophages and the bacterial chromosomes are well marked. The most complete study of this type thus far published is that of Calef and Licciardello (1960) on phage lambda. The assortment of prophage markers among bacteria recombinant for a pair of bacterial genes on either side of the prophage indicates a linear order with the prophage genes lying between those of the bacterium. However, the order of prophage markers on this map is different from that of the vegetative lambda phage. Whereas crosses of vegetative phage give the order "*h-cI-mi*," the order in the lysogenic cell is "*try-h-mi-cI-gal*."

This surprising result is supported by some other data suggesting a singularity in the region between *h* and *cI*. For example, Whitfield and Appleyard (1958) found with doubly lysogenic strains marked at these two loci that one recombinant type was liberated in excess of either parental type. The type preferred depended on the order of lysogenization and on the parental couplings, not on any selective advantages of the markers employed. This seems quite possible if the phage genome splits into two or more pieces at the time of lysogenization and is reassembled later. The results of Calef and Licciardello also allow one to contemplate mechanisms for the transduction of the galactose genes by phage lambda which otherwise would be unthinkable. This point will be amplified in a later section.

If the phage genome is circular rather than linear, the lambda chromosome need not be split into parts, but rather could be cut at a specific point on the circle when it lysogenizes. It is actually very simple (on paper) to insert a circular phage chromosome into a linear bacterial chromosome by reciprocal crossing-over (Fig. 2).

Figure 3 shows the genetic constitutions predicted for the chromosome of doubly lysogenic bacteria and for those carrying the defective, galactose-transducing lambda (Section VIII,A) on the model described here. If induction is imagined to entail a reversal of the process shown in Fig. 2, it is easily seen that one can make many different complete loops from a double lysogen, which would explain the result of Whitfield and Appleyard. The instability of double lysogeny in lambda (Appleyard, 1954b; Arber, 1960), the unstable lysogeny of transducing lambda (Campbell, 1957), and the apparent correlation between loss of transducing lambda and recombination in the *gal* region (Arber, 1958) could all be explained as consequences of the presence of a longer region of duplication.

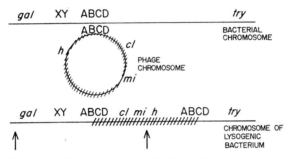

FIG. 2. Possible mechanism of lysogenization by reciprocal crossing-over between a circular phage chromosome and a linear bacterial chromosome. Arrows indicate possible rare points of breaking and joining in the formation of transducing lambda. (See Section VIII.) The genes ABCD are hypothetical and indicate a small region of homology between host and phage. X and Y are unspecified bacterial genes.

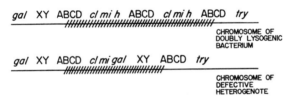

FIG. 3. Genetic constitution of doubly lysogenic bacterium and of defective heterogenote, assuming the mechanism of Fig. 2. The origin of the transducing lambda is described in the text (Section VIII).

On the other hand, this model would predict that it should be easier to lysogenize an already lysogenic strain than a non-lysogenic one, since the former will present a larger region of homology to an entering phage. This prediction is not fulfilled. Lysogeny creates a strong steric hindrance against lysogenization by another phage at the same site (Six, 1961a).

This illustrates well what we pointed out to begin with—that *ad hoc* hypotheses are required to fit most insertion models to the facts. We would assume that the pairing between phage and bacterial chromosomes may have other requirements besides the homology of the regions which pair. This is almost required by the model anyway, because the specificity of the attachment is determined by the immunity region (Kaiser and

Jacob, 1957), which does not coincide with the region of lambda where we wish pairing to take place in this case.

Evidence indicating that the prophage is not inserted into the bacterial chromosome has been presented by Jacob and Wollman (1959b). The most important concerns the non-inducible prophage 18, which is located close to two methionine markers (M1 and M2). Two Hfr strains by which this region was injected early but in opposite directions were isolated, and it was shown that the prophage appears to enter the bacterium after the marker M1 in both cases. They therefore suggest that the prophage may be synapsed parallel with the bacterial chromosome, overlapping the M1 gene, inasmuch as "prophage entrance" would require the entrance of the entire prophage genome. They also observed that the distance between M1 and M2 is not altered by lysogenization of both parents in a cross.

These facts should not be disregarded. They do not, in the reviewer's opinion, necessitate the abandonment of the insertion hypothesis, although it requires some ingenuity to circumvent them. In this review we will discuss mostly insertion hypotheses, not because we strongly favor them, but because we believe alternative ideas have received rather more than their share of attention at the hands of others.

Any model for prophage attachment must ultimately be extended to explain the behavior of homopolylysogenic strains of the lambda type. Strong asymmetries in the patterns of phage liberation and of segregation are seen with such strains, and there are reproducible differences between strains in the type of asymmetry shown (Arber, 1960a). Our picture in Fig. 2 is therefore at best a naïve first approximation to the true situation. It seems permissible at present because no simple explanation of the facts has been provided by alternative hypotheses.

IV. Immunity

A. Superinfection Immunity and Repression

In order that a stable lysogenic culture should exist, it is necessary that the cells of this culture should be unable to support the lytic growth of the phage which other cells have liberated. This conclusion has been verified by experimental superinfection experiments, which have shown, in addition (1) that the lysogenic cells are neither killed or lysed by superinfection with phage of the carried type but that (2) the superinfecting phage are still able to attach to the lysogenic host, and, from genetic data, must also be able to inject their DNA into it. The lysogenic cell is thus immune to superinfection.

Another *sine qua non* for lysogeny is that those genes of the prophage which determine functions involved in growth or maturation of phage or of lysis of the cell—i.e., all those functions which can be characterized as viral—must not express their potentialities in the lysogenic cell. There is much negative evidence that this is indeed the case. The analogy between this repression of viral functions in lysogenic cells and the repression of enzyme synthesis by internal repressors has been brilliantly developed by Jacob (Jacob, 1960; Jacob and Campbell, 1959; Jacob and Monod, 1961).

We have therefore two operational phenomena—the superinfection immunity and the repression of prophage genes—which are similar in that in both cases a phage genome in a cell is prevented from functioning in the same manner that it would upon entering a non-lysogenic cell. One may hypothesize that the two are identical, and that the immunity to superinfection is due to the same repressor(s) which are necessary for the stability of the lysogenic state. A complete proof of this hypothesis will not be possible until the key steps have been defined biochemically, but the available evidence is encouraging.

B. Cytoplasmic Nature

In the first place, both effects are mediated physiologically through the cytoplasm rather than sterically by the attachment of the prophage to the bacterial chromosome. This is indicated by the following facts in the two cases:

1. *Immunity*

As mentioned earlier (Section III,A), the prophage of P2 can occupy various alternative sites on the *E. coli* C chromosome. However, cells carrying P2 in any position are immune (Bertani and Six, 1958; Bertani, 1956). Moreover, the related phage P2 *Hy Dis*, which has a different immunity specificity, does not create immunity against P2, although it can occupy the same sites and interferes with lysogenization by P2 (Cohen, 1959; Six, 1961a).

Equivalent evidence comes from studies of cells recently infected with P22 (Luria *et al.*, 1958; Zinder, 1958) and from merozygotes (Jacob, 1960) and stable partial diploids (Jacob and Monod, 1961) in which one chromosome or one nucleus of a cell has become lysogenized, but its non-lysogenized homolog can segregate from it. This segregation of sensitive progeny from an immune individual implies that immunity is a dominant character. Evidence for its cytoplasmic nature comes also from the persistence of immunity for some time after the genetic segregation has occurred (Luria *et al.*, 1958).

We will refer to the specific cytoplasmic principle involved in immunity as the "immunity substance" (Bertani, 1958).

2. Repression

Bacterial mating is essentially a transfer of part or all of the genetic material of the male parent into the female cell. If the male parent is lysogenic, mating entails the sudden introduction of an established prophage into a non-immune cytoplasm. The dramatic result is that it is induced to multiply vegetatively (Jacob and Wollman, 1954a, 1956b). The implication is that it is ordinarily restrained from entering the autonomous state not by virtue of its physical union with the chromosome but rather by the repressors present in the cytoplasm, which must themselves be synthesized under the direction of the prophage.

C. Genetic Determination of Immunity

In the second place, there is a considerable formal similarity between the genes controlling immunity and the regulator and operator genes which determine the rate of synthesis of an enzyme such as β-galactosidase. For a detailed comparison, the reader is referred to the review of Jacob and Monod (1961). We will concentrate here primarily on the genetic control of immunity.

1. Determination of the Specificity of Immunity

The genetic control of the immunity specificity has been investigated for several phages. Phages independently isolated from nature which are closely enough related to recombine genetically nevertheless frequently differ in their immunity specificity. For example, the two phages lambda and 434 each can form lysogens which are immune to superinfection by the carried phage, but infection of K(λ) by 434 or of K(434) by lambda results in lysis and phage production (Jacob and Wollman, 1956b). If the immunity specificities of two phages are the same, the two are *homo-immune;* if different, *hetero-immune.*

Crosses between hetero-immune phages allow one to localize the determinants of immunity on the genetic map. If one crosses a series of lambda mutants with 434, one finds that some wild type lambda is produced in all cases except for those mutants lying within a single genetic locus (c_I) of the phage. Kaiser and Jacob (1957) backcrossed phage 434 several times with phage lambda and produced a strain which we can call λimm^{434}, in which the immunity determinant of 434 is imbedded in a lambda genome. We will refer to the region containing the c_I locus within which lambda and λimm^{434} differ as the immunity region.

A hetero-immune phage of the P2 type was obtained from infection of *E. coli* B with phage P2. *E. coli* B apparently carries a defective prophage related to P2, and the active hetero-immune phage obtained by recombination with the superinfecting P2 is called P2 *Hy Dis* (Cohen, 1959; Six, 1961a). The rarity of recombination in lytic infection by P2 has discouraged mapping of the immunity determinant. Crosses of *Salmonella* phage P22 with a hetero-immune relative have shown that the immunity determinant is close to and perhaps allelic with the marker v_1 (Zinder, 1958).

2. *Determination of the Ability to Generate or to Respond to the Specific Immunity*

The immunity region thus determines the specificity of immunity, and this immunity has two components. The phage must be able to send a specific message (e.g., a repressor) and also to respond specifically to the same message by failing to grow in an immune cytoplasm. It is not clear why the genes determining these two functions should occupy the same region of the chromosome, but phage present no unique problem in this respect. Bacterial regulator and operator genes can likewise be closely linked (Jacob *et al.*, 1960a). For phage, the implication is that mutations which abolish or alter the ability either to generate or to respond to immunity should be found in the immunity region. Furthermore, mutations altering the *specificity* of immunity should rarely if ever be found, because such mutations would require a simultaneous change in both the regulator and the operator to new forms which again match each other.

Present data support the idea that alterations of immunity specificity do not occur by mutation. We feel justified in making this statement even in the presence of some outstanding exceptions, because of the difficulty in distinguishing between true mutation of a phage and recombination between it and an unknown prophage or defective prophage. That the phage P2 *Hy Dis* was originally classified as an immunity mutant is a good point to bear in mind. More recently, a mutant of lambda has been described (Kellenberger *et al.*, 1961a) which renders the phage not only hetero-immune but also less dense than wild type lambda. On primary isolation, it carried also a small plaque mutation which was separable from the density-immunity alteration by crossing-over. This mutant is therefore complex and could well have a recombinational rather than a mutational origin.

a. *Bacteriophage Lambda.* The mutational pattern of bacteriophage lambda fits well with a regulator-operator picture, and therefore this phage will be discussed first. Several interesting types of mutants are found:

i. *Mutants unable to generate immunity.* Such mutants are the most commonly observed result of mutation in the c_I region. The c_I mutants are still responsive to the immunity generated by wild type lambda and therefore will not grow in lysogenic cells. They are unable to lysogenize. They are called "clear plaque" mutants because whereas there is no lysogenization within the plaques of these mutants, which therefore look clear like the plaques of a virulent phage, the wild type plaques are turbid due to lysogenization of bacterial cells within the plaque.

In lambda, there are two other cistrons closely linked to c_I (one on either side), mutations in which also decrease the ability to lysogenize. These cistrons differ from c_I in several ways: (A) Whereas most c_I mutants show no detectable lysogenization at all, c_{II} and c_{III} mutants exhibit a low but measurable frequency of lysogenization. The lysogens, once formed, are completely stable. (B) Mixed infection with two mutants from different cistrons results in a high frequency of lysogenization. The survivors from such cooperation experiments may be either singly or doubly lysogenic. However, one never finds individuals singly lysogenic for c_I. (C) Kinetic studies show that the c_I function in lysogenization occurs later than the c_{II} or c_{III} functions (Kaiser, 1957).

These data clearly indicate that, of the three cistrons containing clear mutants, all must cooperate in the establishment of the lysogenic condition, but that only the c_I cistron continues to operate in the maintenance of this condition. The c_{II} and c_{III} cistrons perform functions necessary for lysogenization, but not for lysogeny. Their mechanism of action is of interest in its own right, but only the c_I mutants are directly relevant to the problem of immunity; and of these three closely linked cistrons, only c_I lies within the immunity region.

ii. Mutants unable to respond to immunity. These are the so-called "inducing virulent" mutations (Jacob and Wollman, 1954b). The mutation to this state is complex and requires several changes from wild type, one of which is in the immunity region. Inducing virulent mutants are characterized by their ability to grow on lysogenic cells. Whether they have also lost the ability to generate immunity is difficult to test experimentally and has not been decided.

iii. Mutants in which the immunity, but not its specificity, has been altered. A mutant of lambda has been isolated (λ*ind*), lysogens of which are not inducible by ultraviolet. This mutation is located within the immunity region. In double lysogens, non-inducibility is dominant. If a cell of the type K(λ*ind*⁺) which has been induced is superinfected soon afterwards with λ*ind*, lysis is prevented and the survival of cells increases. Most of these survivors are still K(λ*ind*⁺). It thus seems that the *ind* gene produces something which can reverse induction without the *ind* phage's becoming integrated as prophage. The effect is specific for the lambda immunity type. Induced cells of K(λ*ind*⁺*imm*⁴³⁴) or double lysogens K(λ*ind*⁺*imm*^λ)

(λind^+imm^{434}) are not affected by superinfection with λind (Jacob and Campbell, 1959).

This mutant provides additional evidence for the existence of a cytoplasmic immunity substance. One cannot say at present whether the ind mutation alters the substance itself or affects its rate of production. It probably alters the same function which is abolished by the c_I mutations, because a phage which carries both mutations is unable to prevent lysis of K(λind^+) (Jacob and Campbell, 1959).

b. *Bacteriophage P2.* In P2, where the immunity determinants have not been mapped, there occur, in addition to types 1 and 2 (which are called by P2 workers "weak virulent" and "strong virulent," respectively), some intermediate types which show limited growth on lysogenic strains, or which form plaques on singly lysogenic but not on doubly lysogenic strains (Bertani, 1953b, 1958). These are readily explained as involving quantitative alterations in the pattern of response to immunity.

c. *Bacteriophage P22.* The findings with *Salmonella* phage P22 are more difficult to fit into the general scheme constructed for phage lambda. The two phages present certain striking analogies. In both cases, there is a cluster of closely linked cistrons of clear plaque mutants (Levine, 1957; Kaiser, 1957). In both cases, only one of these cistrons is necessary for the maintenance of lysogeny. The others function only in its establishment. In both cases, other mutants are found which form turbid plaques but from which stably lysogenic lines can never be extracted (Lieb, 1953; Zinder, 1958). The function altered in these mutants is apparently not required in order that the infected cell should survive the infection, but is necessary for the establishment or perpetuation of stable lysogeny. These mutants are not closely linked with the c mutants.[1]

In both cases, crosses with related hetero-immune phages allow genetic localization of a determinant of immunity specificity. There is only one important difference between the two systems, but it is one which cannot be ignored. Whereas the immunity determinant of lambda is allelic with the c_I cistron, that of the *Salmonella* phage is close to (perhaps allelic with) the mutant v_1—a mutant of the type which forms turbid plaques but does not lysogenize, and which is unlinked to the clear mutants of P22. This of course does not fit too easily with the idea that the c_I phenotype is the result of the absence of a repressor which determines the immunity specificity. One explanation would be that immunity does not involve one repressor but several, and that hetero-immunity may involve

---

[1] Another mutation of this type has recently been described for phage λ by Kellenberger *et al.* (1961b). It is of especial interest here because its genetic location would make it a possible candidate for the region of homology we have postulated between phage and host (genes ABCD of Fig. 2).

a difference at any one of the immunity "loci." If this is so, one should eventually find two such loci in the same phage.

D. Conclusions

In conclusion, one can therefore say that the immunity is mediated by one or more specific cytoplasmic substances, and that we know something about their genetic control. It seems likely that they function by repressing the synthesis of one or more proteins, but it must be emphasized that this is still only a hypothesis. What we know definitely (from superinfection experiments) is that, when the genome of a phage enters an immune cell, it does not undergo any appreciable amount of either multiplication or decay (Jacob and Wollman, 1953; Bertani, 1954). It sits in the infected cell and is gradually diluted out by growth. Therefore, immunity prevents phage multiplication. We presume this is by repressing some early steps in the lytic cycle. Perhaps the later steps are controlled indirectly, becoming de-repressed during the cycle by sequential induction.

E. Episomes Other than Phage

In the case of the F agent, an apparent incompatibility on the cellular level exists between the autonomous and the integrated state (Jacob *et al.*, 1960b). Whereas F is rapidly transmitted from F+ cells to F− cells, the F factor in Hfr cells is not thus readily transmissible, even though Hfr strains originate as mutants of F+ under conditions where there is ample opportunity for re-infection by F. The presence of an integrated F somehow excludes autonomous F from the same cell line. Likewise, stably colicinogenic strains do not acquire transmissible colicinogeny by contact (Stocker, 1960). It has been suggested that these situations are similar to that with phage, where the lysogenic condition and vegetative phage replication represent alternate states. They represent alternatives at the *cellular* level because the two conditions differ by the presence or absence of repressors which permeate the whole cytoplasm of the cell. If the analogy with phage is valid for F, it would seem that genetic incorporation is a determinative event in initiating physiological changes which suppress the growth of autonomous F. The analogous question for phage remains unsettled (see Section V,B).

Colicinogenic strains are specifically immune to the colicins they produce (Fredericq, 1956c, 1958). As with phage, this immunity does not reflect a loss of receptors for the colicin. If this immunity is due to a repression of gene action, it must exert its effect on some gene of the host which participates in the killing process, because the colicin itself contains no genetic material. The immunity does not always extend to very high concentrations of colicins (Hamon, 1957), but this does not really distinguish it from the phage immunity system where multiplicity effects can also be seen, especially with mutants which generate immunity weakly (L. Bertani, 1961).

V. Lysogenization. Transition from the Vegetative to the Integrated State

A. Infection of Sensitive Cells by Temperate Phage

There are several possible outcomes of the infection of a sensitive cell by a temperate phage (modified from Lwoff, 1953).

(1) The cell dies and produces phage (*productive* response).

(2) The cell dies and does not produce phage (*lethal, abortive* response).

(3) The cell survives and produces a colony of lysogenic cells (*reductive* response).

(4) The cell survives and produces a colony of sensitive cells (*nonlethal, abortive* response).

(5) The cell divides one or more times, producing a group of cells not all of which fall into any one of the above categories (*mixed* response).

Investigators interested in the mechanism of lysogenization are therefore immediately confronted with the question "Why do two genetically identical cells, each infected with a phage particle of the same type, respond differently?" Is there really some gross heterogeneity among cells or phages which we have overlooked? Or is the fate of the cell perhaps determined by statistical fluctuations in the timing of events on the molecular level?

Experimentally, one may ask how changes in the conditions of the experiment may alter the proportion of infected cells giving each type of response. The percentage of cells which survive infection and produce at least some lysogenic progeny (i.e., those giving reductive or mixed responses) is increased by low temperatures (Bertani and Nice, 1954), by high multiplicities of infection (Boyd, 1951), by inhibition of protein synthesis during the latent period, and by exposure to proflavine (L. Bertani, 1957). The temperature and multiplicity effects are large for some phages and small or zero for others, whereas the administration of chemicals is much more effective at certain times during the latent period than at others. The ability to shift the proportions in this way eliminates genetic heterogeneity among either the bacterial or the phage populations as an explanation for the different responses of different cells. No experiment has been devised which allows any decision between a physiological heterogeneity among cells and molecular fluctuations within the infected cell.

What has become clear from the work of many investigators (Lieb, 1953; Luria *et al.*, 1958; Zinder, 1958) is that lysogenization requires the occurrence of at least two events. The first is a physiological decision of the cell whether or not to lyse as a consequence of the primary infection. The second is an interaction between the phage genome and a bacterial nucleus in which the phage is reduced to prophage (*genetic incorporation*).

B. THE DECISION NOT TO LYSE

We have placed the physiological event first because it seems to happen early—before the first cell division following infection. This is shown most simply by a study of clones exhibiting mixed responses. Among the progeny of a single infected cell one frequently finds a mixture of lysogenic and non-lysogenic cells, but rarely if ever a mixture of productive and non-productive responses (Lieb, 1953; Zinder, 1958). There is a period after infection during which exposure to high temperature reduces very much the proportion of lysogenic progeny of those cells giving mixed responses, without changing greatly the proportion of mixed responses itself (Lieb, 1953). The existence of virulent mutants of the type which allow survival but cannot establish stable lysogeny says that the decision not to lyse can occur in the absence of genetic incorporation. If a cell is mixedly infected with more than one genetic type of phage, one can study the early segregation following infection of non-lysogenic and singly lysogenic types, either by plating the culture after a few hours of growth (Zinder, 1958) or by single cell pedigree analysis (Luria *et al.*, 1958). There is general agreement from such studies that the event of genetic incorporation frequently occurs very late, some generations after the primary infection. It is perhaps noteworthy that phage P1 (which, as we have already mentioned, cannot be localized on the genetic map) differs from lambda and P22 in that no mixed responses occur.

All this evidence shows that the decision not to lyse can precede genetic incorporation. It of course does not imply that it is prerequisite to genetic incorporation, or that genetic incorporation has no effect on the probability that the cell will lyse. This last is a question of some importance to which we have as yet no clear answer. If the cell lyses, it is impossible to say whether or not genetic incorporation had occurred within it. The question is not purely academic, however, because one may ask whether any treatment which increases the rate of genetic incorporation will thereby influence the percentage of cells which survive infection rather than lysing. Furthermore, in the case of an episome such as F where vegetative multiplication is not lethal, the exclusion between autonomous and integrated state is hard to understand unless the presence of the integrated element actually represses the autonomous one to a greater extent than one autonomous element represses another in the same cell.

C. Genetic Incorporation

One system has been studied carefully in which genetic incorporation can be followed independently of the decision not to lyse (Six, 1961a). If a strain of *E. coli* C carrying P2 prophage in position II is superinfected with a marked P2 phage, the cells do not lyse because they are immune. In superinfection of an ordinary lysogenic strain [e.g., $C(P2)_I$], the process of genetic incorporation is severely restricted by a steric hindrance due to the prophage already present, but if the preferred location is unoccupied, this effect is abolished. This can be shown by a comparison of the frequency of genetic incorporation of a marked P2 into the three strains $C(P2)_I(P2)_{II}$, $C(P2\ Hy\ Dis)_I(P2)_{II}$, and $C(-)_I(P2)_{II}$. For the first two, the frequency is roughly the same and about 40X lower than that for the third strain. This shows clearly that the steric effect and the immunity are two different things. Steric hindrance reduces genetic incorporation, but immunity prevents phage multiplication. This important distinction has almost never been made in studies of interference between episomes other than phage.

Six found that the probability of genetic incorporation even at an unoccupied site is low (about 0.05 per phage) in an immune cell. Up to multiplicities of about 10, this probability is constant; i.e., the number of cells which have acquired the superinfecting phage as prophage is proportional to the multiplicity of superinfection used. From this we can say that the probability that a phage genome will become incorporated is not influenced by the presence of other phage genomes in the same cell.

Now, we have no reason to suppose that the probability of 0.05 is different for a phage genome in a sensitive rather than in a lysogenic cell. This would lead one to predict that infection of a cell by a single defective phage particle which was unable to multiply would result in a very low probability of genetic incorporation. Unfortunately, no data of this type are available for phage P2, but this is certainly true for phage lambda (Jacob *et al.*, 1957). It is especially well illustrated by the behavior of galactose-transducing lambda. The probability of transduction in single infection is very low, but simultaneous infection by an active lambda phage raises it by a factor of about 20. This "helping" effect of the active lambda is abolished in a lysogenic recipient (Campbell, 1957; Campbell and Balbinder, 1959). We know that the active lambda phage allows the transducing phage to multiply, and it is tempting to suppose that we are here increasing the probability of incorporation mainly by increasing the number of copies of the incorporable genome.

In order to account for the high frequencies of lysogenization observed with many temperate phages under normal or abnormal conditions of infection, one can assume that vegetative multiplication generally precedes lysogenization. The available data on the frequency of recombinant prophage resulting from mixedly infected cells seems to corroborate this (Luria et al., 1958; Bertani, 1958). We are not implying that lysogenization cannot occur in the absence of vegetative multiplication, but only that the frequency of lysogenization by a single, non-multiplying phage element is quite low. Future work will test the validity of this idea.

D. Episomes Other than Phage

Thus, our knowledge of what happens when a phage passes from the autonomous to the integrated state is virtually confined to some information about immunity and lysis. For episomes such as the F agent, which can grow indefinitely in the autonomous state, we have only the bare fact that somehow integration occurs, as a rare sporadic event analogous (from the point of view of population dynamics) to a mutation. The rate of integration has been estimated at 10^{-4} per cell per division (Jacob et al., 1960b). This is sufficiently low so that the vast majority of cells of an F+ culture are F+ rather than Hfr, and indirect selection is necessary for the isolation of new Hfr's. If a phage had as low a rate of integration as F, its temperate nature might be barely noticeable; whereas a non-lethal episome with a high rate of integration might easily be classified simply as a gene. The latter situation is approached with the colicin determinants in *Salmonella* (Stocker, 1960).

Unlike a phage such as lambda, F can become integrated at any point (or at least at a variety of points) along the bacterial chromosome. The difference is probably more apparent than real. Adelberg and Burns (1960) have shown that variants of F can arise which prefer a particular location, and Six (1961b) has found that P2 which has been liberated from a strain carrying it in position II has lost the extreme preference for position I which characterizes wild type P2. Both findings can be explained on the idea that, in the passage from the integrated to the autonomous state, the episome can occasionally pick up a small piece of the bacterial chromosome from near its site of integration. Further evidence on this point will be presented below (Section VIII).

VI. Transition from the Integrated to the Autonomous State

In any lysogenic culture, a few cells occasionally lyse spontaneously each generation so that one always finds some free phage in the culture. With some phages, one can induce the majority of the lysogenic cells to liberate phage by a small dose of irradiation or by any of a number of chemicals. The most convenient working hypothesis is that the primary effect of all these treatments is a disturbance of the balance between repressor and prophage genes.

Levine and Cox (1961) have shown that prior growth of lysogenic cells in the presence of chloramphenicol increases the dose of ultraviolet light required for induction and have presented some evidence that this effect is due to an accumulation of RNA by the chloramphenicol-treated cells. A role of RNA in repression has often been hypothesized, and this would fit very well. There is no direct evidence that this effect is related to repression or immunity, but the fact that the rescue of induced lambda lysogens by λ*ind* (see Section IV,C,2,*a*) is immunity-specific makes this interpretation quite plausible.

Hfr strains occasionally revert spontaneously to F+. The frequency is different for different Hfr's (Lederberg, 1959). The production of colicin can be induced by ultraviolet light (Jacob *et al.*, 1952; Fredericq, 1954c). Quite analogously to the induction of phage production in lysogenic strains, ultraviolet light increases the number of cells in the culture producing colicin rather than merely augmenting the amount of colicin formed by cells which are already producing it (Ozeki *et al.*, 1959). The relationship of colicin production to the state of the colicin factor is uncertain.

When a prophage is introduced into a non-immune cytoplasm, either in mating (Jacob and Wollman, 1956b) or by transduction (Jacob, 1955; Arber, 1960b), it may be induced to develop vegetatively (*transfer induction*). In the case of transduction, the amount of induction is less at low temperatures (Jacob, 1955). The colicinogeny factor probably undergoes something similar to transfer induction, as judged by its spread to all the nuclei of the F− exconjugant in Hfr × F− crosses (Alfoldi *et al.*, 1958).

When the F agent is introduced into an F− cytoplasm by transduction, it is recovered in the autonomous state, even if the donor was Hfr (Arber, 1960b). However, when the integrated F agent enters an F− cell in an Hfr × F− cross, it is transmitted to the progeny in the integrated state (Wollman and Jacob, 1958a).

VII. Curing

For practical reasons, it is desirable to be able to remove an episome from a bacterial strain at will. Methods have been found for effecting this which are still mainly empirical. Cured derivatives of lysogenic cultures can be obtained by (1) superinfection with related hetero-immune (Cohen, 1959) or virulent (Bertani, 1953a) phage or (2) exposure to heavy doses of irradiation (E. Lederberg and J. Lederberg, 1953). F+ strains (Hirota, 1960) and RTF-carrying strains (see Section I,B,3) (Watanabe and

Fukasawa, 1961b) can be cured by exposure to acridine dyes. Cobalt ion is also effective in curing F (Hirota, 1956). Colicinogeny is not cured by acridine dyes, but its transfer is inhibited by them, which may likewise reflect a curing of the autonomous but not of the integrated state (H. Ozeki, S. Smith, and B. Stocker, personal communication). Curing of vegetative phage would result in abortive infection, non-lethal for a temperate stage. This has been reported for chloramphenicol (Ting, 1960) and high temperatures (Lieb, 1953).

VIII. "Gene Pick-Up" by Episomes

One interesting property of episomes is their ability to incorporate small pieces of genetic material from the host into their own structure. We shall refer to this process as "gene pick-up." The importance of gene pick-up lies not in its mechanism (which is unknown, but need not involve any unique properties of episomes), but rather in its consequences. It implies that the ability to pass from the integrated to the autonomous state is not confined to the small group of elements we have discussed thus far but actually pertains to every gene of the bacterial cell.

A. Phage-Mediated Transduction

1. *Special Transduction with Bacteriophage Lambda*

We will first discuss gene pick-up in phage-mediated transduction. The transduction of the galactose genes of *E. coli* by phage lambda has been studied in the greatest detail (Morse *et al.*, 1956a,b). The lambda prophage is located close to the galactose genes, and they are the only genes which this phage can transduce. Transduction does not occur at all with lytically grown lambda, only with lysates prepared by induction of lysogens. The transductants are frequently partially diploid (*heterogenotic*) for the galactose region, and lysates prepared from these transductants are enormously more active in transduction than those prepared from ordinary lysogens. The two types of transduction are called *low frequency transduction* (LFT) and *high frequency transduction* (HFT), respectively.

One most important aspect of the system did not become manifest until careful studies were made on high frequency transduction at very low multiplicities. It was then found that almost all the transductants were not active lysogens but rather defective lysogens. As has been mentioned earlier, the probability of transduction is increased by a factor of 20–40 by simultaneous infection with an active lambda phage (*helping* effect). Most, if not all, of those mixedly infected cells which acquire the galactose genes become doubly lysogenic for the active phage and the transducing phage. This has the technical consequence that, in order to study cells singly infected with the transducing phage, the multiplicities employed must be very low indeed ($<10^{-2}$).

The HFT lysate is a mixture of non-transducing plaque-formers and non-plaque-forming transducers. At multiplicities above 10^{-2} the number of transductions per phage increases because the active phage of the HFT are themselves functioning as helpers. The helping effect creates a technical problem which, when solved, allows one to examine the really interesting object, the defective transducing phage. Its importance in the present context is that similar effects operating in other systems, less technically simple or less thoroughly examined than this one, may lead to erroneous interpretations of results.

The transducing phage has been examined both physically and genetically. The density of transducing lambda is different from that of normal lambda, and independent isolates of transducing lambda differ among themselves (Weigle et al., 1959). Genetically, transducing lambda is missing a large piece from the middle of the known linkage map (Arber, 1958). Finer analysis reveals that independent isolates differ in the exact extent of the missing region, and that the endpoint of the region frequently falls *within* a cistron rather than between cistrons (Campbell, 1959, 1961). The missing regions of all the transducing lambda have a common core, which includes the h marker. Both the physical and the genetic specificity of a particular type of transducing lambda are stable hereditary properties which have never been observed to change during the course either of bacterial or of phage growth.

The fact that the missing piece comes from the middle of the phage genome might lead one to postulate that transducing lambda arises by a genetic interaction similar to crossing-over between homologous regions of phage and host chromosomes. However, no positive evidence has been obtainable to substantiate this viewpoint, and negative evidence enables us to say that the postulated homologies can at best be far from perfect (Campbell, 1960).

As mentioned earlier (Section III,B), the circularity of the phage chromosome and the reversal of phage markers in the prophage seen by Calef and Licciardello allow one to entertain other ideas for the mechanism of origin. One such mechanism is indicated in Fig. 2. Normal induction would produce a lambda genome by a reversal of the lysogenization process, so that essentially one takes from the chromosome a segment whose endpoints fall within the duplicated segment ABCD. In rare abnormal cases, one might take instead a segment such as that indicated by the two arrows, i.e., the deletion of phage material accompanying gene pick-up would be terminal rather than interstitial, as far as the prophage is concerned. It seems quite possible to fit this type of model into the general scheme of genetic and physical speculations concerning circularization (F. Stahl, 1961).

This model would predict (1) that if the Calef and Licciardello experiment is extended to more genes of lambda, the order will be that deduced from circularization of the known map, (2) that any gene between lambda and *gal* should be part of the transducing phage, and (3) that the right hand endpoint of the missing region of transducing lambda (between *h* and *c*) should be identical for all transducing lambda.

Point (1) has not been tested yet. As far as point (2) is concerned, the transduction of *gal* by the phage λimm^{434} creates a problem because this prophage is located much farther away from *gal* than is lambda. However, the data on mapping the location of the hybrid phage (Kaiser and Jacob, 1957) do not really preclude an occasional transposition of the hybrid phage onto the lambda site. Alternatively, one might imagine occasional recombination between the λimm^{434} prophage and some homologous region close to the lambda site immediately preceding induction. Point (3) is correct as far as we know (Arber, 1958).

2. *General Transduction*

The basic feature of lambda transduction is thus the formation, by some still unknown mechanism, of a hybrid genetic structure which is part phage and part bacterium. The defectiveness of the phage, its retention of the ability to generate immunity, its failure to multiply vegetatively, etc. do not seem very basic attributes of the process but rather details of the way in which this particular transduction occurs. One might expect, in general, that in some cases the defect would be so extreme that the persistence of the hybrid structure as prophage would be impossible, whereas in others it might be so mild that the transducing phage could still form plaques.

It is therefore a reasonable conjecture that all transductions require the intermediacy of such hybrid particles, but that in some cases they are very difficult to detect. The available evidence is encouraging. With phages such as P1 and P22 which will transduce any marker after one cycle of lytic growth (*common* transduction), the absence of any obvious correlation between lysogenization and transduction in individual events led early investigators to favor the idea that the phage played a purely mechanical role, by causing the partial degradation of donor DNA and the introduction of fragments of it into recipient cells (Zinder, 1955). However, since the mechanism of lambda transduction has been clarified, many instances of heterogenote formation and high frequency transduction have been reported in common transduction (Adams and Luria,

1958; Luria et al., 1960; Wilson, 1960; Spicer and Datta, 1959; see also Hartman, 1961).

The transduction of the lac gene of E. coli by phage P1 has been particularly illuminating (Luria et al., 1960). If a lac− strain of E. coli is used as recipient, only stable transductions are found. However, if the recipient is either the permanently lac− species Shigella dysenteriae or a mutant of E. coli with a large deletion in the lac region, one then obtains unstable immune strains. The simplest explanation is that, when stable integration is impossible, because of the absence of a homologous region in the recipient, one then sees lysogenization by the hybrid structure. In the normal coli to coli transduction, one would therefore have no way of knowing that any intermediate exists.

The lac transductions by P1 are also of interest because one finds a wider variety of types of transducing particles than have been seen with lambda. A whole array of types such as one might expect to collect by examining many different systems are all found for this one phage and one marker. Some transducing P1 particles retain enough phage functions so that heterogenotes carrying them will produce killing and transducing activity following induction without superinfection, whereas others form highly unstable heterogenotes which show no trace of immunity to P1.

B. F-Mediated Transduction

Gene pick-up by the F agent has also been demonstrated. In an Hfr × F− cross, there is a polarized transfer of genetic material from Hfr to F−, in which the F agent is transferred last. If one selects for rare individuals in which there has been a precocious entrance of the terminal bacterial marker, these are frequently F+ and heterogenotic for the selected marker. This marker behaves subsequently as though it had become part of the F agent; i.e., it is transferred rapidly by direct cellular contact, and cells acquiring it become simultaneously F+ (Jacob and Adelberg, 1959). If one selects in other ways for F+ revertants from Hfr cultures, one also finds that the terminal markers have sometimes become associated with F (Hirota and Sneath, 1961). These modified F agents (which are called F′) have the additional property of a greatly increased rate of formation of the particular type of Hfr strain from which they originated (Adelberg and Burns, 1960).

A variant of F which carries the lac gene has been transmitted into Salmonella and thence to several other genera of the Enterobacteriaceae (Baron et al., 1959a; Falkow and Baron, 1961). Particularly noteworthy is the transfer into Serratia. The two genera have different DNA base ratios, and genetic recombination has seldom been detected in this situation (Lanni, 1959). The DNA of the Serratia which carries the episome contains a physically recognizable fraction with the base ratio of *Salmonella* (Marmur et al., 1961).

In comparing gene pick-up by F (F-mediated transduction or "F-duction") with phage-mediated transduction, one difference should be stressed. In the heterogenotes produced by F-mediated transduction, the F agent is in the autonomous state, whereas in transductional heterogenotes produced with phage, the transducing phage is in the prophage state. This seems to be more than a trivial consequence of the cytopathogenicity accompanying vegetative phage multiplication. There is no evidence to indicate that the F agent forms heterogenotes in the integrated state, although very little data are available on this point.

Hirota and Sneath (1961) have studied the time of entry of a group of genes transferred by an F' factor called F_{13}. A polarized entrance is seen with F entering last. The time of entrance is about 20 minutes instead of the 2 hours required for complete entry in Hfr $\times$ F$-$ crosses. In other words, the block of genes associated with F behaves like a little bacterial chromosome, which strongly suggests that the relationship of these genes to F is the same as that of the bacterial chromosome to F in the Hfr strain. We must not allow our choice of terms to convince us that F has really picked up the genes rather than vice versa.

F' would then behave like a supernumerary bacterial chromosome, and one might expect a continuous gradation of properties from the Hfr chromosome down to the smallest F' particle or to F itself. Of particular interest for such comparative studies would be strains of the type prepared by Clark (1961), where a bacterium with two linkage groups (both of which behave as Hfr) has been synthesized from an Hfr $\times$ Hfr cross. We would predict that if such crosses were performed as to give a large disparity between the sizes of the two linkage groups, then the smaller would behave like F'.

One finds some F$+$ individuals in F' cultures, as well as F$-$ haploids which have lost both F and the supernumerary bacterial genes. In certain matings, a large fraction of F$+$ cells are obtained, indicating that F can become detached from the genes it transduces (Hirota and Sneath, 1961). F', like F, is easily cured by acridine dyes.

Our views of the nature of F' may be illustrated by some more speculative notions concerning the circularization of genetic material. For reasons which need not be detailed here, let us suppose that any genetic element perpetuated in a bacterial cell must be in a circular form. The Hfr chromosome would then be not a linear structure except possibly during conjugation, but rather a circular structure with F located at one point. The F factor itself would be a very small circle, and various F's would constitute progressively larger ones. This is illustrated in Fig. 4.

In the case of temperate phage, we would suppose the situation diagrammed at the bottom of Fig. 4. The only case requiring special comment is the vegetative phage. There is physical evidence that vegetative lambda is linear rather than circular (Meselson and Weigle, 1961), but this applies only to the genetic material of the infecting particle; it would still be linear in this sense if it formed part of a circular structure which contained also some host material. This type of interaction with host material could account for a large fraction of the observations which have justified

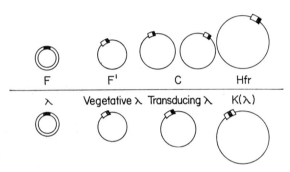

FIG. 4. Diagram of hypothetical circularization of genetic elements. F = fertility factor. F' = transducing fertility factor. C = Clark's strain with two linkage groups. Hfr = bacterial genome with integrated F giving high frequency of recombination. Bacteriophage lambda is shown below at various stages of its life cycle.

the somewhat mystical concept of mating between phage and host (Delbrück and Stent, 1956; Stent, 1958).

C. Existing Episomes as Products of Gene Pick-Up

The resistance transfer factor of *Shigella* might be considered as an episome which had already picked up genes prior to its discovery. RTF can be distinguished from the genes it carries by the fact that the latter can become individually stably integrated into the host genome. Variants of RTF which transmit only some of the four resistance factors originally associated with it have also been found (Watanabe and Fukasawa, 1961b).

The colicin I factor might also be a fertility factor which has picked up the determinant for colicinogeny. If so, the complex can itself mediate the transfer of additional genes, because cells colicinogenic for the two colicins I and E1 transfer both at high frequency, whereas cells carrying E1 alone transfer it at very low frequency. It was also found that the presence of E1 in the donor enhances the frequency of the I-mediated transfer of other bacterial markers. These markers are still transferred very rarely compared with the colicinogeny determinants themselves (Ozeki et al., 1961).

Every episome thus far studied therefore either contains, or can give rise to variants which contain, some genetic material which is not obligatorily episomal. Frequently, it is only by virtue of this material that we are cognizant of the presence of the episome. However, one is really most interested in knowing about that part of the episome which cannot be removed without destroying its episomal properties. With RTF, for example, after the individual determinants of resistance have been lost, something can still remain which is capable of passing from cell to cell at conjugation.

As a matter of nomenclature, we insist that the term "episome" apply not only to the genetic element responsible for episomality but to all associated material. Any other course would be unwarranted at present. Thus the streptomycin gene in RTF, the *lac* gene in P1*dl*, the *h* gene in lambda, etc. each constitute *part* of an episome, even though we know that the episome could still exist if these parts were lost or replaced.

In considering variation of episomes, we do prefer, however, to trace the lineage of the episome only as an episome, which means ultimately to follow whatever part is obligatorily episomal. Thus we consider P1*dl* as a variant of P1 and F-*lac* as a variant of F rather than classifying both as variants of the *lac* gene. All this is of course purely arbitrary, but it is only with such an agreed nomenclature that one can raise the question of the possible relatedness of existing episomes. Since the episome is frequently identified by its nonessential parts, one must consider the possibility that apparently different episomes are really the same except for having picked up, either recently or primordially, different genes from the host. As more variants of known episomes are collected, probably some of the things we have classified as distinct episomes will turn out to be essentially the same.

IX. Partial Diploidy in *Escherichia Coli*

Much of the work on episomes and gene pick-up depends on the formation of partially diploid strains. The discovery of such strains (Lederberg, 1949) antedates the findings with episomes, and there are also recent cases in which no known episome is implicated. The first apparent diploid stock described (called *het*) proved to be hemizygous, or partially diploid, a large fraction of the genes being in the diploid state, but those in a region including the loci for streptomycin resistance and maltose fermentation being haploid.

Het strains exhibit frequent segregation for the heterozygous markers. Some of the segregants are haploid and others homozygous. Haploidy was frequently achieved in a single event, rarely if ever by a progressive diminution of diploidy. If one selected for diploids among the recombinants formed in normal haploid F+ × haploid F+ crosses, such diploids were found at a frequency of about 10^{-3}. These diploids were not all hemizygous for the same region as *het* (Lederberg *et al.*, 1951). A *het* strain may be either F+ or F− (Nelson and Lederberg, 1954).

The heterogenotes originating by gene pick-up show many similarities to these strains. The most obvious difference is the amount of genetic material involved, which is quite small in transduction. Here, too, the amount of genetic material in the diploid state usually remains constant, but partial losses have been reported in some systems (Luria *et al.*, 1960; Watanabe and Fukasawa, 1961b). They probably occur in F-mediated transduction also, and might happen with low frequency in the *het* stocks as well. If there are partial losses from galactose-transducing lambda, they must be quite rare, because both the density of the particle and the amount of phage genetic material are stable hereditary properties of the transducing phage.

Other instances of partial diploidy have been noted in K12 (Matney *et al.*, 1961) and among the progeny of interspecific crosses [Baron *et al.*, (1959a) and Luria and Burrous (1957)]. In these cases the possible relationship of the duplicated piece to the F agent is not clear.

Novick and collaborators (Horiuchi *et al.*, 1962; Novick and Horiuchi, 1961) have obtained, from an Hfr strain of *E. coli*, rare spontaneous "mutants" which synthesize an abnormal amount of β-galactosidase. These proved to be diploid for the *lac* region, but when this diploidy was transmitted to F− recipients, they remained F− and would not transfer *lac* to other strains. The diploidy was transducible by phage P1 and did not extend to neighboring genes outside of the *lac* region.

The amount of enzyme produced by Novick's strains, like that of diploids prepared by F-duction, is three or four times as great as that formed by haploid strains of the same genotype. This suggests the presence of three or four copies of the gene. One might interpret this to mean that one copy is chromosomal and the others cytoplasmic, but the genetic data require that the diploidy have a chromosomal location. Our personal preference would be that these strains have a simple tandem duplication of genetic material in the chromosome itself, and that pairing within this region allows its replication to occur more rapidly than that of the genome as a whole, so that extra copies are cast off into the cytoplasm, which cannot replicate themselves, or at most multiply at a rate slower than that of the cell. We shall return to this type of idea as applied to episomes in general in the discussion (Section XII).

One might also wonder whether these strains might be intermediates in F-mediated transduction, or even whether gene pick-up by F really represents a physical association of F with the transferred genes rather than some kind of helping effect of F on the transfer and/or maintenance of a loose piece of genetic material. The best evidence against the latter hypothesis is that of Hirota and Sneath (1961), who showed not only that F enters at a regular time, but that it exhibits linkage with the genes of the transferred piece. Also, Driskell and Adelberg (1961) showed that F' has a different suicide sensitivity than F. Both findings indicate that F has been itself altered. However, the fact that *non*-transducing F can frequently be recovered from these heterogenotes (Hirota and Sneath, 1961) reminds one of high frequency transduction in the days before we knew that the transducing phage was defective. The possibility that F-mediated transduction constitutes a cooperation between an F particle and a loose gene fragment (or possibly one to which a *part* of F is attached) does not seem too likely at the moment, but it would be premature to discard it altogether.

Campbell [discussion following Novick and Horiuchi (1961) and unpublished data] and Curtiss (1962) have found cases of partial diploidy which resemble Novick's in some respects. In these cases, the cell cannot be cured ot its extra genetic fragment with acridine dyes, under conditions where an F particle in the same cell is removed. The presence of F is necessary for the transfer of the extra fragment to an F− cell, but the transfer occurs at the low rate characteristic of other genes rather than at the high rate characteristic of F-mediated transduction.

X. Transfer of Episomes by Other Episomes

Episomes consist of genetic material and frequently effect the intercellular transfer of other genetic material. It is therefore reasonable to expect that some episomes might be able to transfer other episomes. Prophages can be transduced by other phages, sometimes jointly with linked bacterial markers (Jacob, 1955; Lennox, 1955). The closely linked prophages from double lysogens of the lambda type can be carried together by the same transducing particle (Arber, 1960b), as shown by mixed liberation due to transfer induction. The F agent (Arber, 1960), the resistance transfer factor (Watanabe and Fukasawa, 1961c), and the determinants for several colicins (Demerec *et al.*, 1958; Ozeki and Stocker, 1958; Fredericq, 1959) are likewise transducible by phage. The F agent can of course cause the transfer of prophages when it is in the Hfr form. F-ducing particles (see Section VIII,B) carrying prophage lambda have also been reported (Buttin *et al.*, 1960; Goodgal and Jacob, 1961).

The transfer of one episome by another may reflect a physical connection of the two, so that at least during the time of transfer one episome behaves as part of the other. Alternatively, one episome could merely provide conditions necessary for the transfer of the second, as in the older ideas about transduction where the phage served only as a mechanical vehicle for transfer. It seems reasonable that in some cases of joint transfer during conjugation [e.g., the transfer of other colicin factors concurrently with colicin I (Ozeki et al., 1961)], one episome merely mediates the formation of conjugation bridges between cells, and others then are able to pass through these bridges independently of the first one.

XI. Relationship to Cellular Regulatory Mechanisms

Much of the interest in episomes has arisen from their possible role in initiating changes in the synthetic activities of the cell which carries them. Models to explain developmental processes in terms of episomes are easy to design on paper, and direct evidence is totally absent. The basic fact is simple. When a phage is in the prophage state, its genes do not elicit the phage-specific syntheses they are potentially able to direct. Somehow, these activities are repressed. If the cell is induced, the repression disappears and the phage genes function rapidly. Obviously this provides a mechanism whereby rather nonspecific inducers can initiate highly specific changes in the heredity of a cell.

It is a remarkable fact that, in genetically identical cells, a phage genome may multiply autonomously in one, as a prophage in another, and in a third should be inhibited from multiplying altogether (as a preprophage). Ideas about the mechanism have already been discussed, but the basic importance is independent of mechanism. The same type of control may obviously regulate the multiplication and activity of many genetic elements less obtrusive than phage.

A curious effect of induction of phage lambda on the activity of the galactose genes of *Escherichia coli* K12 has been observed. If a lambda lysogen is induced with ultraviolet, there is a marked increase in the rate of synthesis of the inducible enzymes galactokinase, galactose-1-phosphate uridyl transferase, and uridine diphosphogalactose 4-epimerase. No increase occurs in non-lysogenic cultures exposed to UV, nor to other inducible or constitutive enzymes tested (Yarmolinsky and Wiesmeyer, 1960; Buttin et al., 1960). In partially diploid cultures, only enzymes formed by alleles *cis* to the prophage are increased. The induction of lambda therefore somehow de-represses enzyme formation by a neighboring gene.

XII. General Discussion

Perhaps the greatest importance of the work on episomes to date is to emphasize to what extent the total genetic content of a cell represents an assemblage of components each with some capacity for autonomous multiplication and intercellular transmission. That these potentialities are generally not expressed probably reflects a very efficient regulatory mechanism controlling the multiplication as well as the activity of each element. Where the regulation is perfect, the element remains perpetually integrated, and its potential autonomy is undetectable.

It seems probable that any sharp distinction between episomes and other parts of the genome will break down. In a sense, it has broken down already with the discovery of gene pick-up. Is the lactose gene of *E. coli* to be considered an episome because it can exist in either an autonomous or an integrated state? It is true that it does not become autonomous by itself, but rather by associating itself with P1 or F. But can we be sure that F becomes autonomous by *itself* rather than by forming a part of some as yet undiscovered genetic element of the cell which is assorted independently of the bacterial chromosome?

Similarly we may consider a controlling element in maize which modifies the phenotypic expression of a neighboring gene. In the absence of another controlling element (*activator*), this element will not undergo transposition or cause chromosome breaks (McClintock, 1949, 1955). Under these circumstances it would be classified simply as part of a complex genetic locus concerned with the function it modifies. The qualitative distinction between controlling elements and genes may be replaced eventually by a spectrum of types from the most mobile to the least mobile in a given genetic background.

The existence of gene pick-up seems to render impossible any test of the potentiality of genes in general to "go wild" and pass into a state of unrestrained multiplication as a result of simple mutation. But it provides us with a mechanism which has the same end result. In the author's opinion, gene pick-up is probably due to chromosomal rearrangement rather than to recombination between homologous regions of host and episome, but this remains to be proven.

We have indicated, without delving as deeply as our fancy would dictate, some interpretations of the behavior of episomes based on current speculations about the circularization of genetic material. With our ignorance of the proper rules and restraints for operating with circles, it is inevitable that models constructed at present will have more explanatory than predictive value and will consequently appeal more to their authors than to their critics. It nevertheless would seem to us that a reversible coalescence of two circle such as is illustrated in Fig. 2 (where the circular nature of the bacterial chromosome is not shown) should constitute a very basic mechanism for associating and disjoining genetic material. It is possible that some "episomal" properties would be manifested by any genetic material inserted between the two halves of a duplicated segment as is the prophage of Fig. 2, autonomous multiplication merely reflecting the possibility of copying a small circle more quickly than a larger one. For technical reasons, autonomous multiplication has been demonstrable only for those elements exhibiting transposibility or independent transmissibility. Our range of vision is still quite limited, but it seems obvious that study of episomes will tell us much about the organization of genetic material.

We have devoted what may seem a disproportionate amount of space to phage as compared with other episomes. We actually have much more information about phage than about any of the others. Future work will test the utility of analogies between phages and the other elements mentioned.

Finally, it may not be superfluous to emphasize that the proportion of the genetic properties of an individual which are controlled by episomes in the autonomous state may be much larger than would appear to superficial examination. It is true that almost all known mutant characters in bacteria have been localizable on a single linkage structure, but such a census of mutants is not necessarily equivalent to a census of the genetic material of the cell. The point has already been well made for "cytoplasmic" factors in general (Nanney, 1957).

Acknowledgments

The author is indebted to the many people who have supplied him with ideas, unpublished data, or manuscripts in press. He wishes especially to thank Drs. François Jacob, S. E. Luria, Barbara McClintock, and H. Ozeki for many helpful suggestions concerning the manuscript. Unpublished data from the author's laboratory were from work supported by grant E 2862 United States Public Health Service, National Institute of Allergy and Infectious Diseases.

References

Adams, J., and Luria, S. 1958. Transduction by bacteriophage P1. Abnormal phage function of the transducing particles. *Proc. Natl. Acad. Sci. U.S.* **44**, 590–594.

Adelberg, E., and Burns, S. 1960. Genetic variation in the sex factor of *Escherichia coli*. *J. Bacteriol.* **79**, 321–330.

Akiba, T., Koyama, T., Isshiki, Y., Kimura, S., and Fukushima, T. 1960. Studies on the mechanism of development of multiple drug-resistant strains. *Japan. med. Wochschr.* **1866**, 45–50.

Alfoldi, L., Jacob, F., and Wollman, E. 1957. Zygose létale dans des croisements entre souches colicinogènes et non colicinogènes d'*Escherichia coli*. *Compt. rend. acad. Sci.* **244**, 2974–2977.

Alfoldi, L., Jacob, F., Wollman, E., and Mazé, R. 1958. Sur le déterminisme génétique de la colicinogénie. *Compt. rend. acad. Sci.* **246**, 3531–3533.

Anderson, T., and Mazé, R. 1957. Analyse de la descendance de zygotes formés par conjugaison chez *Escherichia coli* K-12. *Ann. inst. Pasteur* **93**, 194–198.

Appleyard, R. 1954a. Segregation of lambda lysogenicity during bacterial recombination in *Escherichia coli* K-12. *Genetics* **39**, 429–439.
Appleyard, R. 1954b. Segregation of new lysogenic types during growth of a doubly lysogenic strain derived from *Escherichia coli* K-12. *Genetics* **39**, 440–452.
Arber, W. 1958. Transduction des caractères *gal* par le bactériophage lambda. *Arch. Sci. (Geneva)* **11**, 259–338.
Arber, W. 1960a. Polylysogeny for bacteriophage lambda. *Virology* **11**, 250–272.
Arber, W. 1960b. Transduction of chromosomal genes and episomes in *Escherichia coli*. *Virology* **11**, 273–288.
Baron, L., Carey, F., and Spilman, W. 1959a. Characteristics of a high frequency of recombination (Hfr) strain of *Salmonella typhosa* compatible with *Salmonella*, *Shigella*, and *Escherichia* species. *Proc. Natl. Acad. Sci. U.S.* **45**, 1752–1757.
Baron, L., Carey, F., and Spilman, W. 1959b. Genetic recombination between *Escherichia coli* and *Salmonella typhimurium*. *Proc. Natl. Acad. Sci. U.S.* **45**, 976–984.
Belser, W., and Bunting, M. 1956. Studies on a mechanism providing for genetic transfer in *Serratia marcescens*. *J. Bacteriol.* **72**, 582–592.
Bernstein, H. 1958. Fertility factors in *Escherichia coli*. *In* "The Biological Replication of Macromolecules." *Symposia Soc. Exptl. Biol.* **12**, 93–103.
Bertani, G. 1953a. Lysogenic vs. lytic cycle of phage multiplication. *Cold Spring Harbor Symposia Quant. Biol.* **18**, 65–70.
Bertani, G. 1953b. Prophage dosage and immunity in lysogenic bacteria. *Records Genet. Soc. Am.* **22**, 66.
Bertani, G. 1954. Studies on lysogenesis. III. Superinfection of lysogenic *Shigella dysenteriae* with temperate mutants of the carried phage. *J. Bacteriol.* **67**, 696–707.
Bertani, G. 1956. The rôle of phage in bacterial genetics. *Brookhaven Symposia in Biol.* **8**, 50–57.
Bertani, G. 1958. Lysogeny. *Advances in Virus Research* **5**, 151–193.
Bertani, G., and Nice, S. 1954. Studies on lysogenesis. II. The effect of temperature on the lysogenization of *Shigella dysenteriae* with phage P1. *J. Bacteriol.* **67**, 202–209.
Bertani, G., and Six, E. 1958. Inheritance of prophage P2 in bacterial crosses. *Virology* **6**, 357–381.
Bertani, L. 1957. The effect of the inhibition of protein synthesis on the establishment of lysogeny. *Virology* **4**, 53–71.
Bertani, L. 1961. Levels of immunity to superinfection in lysogenic bacteria as affected by prophage genotype. *Virology* **3**, 378–380.
Boice, L., and Luria, S. 1961. Transfer of transducing prophage P1*dl* upon bacterial mating. *Bacteriol. Proc. (Soc. Am. Bacteriologists)* (V80a), 197.
Boyd, J. 1951. Observations on the relationship between symbiotic and lytic bacteriophage. *J. Pathol. Bacteriol.* **63**, 445.
Buttin, G., Jacob, F., and Monod, J. 1960. Synthèse constitutive de galactokinase consécutive au développement des bactériophage λ chez *Escherichia coli* K-12. *Compt. rend. acad. Sci.* **250**, 2471–2473.
Calef, E., and Licciardello, G. 1960. Recombination experiments on prophage host relationships. *Virology* **12**, 81–103.
Campbell, A. 1957. Transduction and segregation in *Escherichia coli* K-12. *Virology* **4**, 366–384.
Campbell, A. 1959. Ordering of genetic sites in bacteriophage lambda by the use of galactose-transducing defective phages. *Virology* **9**, 293–305.
Campbell, A. 1960. On the mechanism of the recombinational event in the formation of tranducing phages. *Virology* **11**, 339–348.

Campbell, A. 1961. Sensitive mutants of bacteriophage lambda. *Virology* **14**, 22–32.
Campbell, A., and Balbinder, E. 1959. Transduction of the galactose region of *Escherichia coli* K-12 by the phages λ and λ-434 hybrid. *Genetics* **44**, 309–319.
Cavalli, L., and Heslot, H. 1949. Recombination in bacteria: outcrossing *Escherichia coli* K-12. *Nature* **164**, 1057–1058.
Clark, A. 1961. Preparation of a strain of *Escherichia coli* K-12 possessing two linkage groups. *Bacteriol. Proc. (Soc. Am. Bacteriologists)* (G105), 98.
Cohen, D. 1959. A variant of phage P2 originating in *Escherichia coli* strain B. *Virology* **7**, 112–126.
Curtiss, R., III. 1962. Genetic studies on a partial diploid of *Escherichia coli* K12. *Bacteriol. Proc. (Am. Soc. Microbiol.)*, p. 55.
de Haan, P. 1954. Genetic recombination in *Escherichia coli* B. I. The transfer of the F agent to *E. coli* B. *Genetica* **27**, 293–300.
de Haan, P. 1955. Genetic recombination in *Escherichia coli* B. II. The influence of experimental conditions on the transfer of unselected markers. *Genetica* **27**, 364–376.
Delbrück, M., and Stent, G. S. 1956. On the mechanism of DNA replication. *In* "The Chemical Basis of Heredity" (W. D. McElroy and B. Glass, eds.), pp. 699–736. Johns Hopkins Press, Baltimore, Maryland.
Demerec, M., Lahr, E. L., Miyake, T., Goldman, I., Balbinder, E., Banic, S., Hashimoto, K., Glanville, E. V., and Gross, J. D. 1958. Bacterial genetics. *Carnegie Inst. Wash. Yearbook* **57**, 390–406.
Demerec, M., Lahr, E. L., Balbinder, E., Miyake, T., Ishidsu, J., Mizobuchi, K., and Mahler, B. 1960. Bacterial genetics. *Carnegie Inst. Wash. Yearbook* **59**, 426–441.
Driskell, P., and Adelberg, E. 1961. Inactivation of the sex factor of *Escherichia coli* K-12 by the decay of incorporated radiophosphorus. *Bacteriol. Proc. (Soc. Am. Bacteriologists)* (P81), 186.
Ephrussi, B., Hottinguer, H., and Chimènes, A. 1949. Action de l'acriflavine sur les levures. *Ann. inst. Pasteur* **76**, 351–367.
Falkow, S., and Baron, L. 1961. An episomic element in a strain of *Salmonella typhosa*. *Bacteriol. Proc. (Soc. Am. Bacteriologists)* (G98), 96.
Fredericq, P. 1953. Recombinants issus du croisement des souches lysogènes et colicinogènes. *Compt. rend. soc. biol.* **147**, 1113–1116.
Fredericq, P. 1954a. Transduction génétique des propriétés colicinogènes chez *Escherichia coli*. *Compt. rend. soc. biol.* **148**, 399–402.
Fredericq, P. 1954b. Intervention du facteur de polarité sexuelle F dans la transduction des propriétés colicinogènes chez *Escherichia coli*. *Compt. rend. soc. biol.* **148**, 746–748.
Fredericq, P. 1954c. Induction de la production de colicine par irradiation ultraviolette de souches colicinogènes d'*Escherichia coli*. *Compt. rend. soc. biol.* **148**, 1276–1280.
Fredericq, P. 1956a. Genetic transfer of colicinogenic properties by *Bact. coli. J. Gen. Microbiol.* **15**, iii.
Fredericq, P. 1956b. Recherches sur la fréquence des souches transductrices des propriétés colicinogènes. *Compt. rend. soc. biol.* **150**, 1036–1039.
Fredericq, P. 1956c. Resistance et immunité aux colicines. *Compt. rend. soc. biol.* **150**, 1514–1517.
Fredericq, P. 1957. Colicins. *Ann. Rev. Microbiol.* **11**, 7–22.
Fredericq, P. 1958. Colicins and colicinogenic factors. *In* "The Biological Replication of Macromolecules." *Symposia Soc. Exptl. Biol.* **12**, 104–122.

Fredericq, P. 1959. Transduction par bactériophage des propriétés colicinogènes chez *Salmonella typhimurium. Compt. rend. soc. biol.* **153,** 357–360.

Fredericq, P., and Betz-Bareau, M. 1953a. Transfert génétique de la propriété colicinogène chez *E. coli. Compt. rend. soc. biol.* **147,** 1110–1112.

Fredericq, P., and Betz-Bareau, M. 1953b. Transfert génétique de la propriété colicinogène en rapport avec la polarité F des parents. *Compt. rend. soc. biol.* **147,** 2043–2045.

Fredericq, P., and Betz-Bareau, M. 1953c. Transfert génétique de la propriété de produire un antibiotique. *Compt. rend. soc. biol.* **147,** 1653–1656.

Fredericq, P., and Betz-Bareau, M. 1956. Influence des diverse propriétés colicinogènes sur la fertilité d'*Escherichia coli. Compt. rend. soc. biol.* **150,** 615–619.

Fredericq, P., and Papavassiliou, J. 1957. Croisement génétique des souches d'*Escherichia coli* marquées par production de colicin I. *Compt. rend. soc. biol.* **151,** 1970–1973.

Goodgal, S., and Jacob, F. 1961. Complementation between lysogenic defectives of *E. coli* K-12(T). *Bacteriol. Proc. (Soc. Am. Bacteriologists)* 28.

Hamon, Y. 1956. Étude générale du transfert des propriétés colicinogènes. *Compt. rend. acad. sci.* **242,** 2064–2066.

Hamon, Y. 1957. Propriétés générales des cultures d'entérobacteriacées rendues colicinogènes par transfert. *Ann. inst. Pasteur* **92,** 363–368.

Harada, K., Suzuki, M., Kameda, M., and Mitsuhashi, S. 1960. On the drug resistance of enteric bacteria. 2. Transmission of the drug-resistance among *Enterobacteriaceae. Japan. J. Exptl. Med.* **30,** 289–299.

Hartman, P. 1961. Methodology in transduction. In *Symposium on Methodology in Basic Genetics* (in press).

Hayes, W. 1953a. Observations on a transmissible agent determining sexual differentation in *Bacterium coli. J. Gen. Microbiol.* **8,** 72–88.

Hayes, W. 1953b. The mechanism of genetic recombination in *Escherichia coli. Cold Spring Harbor Symposia Quant. Biol.* **18,** 75–93.

Hershey, A., and Melechen, N. 1957. Synthesis of phage precursor nucleic acid in the presence of chloramphenicol. *Virology* **3,** 207–236.

Hirota, Y. 1956. Artificial elimination of the F factor in *Bact. coli* K-12. *Nature* **178,** 92.

Hirota, Y. 1960. The effect of acridine dyes on mating type factors in *Escherichia coli. Proc. Natl. Acad. Sci. U.S.* **46,** 57–64.

Hirota, Y., and Sneath, P. 1961. F' and F-mediated transduction in *Escherichia coli.* K-12 *Japanese J. Genetics* **36,** 307–318.

Holloway, B. 1955. Genetic recombination in *Pseudomonas aeruginosa. J. Gen. Microbiol.* **13,** 572–581.

Holloway, B. 1956. Self-fertility in *Pseudomonas aeruginosa. J. Gen. Microbiol.* **15,** 221–224.

Holloway, B., and Jennings, P. 1958. An infectious fertility factor for *Pseudomonas aeruginosa. Nature* **181,** 855–856.

Horiuchi, T., Tomizawa, J., and Novick, A. 1962. Isolation and properties of bacteria capable of high rates of β-galactosidase synthesis. *Biochim. et Biophys. Acta* **55,** 152–163.

Jacob, F. 1955. Transduction of lysogeny in *Escherichia coli. Virology* **1,** 207–220.

Jacob, F. 1960. Genetic control of viral functions. *Harvey Lectures* **1959–60,** 1–39.

Jacob, F., and Adelberg, E. 1959. Transfert de caractères génétiques par incorporation au facteur sexuel d'*Escherichia coli. Compt. rend. acad. Sci.* **249,** 189–191.

Jacob, F., and Campbell, A. 1959. Sur le système de repression assurant l'immunité chez les bactéries lysogènes. *Compt. rend. acad. Sci.* **248**, 3219–3221.
Jacob, F., and Monod, J. 1961. Genetic regulatory mechanisms in the synthesis of proteins. *J. Mol. Biol.* **3**, 318–356.
Jacob, F., and Wollman, E. 1953. Induction of phage development in lysogenic bacteria. *Cold Spring Harbor Symposia Quant. Biol.* **18**, 101–121.
Jacob, F., and Wollman, E. 1954a. Induction spontanée du developpement du bacteriophage lambda au cours de la recombinaison génétique chez *Escherichia coli*. *Compt. rend. acad. sci.* **239**, 317–319.
Jacob, F., and Wollman, E. 1954b. Étude génétique d'un bactériophage temperé d'*Escherichia coli*. 1-Le système génétique du bactériophage λ. *Ann. inst. Pasteur* **87**, 653.
Jacob, F., and Wollman, E. 1956a. Recombinaison génétique et mutants de fertilité chez *Escherichia coli*. *Compt. rend. acad. Sci.* **242**, 303–306.
Jacob, F., and Wollman, E. 1956b. Sur les processus de conjugaison et de recombinaison chez *Escherichia coli*. I-L'induction par conjugaison ou induction zygotique. *Ann. inst. Pasteur* **91**, 486–510.
Jacob, F., and Wollman, E. 1956a. Analyze des groupes de liaison génétique de differentes souches donatrices d'*Escherichia coli* K-12. *Compt. rend. acad. sci.* **245**, 1840–1843.
Jacob, F., and Wollman, E. 1957b. Genetic aspects of lysogeny. *In* "The Chemical Basis of Heredity" (W. D. McElroy and B. Glass, eds.), pp. 468–500. Johns Hopkins Press, Baltimore, Maryland.
Jacob, F., and Wollman, E. 1958a. Genetic and physical determinations of chromosomal segments in *Escherichia coli*. *In* "The Biological Replication of Macromolecules." *Symposia Soc. Exptl. Biol.* **12**, 75–92.
Jacob, F., and Wollman, E. 1958b. Les épisomes, éléments génétiques ajoutes. *Compt. rend. acad. sci.* **247**, 154–156.
Jacob, F., and Wollman, E. 1959a. Lysogeny. *In* "The Viruses" (F. Burnet and W. Stanley, eds.), pp. 319–352. Academic Press, New York.
Jacob, F., and Wollman, E. 1959b. The relationship between the prophage and the bacterial chromosome in lysogenic bacteria. "Recent Progress in Microbiology," *7th Intern. Congr. Microbiol.*, pp. 15–30. Almquist and Wiksells, Uppsala.
Jacob, F., and Wollman, E. 1961. "Sexuality and the Genetics of Bacteria." Academic Press, New York
Jacob, F., Siminovitch, L., and Wollman, E. 1952. Sur la biosynthèse d'une colicine et sur son mode d'action. *Ann. inst. Pasteur* **83**, 295–315.
Jacob, F., Fuerst, C., and Wollman, E. 1957. Recherches sur les bactéries lysogènes défectives. II-Les types physiologiques liés aux mutations du prophage. *Ann. inst. Pasteur* **93**, 724–753.
Jacob, F., Perrin, D., Sanchez, C., and Monod, J. 1960a. L'opéron: groupe de gènes à expression coordonée par un operateur. *Compt. rend. acad. sci.* **250**, 1727–1729.
Jacob, F., Schaeffer, P., and Wollman, E. 1960b. Episomic elements in bacteria. *In* "Microbial Genetics." *Symposium Soc. Gen. Microbiol. 10th* pp. 67–91.
Kaiser, A. 1957. Mutations in a temperate bacteriophage affecting its ability to lysogenize *Escherichia coli*. *Virology* **3**, 42–61.
Kaiser, A., and Jacob, F. 1957. Recombination between related temperate bacteriophages and the genetic control of immunity and prophage localization. *Virology* **4**, 509–521.
Kellenberger, G., Zichichi, M., and Weigle, J. 1961a. Exchange of DNA in the recombination of bacteriophage λ. *Proc. Natl. Acad. Sci. U.S.* **47**, 869–878.

Kellenberger, G., Zichichi, M., and Weigle, J. 1961b. A mutation affecting the DNA content of bacteriophage lambda and its lysogenizing properties. *J. Mol. Biol.* **3**, 399–408.

Kozloff, L. 1960. Biochemistry of viruses. *Ann. Rev. Biochem.* **29**, 475–502.

Lanni, F. 1959. Base composition of microbial *DNA* in relation to genetic homology. *Bacteriol. Proc. (Soc. Am. Bacteriologists)* (G58), 45–46.

Lavallé, R., and Jacob, F. 1961. Sur la sensibilité des épisomes sexuel et colicinogène d'*Escherichia coli* K 12 à la désintégration du radiophosphore. *Compt. rend. acad. sci.* **252**, 1678–1680.

Lederberg, E., Lederberg, M., and Lederberg, J. 1953. Genetic studies of lysogenicity in *Escherichia coli*. *Genetics* **38**, 51–64.

Lederberg, J. 1949. Aberrant heterozygotes is *Escherichia coli*. *Proc. Natl. Acad. Sci. U.S.* **35**, 178–184.

Lederberg, J. 1952. Cell genetics and hereditary symbiosis. *Physiol. Revs.* **32**, 403–430.

Lederberg, J. 1956. Conjugal pairing in *Escherichia coli*. *J. Bacteriol.* **71**, 497–498.

Lederberg, J. 1957. Sibling recombinants in zygote pedigrees of *Escherichia coli*. *Proc. Natl. Acad. Sci. U.S.* **43**, 1060–1065.

Lederberg, J. 1959. Bacterial reproduction. *Harvey Lectures* **1957–58**, 69–82.

Lederberg, J., and Lederberg, E. 1956. Infection and heredity. *In* "Cellular Mechanisms of Differentiation and Growth" (D. Rudnick, ed.), pp. 101–124. Princeton Univ. Press, Princeton, New Jersey.

Lederberg, J., and Tatum, E. L. 1946. Novel genotypes in mixed cultures of biochemical mutants of bacteria. *Cold Spring Harbor Symposia Quant. Biol.* **11**, 113–114.

Lederberg, J., Lederberg, E., Zinder, N., and Lively, E. 1951. Recombination analysis of bacterial heredity. *Cold Spring Harbor Symposia Quant. Biol.* **16**, 413–443.

Lederberg, J., Cavalli, L., and Lederberg, E. 1952. Sex compatibility on *Escherichia coli*. *Genetics* **37**, 720–730.

Lennox, E. 1955. Transduction of linked genetic characters of the host by bacteriophage P1. *Virology* **1**, 190–206.

Levine, M. 1957. Mutations in the temperate phage P22 and lysogeny in *Salmonella*. *Virology* **3**, 22–41.

Levine, M., and Cox, E. 1961. Repression of induction of phage production in lysogenic bacteria. *Bacteriol. Proc. (Soc. Am. Bacteriologists)* (V73), 163.

Levinthal, C. 1959. Bacteriophage genetics. *In* "The Viruses" (F. Burnet and W. Stanley, eds.), pp. 281–318. Academic Press, New York.

Li, K., Barksdale, L., and Garmise, L. 1961. Phenotypic alterations associated with the bacteriophage carrier state of *Shigella dysenteriae*. *J. Gen. Microbiol.* **24**, 355–367.

Lieb, M. 1953. The establishment of lysogenicity in *Escherichia coli*. *J. Bacteriol.* **65**, 642–651.

Lieb, M., Weigle, J., and Kellenberger, E. 1955. A study of hybrids between two strains of *Escherichia coli*. *J. Bacteriol.* **69**, 468–471.

Luria, S., and Burrous, J. 1957. Hybridization between *Escherichia coli* and *Shigella*. *J. Bacteriol.* **74**, 461–476.

Luria, S., Fraser, D., Adams, J., and Burrous, J. 1958. Lysogenization, transduction, and genetic recombination in bacteria. *Cold Spring Harbor Symposia Quant. Biol.* **23**, 71–82.

Luria, S., Adams, J., and Ting, R. 1960. Transduction of lactose-utilizing ability among strains of *E. coli* and *S. dysenteriae* and the properties of the transducing phage particles. *Virology* **12**, 348–390.

Lwoff, A. 1953. Lysogeny. *Bacteriol. Revs.* **17**, 269–337.

McClintock, B. 1949. Mutable loci in maize. *Carnegie Inst. Wash. Yearbook* **48**, 142–154.
McClintock, B. 1955. Controlled mutation in maize. *Carnegie Inst. Wash. Yearbook* **54**, 245–255.
McClintock, B. 1956. Controlling elements and the gene. *Cold Spring Harbor Symposia Quant. Biol.* **21**, 197–216.
McClintock, B. 1961. Some parallels between gene control systems in maize and in bacteria. *Am. Naturalist* **95**, 265–277.
Marmur, J., Rownd, R., Falkow, S., Baron, L., Shildkraut, C., and Doty, P. 1961. The nature of intergeneric episomal infection. *Proc. Natl. Acad. Sci. U.S.* **47**, 972–979.
Matney, T., McDonald, W., and Goldschmidt, E. 1961. Evidence of persistent diploidy in an Hfr mating system in *Escherichia coli* K-12. *Bacteriol. Proc. (Soc. Am. Bacteriologists)* (G99), 96.
Meselson, M., and Weigle, J. 1961. Chromosome breakage accompanying genetic recombination in bacteriophage. *Proc. Natl. Acad. Sci. U.S.* **47**, 857–868.
Mitsuhashi, S., Harada, K., and Hashimoto, K. 1960a. Multiple resistance of enteric bacteria and transmission of drug-resistance to other strains by mixed cultivation. *Japan. J. Exptl. Med.* **30**, 179–184.
Mitsuhashi, S., Harada, K., Kameda, M., Suzuki, M., and Egawa, R. 1960b. On the drug resistance of enteric bacteria. 3. Transmission of the drug resistance from *Shigella* to F− or Hfr strains of *E. coli* K-12. *Japan. J. Exptl. Med.* **30**, 301–306.
Miyake, T., and Demerec, M. 1959. *Salmonella-Escherichia* hybrids. *Nature* **183**, 1586.
Morse, M., Lederberg, E., and Lederberg, J. 1956a. Transduction in *Escherichia coli* K-12. *Genetics* **41**, 121–156.
Morse, M., Lederberg, E., and Lederberg, J. 1956b. Transductional heterogenotes in *Escherichia coli*. *Genetics* **41**, 758–779.
Nanney, D. 1957. The role of the cytoplasm in heredity. *In* "The Chemical Basis of Heredity" (W. D. McElroy and B. Glass, eds.), pp. 134–166. Johns Hopkins Press, Baltimore, Maryland.
Nelson, T., and Lederberg, J. 1954. Postzygotic elimination of genetic factors in *Escherichia coli*. *Proc. Natl. Acad. Sci. U.S.* **40**, 415–419.
Novick, A., and Horiuchi, T. 1961. Hyper-production of β-galactosidase by *Escherichia coli* bacteria. *Cold Spring Harbor Symposia Quant. Biol.* **26**, 239–245.
Ochiai, K., Yamanaka, T., Kimura, K., and Sawada, O. 1959. Studies on inheritance of drug resistance between *Shigella* strains and *Escherichia coli* strains. *Japan. med. Wochschr.* **1861**, 34–46.
Ørskov, F., and Ørskov, I. 1961. The fertility of *Escherichia coli* antigen test strains in crosses with K12. *Acta Pathol. Microbiol. Scand.* **51**, 280–290.
Ørskov, F., Ørskov, I., and Kauffmann, F. 1961. The fertility of *Salmonella* strains determined in mating experiments with *Escherichia* strains. *Acta. Pathol. Microbiol. Scand.* **51**, 291.
Ozeki, H., and Stocker, B. 1958. Phage-mediated transduction of colicinogeny in *Salmonella typhimurium*. *Heredity* **12**, 525–526.
Ozeki, H., Stocker, B., and deMargerie, H. 1959. Production of colicine by single bacteria. *Nature* **184**, 337–339.
Ozeki, H., Howarth, S., and Clowes, R. 1961. Colicine factors as fertility factors in bacteria. *Nature* **190**, 986–989.
Richter, A. 1957. Complementary determinants of an Hfr phenotype in *E. coli* K-12. *Genetics* **42**, 391.

Richter, A. 1961. Attachment to wild type F factor to a specific chromosomal region in a variant strain of *Escherichia coli* K12: the phenomenon of episomic alternation. *Genet. Research, Camb.* **2**, 335–345.

Roper, J. 1958. Nucleo-cytoplasmic interactions in *Aspergillus nidulans*. *Cold Spring Harbor Symposia Quant. Biol.* **23**, 141–154.

Six, E. 1961a. Inheritance of prophage P2 in superinfection experiments. *Virology* **14**, 220–233.

Six, E. 1961b. Attachment site specificity of temperate coli phage P2. *Bacteriol. Proc. (Soc. Am. Bacteriologists)* (V72), 162.

Skaar, P., and Garen, A. 1956. The orientation and extent of gene transfer in *Escherichia coli*. *Proc. Natl. Acad. Sci. U.S.* **42**, 619–624.

Skaar, P., Richter, A., and Lederberg, J. 1957. Correlated selection for motility and sex incompatibility in *Escherichia coli* K-12. *Proc. Natl. Acad. Sci. U.S.* **43**, 329–333.

Spicer, C., and Datta, N. 1959. Reversion of transduced antigenic characters in *Salmonella typhimurium*. *J. Gen. Microbiol.* **20**, 136–143.

Stahl, F. 1961. A chain model for chromosomes. *J. Chimie Physique* No 2392, 1072–1077.

Stent, G. 1958. Mating in the reproduction of bacterial viruses. *Advances in Virus Research* **5**, 95–149.

Stocker, B. 1960. Introduction. Microorganisms in genetics. *In* "Microbial Genetics." *Symposium Soc. Gen. Microbiol.* 10th pp. 1–12.

Taylor, A., and Adelberg, E. 1960. Linkage analysis with very high frequency males of *Escherichia coli*. *Genetics* **45**, 1233–1243.

Thomas, R. 1959. Effects of chloramphenicol on genetic replication in bacteriophage. *Virology* **9**, 275–289.

Ting, R. 1960. A curing effect of chloramphenicol on bacteria infected with bacteriophage. *Virology* **12**, 68–80.

Tomizawa, J. 1958. Sensitivity of phage precursor nucleic acid synthesized in the presence of chloramphenicol to ultraviolet irradiation. *Virology* **6**, 55–80.

Watanabe, T., and Fukasawa, T. 1960a. Episomic resistance factors in Enterobacteriaceae. I. Transfer of resistance factors by conjugation among Enterobacteriaceae. *Medicine and Biol. (Japan)* **56**, 56–59.

Watanabe, T., and Fukasawa, T. 1960b. Episomic resistance factors in Enterobacteriaceae. II. Transduction of resistance factors in *Escherichia coli* and *Salmonella typhimurium*. *Medicine and Biol. (Japan)* **56**, 64–67.

Watanabe, T., and Fukasawa, T. 1960c. Episomic resistance factors in Enterobacteriaceae. III. Elimination of resistance factors by treatment with acridine dyes. *Medicine and Biol. (Japan)* **56**, 71–74.

Watanabe, T., and Fukasawa, T. 1960d. Episomic resistance factors in Enterobacteriaceae. IV. Kinetics of transfer of resistance factors. *Medicine and Biol. (Japan)* **56**, 98–100.

Watanabe, T., and Fukasawa, T. 1960e. Episomic resistance factors in Enterobacteriaceae. V. Transfer of resistance factors mediated by a new episome. *Medicine and Biol. (Japan)* **56**, 201–206.

Watanabe, T., and Fukasawa, T. 1960f. Episomic resistance factors in Enterobacteriaceae. VI. Suppression of sex factor of *Escherichia coli* strain K-12 by resistance transfer factor. *Medicine and Biol. (Japan)* **57**, 30–33.

Watanabe, T., and Fukasawa, T. 1960g. Episomic resistance factors in Enterobacteriaceae. VII. Suppression of transfer of resistance transfer factor from resistant *Shigella* to *Escherichia coli* strain K-12 by the presence of F-factor in recipients. *Medicine and Biol. (Japan)* **57**, 169–172.

Watanabe, T., and Fukasawa, T. 1960h. "Resistance transfer factor." An episome in *Enterobacteriaceae*. *Biochem. Biophys. Research Communs*. **3**, 660–665.
Watanabe, T., and Fukasawa, T. 1961a. Episome-mediated transfer of drug resistance in *Enterobacteriaceae*. I. Transfer of resistance factors by conjugation. *J. Bacteriol.* **81**, 669–678.
Watanabe, T., and Fukasawa, T. 1961b. Episome-mediated transfer of drug resistance in *Enterobacteriaceae*. II. Elimination of resistance factors with acridine dyes. *J. Bacteriol.* **81**, 679–683.
Watanabe, T., and Fukasawa, T. 1961c. Episome-mediated transfer of drug resistance in *Enterobacteriaceae*. III. Transduction of resistance factors. *J. Bacteriol.* **82**, 202–209.
Watanabe, T., and Fukasawa, T. 1961d. Episomic resistance factors in *Enterobacteriaceae*. VIII. Kinetics of transfer of resistance factors studied by interrupted conjugation with phage T6. *Medicine and Biol. (Japan)* **58**, 202–206.
Watanabe, T., and Fukasawa, T. 1961e. Episomic resistance factors in *Enterobacteriaceae*. IX. Interaction between resistance factors and F' in *Escherichia coli* K-12. *Medicine and Biol. (Japan)* **59**, 42–47.
Watanabe, T., and Fukasawa, T. 1961f. Episomic resistance factors in *Enterobacteriaceae*. X. Relationship between resistance factors and resistance transfer factor. *Medicine and Biol. (Japan)* **59**, 115–118.
Watanabe, T., and Fukasawa, T. 1961g. Episomic resistance factors in *Enterobacteriaceae*. XI. High frequency resistance transfer system (HFRT) *Medicine and Biol. (Japan)* **59**, 153–157.
Watanabe, T., and Fukasawa, T. 1961h. Episomic resistance factors in *Enterobacteriaceae*. XII. Chromosomal attachment of resistance transfer factor in *Escherichia coli* strain K-12. *Medicine and Biol. (Japan)* **59**, 180–184.
Watanabe, T., and Fukasawa, T. 1962. Episomic resistance factors in *Enterobacteriaceae*. IV. Interactions between resistance transfer factor and F-factor in *Escherichia coli* K-12. (Submitted to *J. Bacteriol.*)
Weigle, J., Meselson, M., and Paigen, K. 1959. Density alterations associated with transducing ability in the bacteriophage lambda. *J. Mol. Biol.* **1**, 379–386.
Whitfield, J., and Appleyard, R. 1958. Recombination and phenotypic mixing during phage growth in strains of *Escherichia coli* doubly lysogenic for coliphage LAMBDA. *Virology* **5**, 275–290.
Wilson, D. 1960. The effects of ultraviolet light and ionizing radiation on the transduction of *Escherichia coli* by phage P1. *Virology* **11**, 533–546.
Wollman, E., and Jacob, F. 1955. Sur le mécanisme de transfert de matériel génétique au cours de la recombinaison chez *Escherichia coli* K-12. *Compt. rend. acad. sci.* **240**, 2449–2451.
Wollman, E., and Jacob, F. 1958a. Sur le déterminisme génétique des types sexuels chez *Escherichia coli* K-12. *Compt. rend. acad. sci.* **247**, 536–539.
Wollman, E., and Jacob, F. 1958b. Sur les processus de conjugaison et de recombinaison chez *Escherichia coli*. V. Le mécanisme du transfert de matériel génétique. *Ann. inst. Pasteur* **95**, 641–646.
Yarmolinsky, M., and Wiesmeyer, H. 1960. Regulation by coliphage lambda of the expression of the capacity to synthesize a sequence of host enzymes. *Proc. Natl. Acad. Sci. U.S.* **46**, 1626–1645.
Zinder, N. 1955. Bacterial transduction. *J. Cellular Comp. Physiol.* **45**, (Suppl. 2), 23–49.
Zinder, N. 1958. Lysogenization and superinfection immunity in *Salmonella*. *Virology* **5**, 291–326.
Zinder, N. 1960a. Hybrids of *Escherichia* and *Salmonella*. *Science* **131**, 813–815.
Zinder, N. 1960b. Sexuality and mating in *Salmonella*. *Science* **131**, 924–926.

THE CYTOGENETICS OF Oenothera

Ralph E. Cleland*

Department of Botany, Indiana University, Bloomington, Indiana

	Page
I. Introduction	147
II. Outline of Genetic Behavior	148
III. Early Cytological Work	152
IV. The Relation of Cytological to Genetic Behavior	154
V. The Cause of Circle Formation in *Oenothera*	156
VI. Chromosome Structure	160
A. Size	160
B. Morphology	163
VII. Early Meiotic Stages	164
A. Leptophase	165
B. Zygophase	165
C. Pachyphase and Diplophase	165
VIII. Factors Influencing Genetic Analyses	167
A. Megaspore Competition and Development of Female Gametophyte	167
B. Pollen Tube Competition	169
IX. Factor Analysis and the Location of Genes	170
X. Crossing-Over	172
XI. Position Effect	180
XII. Gene Conversion	183
XIII. The Nature of de Vries' Mutants	188
XIV. Induced Mutations	202
XV. Incompatibility	204
XVI. Plastid Behavior	208
XVII. Evolutionary Considerations	215
References	229

I. Introduction

According to Munz, who has published a series of taxonomic revisions on the genus (1928, 1949), *Oenothera* comprises fifteen subgenera, of which only the largest (subgenus *Euoenothera* of Munz, *Onagra* of much genetic literature) has been extensively studied cytogenetically. This is not the *Euoenothera* of Schwemmle, which according to Munz should be called *Raimannia*. The present review will deal exclusively with *Euoenothera* (*Onagra*) as found in North America or as represented by European derivatives of North American origin.

* Original work reported here was supported by grants from the Rockefeller Foundation.

The genus is native to the western hemisphere but has been extensively introduced into other areas of the world. The first reference to *Oenothera* in Europe appears to be that of Prosperus Alpinus of Padua (1627). It has become widely distributed throughout Europe where it constitutes an important element in the flora. While some of the races found in Europe have been derived directly from North America, most European forms have probably arisen *de novo* in Europe as a result of hybridization and thus are not represented in the western world. This is undoubtedly true of *O. lamarckiana, biennis,* and *suaveolens* of de Vries among others.

II. Outline of Genetic Behavior

Hugo de Vries first brought *Oenothera* into scientific prominence when he used it as the chief support for his Mutation Theory of Evolution. Although it turned out later that de Vries' "mutants" were not what he thought they were, the genus has nevertheless retained its central scientific interest because of the unusual hereditary behavior which de Vries showed it to have.

Although de Vries did not call attention to Mendel's work until 1900, and presumably knew nothing about this work before that time, it is evident (1900a,b) that he had already carried out an extensive breeding program using a variety of plants, and was already aware of the behavior now termed "Mendelian." It is also clear that he had found in *Oenothera* a type of behavior that did not conform with the Mendelian scheme, thus causing him to doubt the universality of the latter. For example he found cases (de Vries, 1900b) where true-breeding Oenotheras, when crossed, produced segregating progeny, and where hybrids thus produced bred true—the reverse of what was to be expected on the basis of Mendelian behavior.

De Vries' attempt to explain this behavior and to reconcile it with the behavior of other organisms proved inadequate, partly because it was based on his earlier theory of Intracellular Pangenesis (1889) (see p. 188) and partly because he was unwilling to concede the possibility of his material being genetically heterozygous. It remained for Renner (1917a,b,c, 1919, 1925, etc.) and his school to analyze the breeding behavior in *Oenothera* and to apply the correct genetic explanation to it. Bartlett and his co-workers (1916) independently arrived at conclusions which bore some similarity to those of Renner, as did Muller (1917). Although the principal features of ordinary *Oenothera* genetics as presented by Renner are well known, they will be briefly summarized as background for what follows.

1. Renner found that most Oenotheras behave as though they possess

but a single linkage group, for which they are highly heterozygous—this in spite of the fact that they have a diploid number of 14 chromosomes and would be expected to have 7 linkage groups.

2. A plant which is heterozygous for a single linkage group will theoretically be capable of producing two and only two kinds of gametes, and these will be identical genetically with the two that united to form the plant. This is the situation in most Euoenotheras.

3. Each haploid set of genes, or genome, possesses one or more lethals which prevent its existence in double dose. Lethals in *Oenothera* are of two kinds—zygotic and gametophytic (Fig. 1). Zygotic lethals cause the death

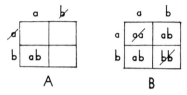

FIG. 1. Action of lethals: (*left*) gametophytic; (*right*) zygotic. Reprinted from *Scientific Monthly* by permission.

of zygotes receiving the same genome through sperm and egg; gametophytic lethals kill or inhibit the male or the female gametophyte, as the case may be. Most Oenotheras have a balanced lethal situation, i.e., both of their genomes possess lethals, although at different loci. In the case of gametophytic lethals, one genome has an "egg lethal" that kills or slows the growth of the female gametophyte (embryo sac and its contents); the other genome has a "pollen lethal" that kills or slows the development of the male gametophyte or the growth of the pollen tube. Thus, a plant with a balanced set of gametophytic lethals produces only one kind of functional egg and one kind of sperm. Plants possessing only zygotic lethals produce two kinds of functional eggs and sperm, but half the zygotes resulting from self pollination will die because they receive one or the other of these lethals in double dose.

Some Oenotheras have intermediate situations in that only one of the genomes may possess a gametophytic lethal. This will result in the production of two kinds of functional eggs, and only one kind of sperm, or vice versa. In still others, the so-called lethals are only partially effective, so that one genome will be carried through the egg less frequently than the other, or only rarely; the other may be transmitted with reduced frequency through the sperm. This suggests that at least in some cases "gametophytic lethals" are not lethals at all, but they merely slow down the development of the embryo sac (on the female side) or the growth of the pollen tube (on the male side), thus allowing the other genome to win out in the competition to produce or to fertilize the eggs.

4. In an inbred line, as a result of the inclusion of all genes into a single linkage group, and the presence of the lethals, a plant can produce only one kind of offspring—all its progeny will be exact duplicates of itself, and of each other (Fig. 2).

5. In nature, most Oenotheras with genes in a single linkage group and possessing lethals are normally self-pollinating. Consequently, each line is a true-breeding line, and is isolated fairly effectively from other lines because the self-pollinating habit ordinarily prevents outcrossing.

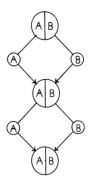

FIG. 2. In selfed line, complex-heterozygotes breed true. Reprinted from *Cytologia*.

6. The linkage of all genes into a single group, coupled with balanced lethals and self-pollination, also means that a given genome becomes an entity of indefinite duration. The entire set of genes received through the egg is segregated in meiosis from the entire set received through the sperm. Self pollination prevents combination with genomes from other lines, and the lethals prevent the existence of genomes in double dose. Only the reunion of the two genomes present in the parent plant is possible. The genomes in a strain are thus passed on intact from one generation to the next indefinitely. Renner emphasized this point and gave the name "complex" to a genome which is thus transmitted intact for an indefinite period. Such a genome is now known as a "Renner complex." Most races are "complex-heterozygotes" since they possess two different complexes. Renner gave Latin names to all complexes with which he worked. The present reviewer uses "alpha" and "beta" to designate the egg and the sperm complexes, respectively, of a given race. Thus, the race "*Platteville*" has the complexes alpha *Platteville* and beta *Platteville*.[1] (See facing page for footnote.)

We can now see why de Vries' races bred true. They were not pure lines, but permanent heterozygotes. It is also clear why outcrosses between true-breeding lines often produce segregating progeny in the F_1. One may expect one, two, or four classes of progeny following an outcross, depending on whether one or both parents produce one or two kinds of functional sperm or egg.

Renner (1917b) ascribed the origin of complex-heterozygotes to hybridization, as a result of which initially unrelated genomes have been brought into more or less permanent association.

7. Renner (1928) found another extraordinary situation: hybrids that have been derived by the crossing of true-breeding complex-heterozygotes may have two or more independent linkage groups instead of one. Genes that were linked in the parents may be independent in the hybrid. A given hybrid always has the same number and kind of linkage groups, but different hybrids will differ in respect to the number and composition of these groups. From this it is clear that the linking of all genes into a single group, where this occurs, as in most wild races, must have some other physical basis than that responsible for ordinary linkage. Renner was inclined toward the assumption that the genes making up a complex are distributed among the various chromosomes and that some mechanism exists which brings about the regular disjunction of the entire paternal set of chromosomes from the maternal set in most races, but which allows the chromosomes to be variously independent in hybrids between these races. This mechanism was revealed by cytological studies beginning in 1922 (see below).

8. Renner further pointed out that some Oenotheras do not show the peculiarities outlined above. They are relatively homozygous, do not have lethals or self pollination. They are "complex-homozygotes," possessing a single complex in double dose.

Extensive studies of the North American Euoenotheras have shown that practically all races found from the Rocky Mountains eastward are complex-heterozygotes, with balanced lethals and self pollination. By contrast, the Oenotheras found in the southwestern and western parts of the United States and northern Mexico are largely complex-homozygotes, open pollinated and free of lethals.

[1] Since the term "race" is variously used by various investigators, it is necessary to explain the sense in which the term is employed in this treatise. It is applied to a line of descent which is isolated from other lines in nature because of its self-pollinating character, and which breeds true because of the presence of large circles and balanced lethals. The hundreds or thousands of such races differ from each other in phenotype (slightly to markedly) and/or in the arrangement of their chromosome segments.

We are using the term also for pedigrees derived from populations of open-pollinated, alethal Oenotheras, such as characterize California and adjacent areas. There is less justification for this usage, but it is adopted as a matter of convenience.

A race ordinarily carries the name of the locality where it was first found.

It should be mentioned at this point that Shull (1923b), working with
O. lamarckiana and its derivatives, also found that most genes are confined to a single linkage group. He actually recognized three linkage
O. groups in *lamarckiana* (Shull, 1925), but two of them were small, most
genes belonging to a single large group. Shull assumed that each of these
groups was related to an individual chromosome pair, as is true in other
organisms.

To summarize this situation, most Oenotheras are "complex-heterozygotes," as Renner called them. Each race has two different complexes
or permanent genomes. This results from (a) the inclusion of all genes
within a single linkage group for which the plant is heterozygous, (b)
the presence of one or more lethals in each genome, and (c) self pollination. Hybrids derived from the crossing of true-breeding complex-heterozygotes, however, may have two or more linkage groups. Genes linked
n the parents are not necessarily linked in the hybrids.

III. Early Cytological Work

The physical basis for the genetic peculiarities revealed by the Renner
school became evident when the chromosome behavior of this group was

FIG. 3. Ring of 14 chromosomes in diakinesis. Reprinted from *Jahrb. wiss. Botan.*

FIG. 4. Ring of 14 chromosomes, side view of first metaphase. Reprinted from *Botanical Gazette* by permission.

analyzed. Cytological work on *Oenothera* began in 1907 with papers by
Gates (1907a,b), Lutz (1907), and Geerts (1907); and in the following
years these authors, as well as Davis (1909, 1910) and others, produced a

considerable number of papers which showed that, while the chromosomes behave normally during meiosis in some Oenotheras, in most they fail to form pairs in the usual fashion and seem to behave in a highly irregular manner. It was not until 1922 and 1923, however, that it was found that the apparent irregularity in behavior was in reality a highly regular process. Cleland found in 1922 that 4 of the 14 chromosomes in a strain of *O. franciscana* fail to pair normally in meiosis, but instead form a closed circle of 4 (⊙ 4), the other chromosomes forming pairs. The complete regularity of this behavior, every cell showing ⊙ 4 and 5 pairs, suggested that other forms, in which chromosome behavior had been described as highly irregular, might also show regularity in their behavior. Investigation of a variety of materials soon showed this to be the case (Cleland, 1923, 1925). One form ("*O. franciscana sulfurea*") showed ⊙12 and one pair, and *O. muricata* showed all 14 chromosomes in a single circle; de Vries' *lamarckiana* was found to have ⊙ 12, 1 pair; his *biennis* had ⊙ 6, ⊙ 8. Subsequent work showed that most races found in nature have

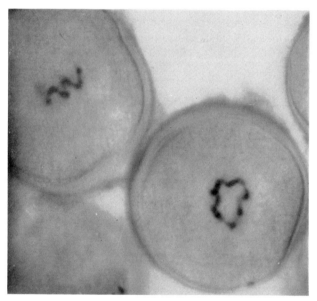

FIG. 5. Photomicrograph of first metaphase. *Above*, side view; *below*, polar view.

⊙ 14 (Figs. 3 and 5). This is the all but universal configuration in Euoenotheras from the Rocky Mountains eastward.

But this was not the only peculiarity exhibited by *Oenothera* chromosomes. Cleland (1923) also found that the circles remain intact into metaphase, and become arranged across the spindle with adjacent chromosomes directed toward opposite poles (Figs. 4 and 5). Since all chromosomes have median centromeres and all are of essentially the same size, the chain takes on a regular zigzag appearance when seen from the side in metaphase. The result of this is that adjacent chromosomes are regularly separated to opposite poles in anaphase I. The presence of circles in *Oenothera* was a new phenomenon when first discovered, the only previous instance of multiple association being that of a ⊙ 4 in tetraploid *Primula kewensis*, reported by Digby in 1912. The separation of adjacent chromosomes to opposite poles was an entirely new phenomenon.

One should not leave the impression that this behavior takes place with 100% regularity. The closed circle may open into a chain and may occasionally become shattered into two or more pieces, especially under unfavorable environmental conditions. Furthermore, the zigzag arrangement fails sometimes to become completely established, so that two adjacent chromosomes may go to one pole, with two adjacent chromosomes elsewhere in the circle going to the same or opposite pole. Such an irregularity results in non-disjunction and probable failure of the ensuing gametes. Some irregularities will produce 8-chromosome gametes, which may result in the appearance in the next generation of trisomics. On the whole, however, the behavior of the chromosomes in meiosis is remarkably regular.

IV. The Relation of Cytological to Genetic Behavior

The presence of extensive chromosome linkage in *Oenothera* naturally led to the suggestion by Cleland (1923) and Oehlkers (1924, 1926a) that this

FIG. 6. Diagram to show segregation of paternal and maternal complexes in first meiotic anaphase. Reprinted from *Scientific Monthly* by permission.

might constitute the physical basis for the extensive genetic linkage which Renner and Shull had found to exist in various races. If one assumed that the chromosomes have fixed positions in the circle, paternal and maternal chromosomes alternating, the regular separation of adjacent chromosomes to opposite poles would then carry all paternal chromosomes and genes to one pole, all maternal chromosomes and genes to the other, so that, where a circle of 14 existed, only two kinds of gamete would be produced, identical genetically with the two that united to form the plant (Fig. 6).

If this were the case, however, how would one explain the presence of two or more linkage groups in many hybrids? Obviously this would require other configurations than ⊙ 14 to be present in such hybrids. Theoretically, assuming that a circle always has an equal number of paternal and maternal chromosomes in it, and therefore has an even number of chromosomes, 15 arrangements of 14 chromosomes into circles and pairs are possible (Table 1). An analysis of a variety of hybrids by Oehlkers (1926b) and Cleland (1928) soon showed that many hybrids do have configurations other than ⊙ 14 or 7 pairs. It was not too long before all 15 possible arrangements were found among the many hybrids studied by the writer. Each hybrid was found to have a specific chromosome configuration which it could be depended upon to show. What was necessary at this point, therefore, was to study a large number of hybrids cytologically and genetically, to determine the chromosome configuration of each, and to discover whether there was a correspondence between the number of chromosome groups present in a hybrid and the number of linkage groups that it possessed.

TABLE 1
Possible Configurations of 14 Chromosomes into Circles and Pairs

⊙ 14	⊙ 10, 2 pairs
⊙ 10, ⊙ 4	⊙ 6, ⊙ 4, 2 pairs
⊙ 8, ⊙ 6	⊙ 8, 3 pairs
⊙ 6, ⊙ 4, ⊙ 4	⊙ 4, ⊙ 4, 3 pairs
⊙ 12, 1 pair	⊙ 6, 4 pairs
⊙ 8, ⊙ 4, 1 pair	⊙ 4, 5 pairs
⊙ 6, ⊙ 6, 1 pair	7 pairs
⊙ 4, ⊙ 4, ⊙ 4, 1 pair	

The first test of this hypothesis was performed by Oehlkers (1926b), using the cross *suaveolens* (⊙ 12) × *strigosa* (⊙ 14). Twin hybrids were produced, of which the complex-combination *albicans.stringens* with ⊙ 12, 1 pair, bred true, whereas *flavens.stringens* with 7 pairs (should have had ⊙ 4, 5 pairs) showed independent assortment of at least three factors. Much more extended tests were made by Cleland and Oehlkers (1930) and Renner and Cleland (1933). In these tests, about fifty different hybrid combinations were studied with respect to their chromosome configurations and breeding behavior. The results showed in a striking way that those hybrids that had large circles tended to breed true, whereas those with more than one chromosome group tended to show splitting, the complexity of splitting increasing with the number of separate chromosome groups. Only one gene failed to fit into this picture, namely the flower size factor (Co), which was inherited independently in all cases where it was present in heterozygous condition, no matter what the chromosome configuration. This anomalous behavior is also characteristic of *brevistylis* (*br*), a gene for greatly shortened styles. The behavior of these two genes, which according to Emerson and Sturtevant (1932) are situated in the same chromosome, about 15 crossover units apart, has been postulated (Cleland and Brittingham, 1934) to be the result of the terminal position of these genes in their chromosome, crossing-over between them and the rest of the complex reaching approximately 50%. These exceptional cases, therefore, do not invalidate the conclusion that chromosome cohesion (Gates' term) produces linkage of the genes present in the chromosomes thus brought together. Where all chromosomes belong to a single circle, all of the genes belong to a single linkage group. Where more than one chromosome group is present, more than one linkage group is also present; the number and size of the linkage groups depends on the number and size of the chromosome groups present. Thus the peculiar genetic behavior, not only in the various races, but also in their hybrids, depends upon the type of chromosome behavior found in these plants.

V. The Cause of Circle Formation in Oenothera

At first it was not clear why the chromosomes in *Oenothera* tend to form circles instead of pairs in meiosis, and various hypotheses were advanced to account for this phenomenon. The first clue to what ultimately proved to be the correct explanation came from Belling (1927; Belling and Blakeslee, 1926), who advanced the hypothesis of segmental interchange to account for the presence of a circle of 4 which he found in a hybrid between two perfectly pairing races of *Datura*. Belling suggested (1) that synapsis is a process of attraction, not between chromosomes as a whole, but between homologous segments or even loci; (2) if two non-homologous chromosomes should exchange segments and the interchange chromosomes thus produced should become associated with non-interchange chromosomes, synapsis would produce a circle of 4 instead of two independent pairs (Fig. 7). Circle formation results, therefore, from a mutual exchange of segments between non-homologous chromosomes followed by synapsis or side-by-side pairing of homologous segments in an interchange heterozygote.

Belling suggested that segmental interchange (or reciprocal translocation, as it is now more generally called) might account for the circles in *Oenothera*. This suggestion was accepted by Darlington (1929, 1931) largely on theoretical grounds, and was put to experimental test independently by Emerson and Sturtevant (1931) and Cleland and Blakeslee (1931), who realized that if the hypothesis is correct, each complex must have its own specific arrangement of chromosome segments, and it should be possible to determine the segmental arrangements of given complexes in terms of a standard complex and thus reach a point where it would be possible to predict in advance what chromosome configuration would be found in a given hybrid combination when complexes whose segmental arrangements had been analyzed were combined. Both groups of authors succeeded in making such predictions, which were subsequently verified when the hybrids in question were obtained and studied, and thus demonstrated the correctness of the segmental interchange hypothesis of Belling in the case of *Oenothera*.

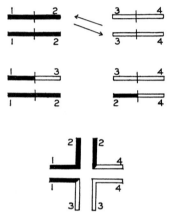

Fig. 7. Diagram to show formation of circle of four, as a result of reciprocal translocation. Reprinted from *Scientific Monthly* by permission.

This must mean, of course, that interchanges have been occurring and accumulating at an unprecedented rate in *Oenothera*, since circles of 14, each representing a past history of at least 6 interchanges, are the commonest configuration in North America. The reviewer and his students have completely analyzed the segmental arrangements of approximately 385 complexes [see Cleland (1958) for a list of most of them], mostly from wild races in North America. Catcheside (1940), Renner (1942), and others have made similar analyses of other material. More than 160 different segmental arrangements have been found. Each complex possesses one and only one segmental arrangement out of the 135,135 possible arrangements of 14 ends into sets of 7 chromosomes. A completely consistent system has been built up, in which each of the hundreds of complexes thus analyzed gives the correct configuration with each of the other complexes with which it has been combined. In scores of cases the configuration of a hybrid complex-combination has been worked out on paper before the combination has been made and tested. In every such case the predicted configuration has been obtained. Hundreds of additional predictions can be made for combinations which have not as yet been grown. Thus, it is completely demonstrated that each complex has its own specific arrangement of end segments, which characterizes it, and from which it does not depart except as a result of further interchange. It may be added that analysis of these segmental arrangements has been made without the presence of distinguishing morphological characteristics on the chromosomes. All chromosomes in all Euoenotheras look so much alike in meiosis that they cannot be told apart. Some workers, including Emerson and Sturtevant (1931) and especially Renner (1942, 1949), have used genetic markers to assist in the determinations of segmental arrangement; but Cleland and his students have made a practice of using only the chromosome configurations found in the various complex-combinations as the data upon which to base such analyses.

The complex that was chosen as the standard, in terms of which all other segmental arrangements have been determined, was ^{h}*hookeri* (haplo-*hookeri*), the complex present in double dose in the complex-homozygous *hookeri* of de Vries. Numbering the segments (probably whole arms) of the chromosomes, and representing each chromosome by two numerals connected by a dot, this complex was given the formula

$$1\cdot 2 \quad 3\cdot 4 \quad 5\cdot 6 \quad 7\cdot 8 \quad 9\cdot 10 \quad 11\cdot 12 \quad 13\cdot 14$$

This complex was made the standard because it was thought that de Vries' *hookeri*, having none of the peculiarities that characterize most Oenotheras, was probably primitive, and might therefore have the original arrangement of ends, from which all other arrangements have descended.

Later work showed that the original arrangement was probably one that is removed by one interchange from that of ^{h}*hookeri*, namely

$$1\cdot 2 \quad 3\cdot 4 \quad 5\cdot 6 \quad 7\cdot 10 \quad 9\cdot 8 \quad 11\cdot 12 \quad 13\cdot 14$$

This arrangement is the commonest among the races of *hookeri* found in California and adjacent areas. Furthermore, the seven chromosomes making up its formula are found more commonly than any other chromosomes in the North American Euoenotheras (Fig. 8), indicating that they are probably relict chromosomes which have escaped interchange in the evolution of the complexes of which they are a part.

At this point, it may be of interest to note that reciprocal translocations have not been limited to certain of the 14 ends present in a complement of chromosomes. Every one of the 14 ends has been found associated with every other end among the chromosomes so far analyzed. There are 91 possible associations of the 14 ends by twos—1·2, 1·3, 1·4, etc. . . .

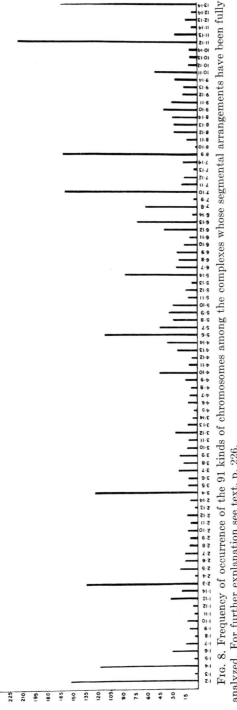

FIG. 8. Frequency of occurrence of the 91 kinds of chromosomes among the complexes whose segmental arrangements have been fully analyzed. For further explanation see text, p. 226.

2·3, 2·4, etc., i.e., 91 different chromosomes from the standpoint of segmental arrangement are possible in *Oenothera*. Every one of these associations has been found among the complexes whose segmental arrangements have been fully determined (Fig. 8). Thus, there has been a wholesale shuffling of ends in the course of evolution.

VI. Chromosome Structure

The prevalence of translocations which have brought about the widespread occurrence of structural hybridity naturally tends to focus interest upon chromosome structure in *Oenothera*. Two questions have been of principal interest: (1) Are the chromosomes of a complement similar or identical in size and basic morphology, or have reciprocal translocations brought about alterations in structure? (2) What is the nature of the clear-cut morphological distinction that is observed between the proximal and distal regions of the same chromosome?

A. Size

With regard to the question of size, divergent points of view have been expressed. Some workers have found what they have interpreted as significant differences in size and structure among the chromosomes of a genome—others have failed to find such differences.

In the first place, it is generally agreed that such differences as exist are reduced to a minimum during diakinesis and the first meiotic metaphase, where it is ordinarily not possible to distinguish one chromosome from another. Most attempts to analyze chromosome structure have been based upon studies of somatic cells (root tips, anther walls, etc.). Among the workers who have found what they have interpreted as significant size or other differences within a single genome are van Overeem (1922), Darlington (1931), Wisniewska (1935), Marquardt (1937), and Bhaduri (1940). These differences include slight variations in length or in position of centromere, or the presence or absence of terminal knobs, etc.

It is interesting to note that there has not always been agreement between investigators who have worked with the same material. For instance, van Overeem (1922) classified *O. lamarckiana* chromosomes into four groups: (1) two long pairs, strongly bent in the middle, (2) two long pairs, slightly bent in the middle, (3) one middle-sized pair, and (4) two short pairs. (Van Overeem, of course, assumed that the chromosomes form pairs in *lamarckiana*.) Lewitsky (1931, pp. 128–130), on the other hand, found it impossible to classify the chromosomes of *lamarckiana* into groups, the shortest being three-quarters the length of the longest, with the others ranging imperceptibly between. He was not inclined to consider size differences in *Oenothera* as significant.

To cite another instance, the two published accounts of chromosome size and structure in *O. hookeri* (Wisniewska, 1935; Marquardt, 1937) differ considerably although the material used was presumably identical. Wisniewska recognized two size classes in the haploid set, one chromosome being larger, the others smaller but equal to each other. Marquardt distinguished two large, one doubtfully large, two median, one doubtfully small, and one small chromosome. Wisniewska emphasized the presence or absence of terminal knobs, one small chromosome having no knobs, five chromosomes having a knob at one end (one of them with a satellite at the other end), the large chromosome having a knob at both ends. Marquardt considered knobs to be inconstant in appearance. According to Wisniewska the centromeres are subterminal in one chromosome, submedian in another, and median in five chromosomes; Marquardt mentions the position of the centromere in the case of five chromosomes, one of them submedian, the others median. His figures suggest that the other two chromosomes also have median insertions. Thus, the two observers came to quite different conclusions with regard to the same material.

Discrepancies such as these are not surprising when one realizes that an *Oenothera* chromosome at late prophase of mitosis averages only about 4 μ in length after Navashin fixation. In such small chromosomes, very minor differences in absolute length become sizable differences on a percentage basis. Such factors as differential condensation and the effects of fixing fluids and of the staining, dehydrating, and clearing processes produce differences in length which become relatively more important as chromosome size decreases. One must demonstrate that recognizable differences exist of an order of magnitude clearly greater than those brought about by differential condensation or handling before he can claim that these differences are significant.

The reviewer considers that significant size differences within genomes and between different genomes have not been convincingly demonstrated. He has encountered a few cases, mostly in forms with paired chromosomes, where one chromosome is noticeably smaller or larger than the rest, but these are exceptions to the rule that all chromosomes within a genome are essentially equal in size and equal-armed, and that different genomes do not differ significantly in the size or morphology of their chromosomes. Specific evidence, by Gardella, published so far only in abstract form (1953), tends to confirm this impression. She made detailed measurements of prophase, metaphase, and anaphase chromosomes in root tip and pollen tube mitoses in two all-pairing races and two with a circle of 14. She finds no recognizable size classes in any of this material. Chromosomes are essentially isobrachial and of uniform size in all races studied, whether 7-paired or circle-bearing races with long histories of interchange behind them. The range of variation in chromosome length between races is no greater than that found between the cells of a single race. The variation in size of chromosome and position of the centromere among the chromosomes of a race is no greater than that found in the satellite chromosomes alone. Variation among the chromosomes in a single anaphase complement is of the same order of magnitude as that found between sister chromosomes in the two halves of the anaphase cell. The differences between the chromosomes of a cell are small enough that it is not possible to recognize individual chromosomes, except, of course, for the satellite chromosome.

This is a matter of considerable interest and importance. In the course of its evolution, Oenothera has experienced a relatively high frequency of reciprocal translocation. Many chromosomes in many complexes owe their present arrangement of end segments to a succession of interchanges. One might expect that a chromosome that had suffered two or three or more interchanges would have become considerably altered in its morphology as a result. Apparently, however, most of the chromosomes found in nature still retain an unaltered morphology, in spite of the interchanges that have occurred in their evolution. Consequently, we find that the chromosomes of circle-bearing races are as uniform as those in all-pairing forms with the original arrangement of ends; in fact, they are, on the average, more uniform. Apparently, those interchanges that have occurred, or at least have survived in the evolution of the group, have involved segments or arms of equal length and have probably occurred for the most part in the region of the centromere. As a result, the interchange chromosomes have not been altered morphologically by the addition or subtraction of material.

This has been an important factor in the success of the circle-bearing Oenotheras. In other circle-bearing organisms, such as Rhoeo, whose chromosomes vary markedly in total length and in the length of arms, so that the centromeres within the circle are unevenly spaced in metaphase, a high degree of irregularity in disjunction results. In Oenothera, however, the essential equality in size of the circle chromosomes and the median position of the centromeres has resulted in the centromeres being so evenly spaced that the forces operating between them are equalized, thus permitting the regular separation of adjacent chromosomes to opposite poles. Whatever differences there are in chromosome size and in the position of the centromere in the circle chromosomes are of insufficient magnitude to cause much disturbance in the regular separation of paternal and maternal chromosomes.

While the maintenance of isobrachy, and of equality in size, would seem to be essential to the permanent survival of circle chromosomes, the same is not true of chromosomes that are found in paired forms. It makes no difference in the success of disjunction whether paired chromosomes are larger or smaller, or whether all the chromosomes are of the same size. Inequality of size, if it developed, would not affect survival in a paired form; but it would have a deleterious effect in a circle-bearing form if it were of sufficient magnitude to render the separation of adjacent chromosomes more difficult. One would not expect, however, that an all-pairing form with chromosomes of different size and morphology would be able to give rise in the evolutionary process to successful circle-bearing races.

The situation as here pictured is not in line with the concepts that Darlington has developed regarding chromosome structure in *Oenothera* (1931, 1933, 1936). Darlington reasons that breaks do not necessarily take place at the region of the centromere and that it is too much to expect that they will be confined to this area. Since breaks, therefore, may occur elsewhere, many interchange chromosomes will be found to be longer or shorter than the original. Darlington claims that such differences exist. In the opinion of the writer, however, such differences as Darlington has described (1931) are of the same order of magnitude as those resulting from the vicissitudes of cytological handling. With regard to the likelihood that exchanges are more apt to occur at the centromere than elsewhere, the evidence indicates that the region contiguous to the centromere is heterochromatic, and it is reasonable to assume that *Oenothera* chromosomes are therefore more easily broken close to the centromere region than elsewhere. For additional discussion of Darlington's ideas regarding chromosome structure, see p. 179.

B. Morphology

A second question relating to chromosome structure involves the basis for the morphological differentiation that can be seen along the length of the chromosome. Marquardt (1937) and Seitz (1935) claim that the central region of the *Oenothera* chromosome is largely composed of heterochromatin, the centromere occupying a more or less median position in this region. The distal portions of the chromosome, on the other hand, are euchromatic. According to Japha (1939) the euchromatic region represents in pachyphase (see Section VII) about three-quarters of the total length of the chromosome; more rapid shrinkage of the euchromatic segments, however, reduces the discrepancy until by the end of diakinesis, euchromatin and heterochromatin are approximately equal; and, by metaphase, the euchromatic portion is reduced to a small region at either end of the heterochromatin. According to these authors, the distinction between heterochromatin and euchromatin is clear at all stages from presynapsis to metaphase. The heterochromatic regions exist in the premeiotic interphase in the form of prochromosomes. According to Marquardt, homologous prochromosomes pair in leptophase. During zygophase and pachyphase, the heterochromatic regions are considerably more condensed than the rest of the chromosomes and tend to be concentrated toward one side of the nucleus. During diplophase the distinction remains clear, the end segments being decidedly thinner than the central heterochromatin.

There can be no doubt that, in the mature somatic chromosome, a clear morphological distinction exists between the central portion and the distal portions of the chromosome. Gardella (1953) has found that mid- or late prophase chromosomes in root tips and in the pollen tube division have a seemingly thicker central portion and seemingly thinner distal portions. This difference, however, is at least in part a result of the fact that the chromatids are closely associated in the central region whereas in the distal regions they are separated, often quite widely, from each other, thus making the central portion appear thicker than the distal regions. This does not mean, however, that heterochromatin is wholly lacking in the central region. The fact that the region about the centromere remains condensed in metabolic nuclei, forming prochromosomes, indicates that this region, at least, is largely heterochromatic; and it is probable that blocks of heterochromatin are scattered along the remainder of the central segment. Much of the proximal portion, however, must be euchromatic, since it is necessary to assign the genes of the Renner complexes in large part to this region of the chromosome. The pairing segments appear to be euchromatic except for the terminal knobs that characterize most *Oenothera* chromosomes, and which may contain heterochromatin.

VII. Early Meiotic Stages

The study of early stages of meiosis in *Oenothera* is beset with many difficulties. The nuclei are small; the sporocytes are surrounded by heavy cutinized anther walls which impede the entrance of fixing fluids and make adequate smear preparations difficult to achieve. It is not surprising, therefore, that interpretation has often run far ahead of observational evidence. All accounts of early meiosis in *Oenothera* should be received with considerable skepticism. Because the terms "leptotene," "zygotene," etc. are adjectives and should not be used as nouns and since "leptonema," "zygonema," etc., although nouns, refer to the chromosomes rather than the stage, we shall designate the respective stages by the terms "leptophase," "zygophase," etc.

A. Leptophase

According to Marquardt (1937) the prochromosomes, which in the pre-meiotic interphase are split (in opposition to Darlington's "precocity" theory), pair with their homologs in the leptophase, thus reducing their number from 14 to 7. This claim raises serious doubt as to whether leptophase has not been confused with the zygophase or later stages in this case. This is not the only instance in which stages in early meiosis in *Oenothera* have been confused. Weier (1930) and Wisniewska (1935) have probably mistaken early or even mid-diplophase stages for leptophase or zygophase (see, for instance, Wisniewska's Fig. 14 which is labeled "zygotene," but is almost surely late diplophase).

B. Zygophase

It is possible that synapsis in *Oenothera* is of the polarized type. Catcheside (1932) and Japha (1939) both suggest this, as does Darlington (1931, p. 431). Whether synapsis involves the chromosomes throughout their entire length, or only at the ends, is a matter of controversy. Marquardt considers that, at least in the case of chromosomes which are homologous from end to end and will form pairs in diakinesis, synapsis occurs throughout the entire length, in fact beginning in the central (supposedly heterochromatic) region. Catcheside (1931) suggests that the chromosomes which form pairs synapse throughout their length, but that circle chromosomes synapse only at the ends. Weier (1930) has made a similar claim. The reviewer favors the view that true synapsis (an association capable of producing chiasmata) is confined to the pairing ends, and that the central regions either fail to pair (in circle chromosomes), or, in chromosomes which will form pairs, are brought into loose association because of the synapsis that occurs at both ends.

C. Pachyphase and Diplophase

It is somewhat easier to obtain adequate fixation of these stages, and the various workers, therefore, tend to agree in the general picture presented, although there are noteworthy differences in interpretation. Marquardt (1937) pictures the chromosomes in this period as definitely paired throughout their entire length if they are to form pairs. In the case of circle chromosomes, he claims to have evidence of side-by-side pairing for considerable distances back from the ends, a given chromosome being paired with two different chromosomes at its two ends. He finds a clear distinction between heterochromatic and euchromatic regions in pachyphase and diplophase. Cleland (1924, 1926a,b), on the other hand, has considered the threads at this stage as primarily unpaired. This opinion was shared by Darlington (1931); and Catcheside (1931) also accepted it, so far as circle chromosomes are concerned. Weier (1930) has claimed that *hookeri* chromosomes are paired throughout their entire length at this stage, whereas circle chromosomes in *lamarckiana* are unpaired proximally. Håkansson (1926) found doubleness in pachyphase but considered this to be in preparation for the second meiotic division, and not an evidence of bivalence. Cleland (unpublished) has recently found many cases where side-by-side pairing occurs in diplophase, but only for a short distance back from the tip. He fails to find clear-cut evidence that the unpaired regions are heterochromatic. Thus a variety of opinions has been held regarding the nature of the pachynema and diplonema.

The question as to whether chiasmata are produced along the entire length of the chromosomes or only near the ends has been variously answered by various investigators. Marquardt (1937) and Japha (1939) agree that chiasmata are not ordinarily formed in the central ("heterochromatic") portions of the chromosomes although these regions tend to be closely synapsed up to the beginning of diplophase. The euchromatic distal segments, on the other hand, form chiasmata, ranging in number, according to Marquardt, from two to five per bivalent in the case of chromosomes which will form pairs [also see Catcheside (1931), who finds a similar number in paired chromosomes of a *lamarckiana* derivative].

In most cases, such chiasmata as are formed tend to terminalize completely. Occasionally, however, a minute lump of material on a thread connecting two chromosomes at metaphase suggests that a chiasma has not completely terminalized, although such lumps are by no means as frequent as some writers would suggest. It is not even certain that such lumps actually represent incomplete terminalization, since it is very common to find terminal knobs in both somatic prophase and late diplophase (heterochromatin?), retention of which into metaphase would suffice to produce the lump. Marquardt and Japha, however, interpret these cases as incompletely terminalized chiasmata which have originated in the distal segments of the chromosome. Darlington goes so far as to claim (1931, Figs. 16, 17, 19, Plate II, Figs. 9–16) that interstitial chiasmata may be formed in the central portion of the chromosome, resulting from crossing-over between homologous segments in otherwise non-homologous chromosomes. The writer has been unable to find convincing evidence of the occurrence of such chiasmata and considers Darlington's figures as examples of overlapping chromosomes.

With respect to the earlier stages of meiosis in *Oenothera*, therefore, one is forced to conclude that our knowledge of these stages is in a very chaotic condition. This is in part owing to the difficulty of obtaining adequate fixation, partly to the small size of the *Oenothera* nucleus and chromosomes. It will be noted, for instance, that Marquardt's figures are drawn mostly at a magnification of 3800, some at 4800. When one notes the extreme delicacy of the threads as drawn in his Figs. 7 or 8 (Marquardt, 1937) and considers that the ordinary light microscope begins to lose definition at magnifications above 1500, one realizes how difficult it is to be sure of the finer details of structure and organization, and how easy it is for various authors to interpret certain stages differently. Our knowledge of early meiosis, therefore, cannot be said to rest on a basis of solid fact. Such evidence as there is suggests that synapsis may be polarized, that it occurs primarily at the ends of the chromosomes, the middle segments remaining unsynapsed, at least in circle chromosomes, and probably also in chromosomes that form pairs.

VIII. Factors Influencing Genetic Analyses

In the relatively few Euoenotheras which do not possess circles and lethals and in which open pollination occurs (such as *hookeri*) there is little to interfere with normal Mendelian behavior. In the great majority of races, however, the peculiar cytological behavior tends to disturb typical Mendelian ratios so that ordinary genetic analysis becomes a somewhat difficult matter. The chief factors which tend to modify typical Mendelian behavior include chromosome catenation leading to linkage of all genes to the lethals, megaspore competition, pollen tube competition, and plastid imcompatibility.

The effects of chromosome catenation and balanced lethals have been considered above. The plastid situation will be discussed in a later section. The effects of megaspore and pollen tube competition will be briefly mentioned at this point.

A. Megaspore Competition and Development of Female Gametophyte

To the other well-known peculiarities of *Oenothera* may be added two, which are characteristic, however, not of this genus alone, but of the family to which it belongs—the Onagraceae. These are the tendency for the embryo sac to develop from the spore nearest to the micropyle, in contrast to most plants, in which it regularly develops from the chalazal spore; and the production of a 4-nucleate embryo sac, as opposed to the more usual 8-nucleate structure. There are no antipodals, the embryo sac containing one polar body and two synergids in addition to the egg.

These facts, as they relate to *Oenothera*, were first set forth by Geerts (1909). Renner (1921a) showed that there is a tendency for Oenotheras to differ among themselves in respect to the spore which will develop into the embryo sac, depending upon whether they are complex-homozygotes, or isogamous or heterogamous complex-heterozygotes.[2] In homozygotes, such as *hookeri*, the embryo sac seemed to develop regularly from the micropylar spore; in isogamous heterozygotes (those which lack gametophytic lethals, e.g., *lamarckiana*) the same occurred. In heterogamous forms, however (those which have balanced gametophytic lethals, e.g., *muricata*), the chalazal spore was found frequently to develop instead of the micropylar, and in a fair number of cases, both the micropylar and the chalazal spores were found to be developing, indicating a competition between them. For example, Renner's interpretation of the situation in *muricata* was as follows: Almost all of the functional eggs carry *rigens;* the pollen complex (*curvans*) rarely succeeds in the egg. When *rigens* finds itself in the micropylar spore, this cell develops into the embryo sac without difficulty; when *rigens* is found in the chalazal spore, however, *curvans* thus being in the micropylar spore, the latter has the advantage of position, but the former has the advantage of innate ability to favor growth and development of the female gametophyte. As a result, both spores may begin to grow, in which case, *rigens*, even if it is in the chalazal spore, will succeed in most of the ovules, but in a small percentage of cases *curvans* will win out if it is in the micropylar spore. This behavior is known as the "Renner effect." In contrast with this situation, isogamous races possess complexes which do not differ greatly in their ability to induce development of the embryo sac; hence, the micropylar spore will develop, no matter which complex it carries. In the case of homozygous races, there is only one type of complex present, and seemingly nothing to prevent the micropylar spore from developing.

Renner found, however, that the situation became more complicated than this when hybrid combinations were considered. He showed (Renner, 1921a) that the ability of a complex to compete in spore development depends to some extent upon what complex it is associated with in the plant. Thus, *albicans* succeeds in a fair number of cases in competition with *rubens* in de Vries' *biennis;* in competition with *hookeri*, however, it is rarely successful, even though the hybrid combination *albicans.hookeri* is derived from the cross *biennis* × *hookeri* so that *albicans* has the advantage of being in its own cytoplasm, whereas *hookeri* is in a foreign cytoplasm.

In later papers, he and some of his students (Hoeppener and Renner, 1929; Langendorf, 1930; Rudloff, 1929, 1931) uncovered situations which

---

[2] "Isogamy" as used by de Vries, is the condition where reciprocal crosses each produce twin hybrids: "heterogamy" is the condition where reciprocal hybrids are unlike, but tend to be uniform. Expressed in terms of Renner's analysis, isogamous races transmit both complexes through both sperm and egg; heterogamous races transmit one complex through the egg, the other through the sperm.

they were at first inclined to interpret in terms of polarization of the first meiotic spindle in the ovule, i.e., orientation in such a way that one complex always or usually goes to the micropylar end. This suggestion was based on the discovery that certain heterogamous combinations seemed to show no competition, only the micropylar spore developing, although one complex predominated in the eggs, as for instance in *rubricaulis* (Rudloff, 1929).

More extensive work, however, has not supported the idea of polarization. It now appears that megaspore competition may occur in all kinds of *Oenothera*, complex-homozygotes and complex-heterozygotes. There seems to be little relation between this behavior and the type of plant in which it is found. Renner (1940a) is inclined to accept the view that ability or failure of a complex to be transmitted through the egg is primarily a result of its genetic adequacy or inadequacy with respect to embryo sac development. When a decided difference in adequacy exists between associated complexes, the more adequate complex will succeed in the micropylar position, and in some cases may succeed through competition if it is in the chalazal position. However, the mere fact that it is in the chalazal position may not be the sole contributing cause of competition between spores, since it is now known that competition between chalazal and micropylar spores may occur in homozygous races, and certain heterogamous races show little or no competition. We may accept the conclusion, therefore, that complexes differ in their effect on the development of the female gametophyte. Some may exert a deleterious effect, ranging all the way from complete inhibition to various degrees of inability to compete with certain other complexes. As a result of these differences, ratios of appearance of certain complexes or genes in the eggs may vary greatly from cross to cross.

B. Pollen Tube Competition

In 1917, Correns, working on *Melandrium*, and Renner, working on *Oenothera*, independently showed that different classes of pollen tube may grow through the style at different rates. In *Melandrium*, the faster growing pollen tubes tend to give rise to female plants, and vice versa. Renner (1917b) found in *Oenothera* that pollen tubes carrying one complex may tend to grow faster than those containing the competing complex—e.g., *velans* is carried to the ovules faster than *curvans* when these complexes are competing in styles of *biennis* of de Vries. In 1919 Renner presented additional evidence for this type of behavior in *Oenothera*, and emphasized the fact that the ability of two complexes to compete in pollen tube growth depends at least in part upon the kind of style in which they find themselves. For instance, when pollen from the hybrid *velans.curvans* is placed on *muricata* stigmas, *velans* reaches and fertilizes four times as many eggs as *curvans;* in styles of the hybrid *rigens.velans*, using the same pollen, the ratio is 12:1; in *biennis* styles, it is 30:1; in *lamarckiana* styles, no *curvans* succeeds in competition with *velans*.

The ratios actually obtained, however, may depend also on richness of pollination. When, for instance, *biennis* is pollinated richly with *lamarckiana* pollen, many more *velans* tubes reach and fertilize the eggs than *gaudens;* when, however, pollination is sparse, so that the number of ovules to be fertilized exceeds the number of pollen grains, equal numbers of eggs are fertilized by *velans* and *gaudens* sperm. Evidently, *velans* tubes outgrow the *gaudens*, and where there is an excess of pollen, most of the eggs are fertilized by *velans* before the *gaudens* tubes reach the ovules; where pollination is scanty, however, *gaudens* may fertilize a larger proportion of the eggs.

It is very common, therefore, to find that pollen tubes carrying associated complexes grow at different rates, thus modifying the genetic ratios expected in the offspring. This may even result from the presence of a single pair of heterozygous genes. Thus, R (red midrib) pollen tubes grow faster than r-tubes (Heribert-Nilsson, 1920; Renner, 1921b).

The combined effect, therefore, of differing abilities of the various complexes to induce megaspore development, and to promote pollen tube growth, is to bring about widespread departure from normal Mendelian ratios and to increase the difficulty of genetic analyses.

IX. Factor Analysis and the Location of Genes

In spite of the handicaps thus created, considerable progress has been made in investigating the genetic composition of a number of complexes. Initial efforts of course were directed toward the analysis of individual genomes in order to determine the distinguishing characteristics of each complex. As early as 1925 Renner had recognized a considerable number of genes in each of the complexes with which he was concerned and had analyzed their behavior. This process of "factor analysis" was continued in later papers by himself and students until many complexes were analyzed in considerable detail. The same process, applied by the writer and his students, as well as by Emerson and Sturtevant, to North American material, has characterized many additional complexes.

This analysis, in and by itself, does not locate genes other than to determine the complex to which they belong. With the help of cytological studies, however, it has been possible to carry the analysis somewhat farther. While nothing approaching a real genetic map has been achieved in *Oenothera*, certain genes have been located within certain chromosomes, or even in a few cases within a single arm of a chromosome.

The methods by which genes are located in specific chromosomes or arms of chromosomes are illustrated by the following: (a) In a complex-combination possessing a large circle with balanced lethals, plus a pair, any heterozygous genes which segregate independently of the lethals are probably situated in the pair. (b) In a complex-combination possessing a small lethal-free circle and several pairs (e.g., ⊙ 4, 5 pairs, or ⊙ 6, 4 pairs), a recessive gene may produce its effect only in individuals which lack the circle, thus indicating that it belongs to a circle chromosome. Other recessive genes may express themselves independently of the presence or absence of the circle, thus indicating that they lie in one of the pairs.

(c) A comparison of the behavior of a gene belonging to a certain complex in different combinations of that complex, having different chromosome configurations, serves often to restrict the possible positions of this gene to a single chromosome or arm of a chromosome. For instance (Emerson and Sturtevant, 1931), P and R are independent in *lamarckiana*: since R is in the pair, P must be in a circle chromosome. Since P is found only in combinations with *velans*, not with *gaudens*, when *lamarckiana* is crossed to races which do not themselves contribute P to the hybrid, P must belong to a *velans* chromosome, not to a *gaudens* chromosome. In *flavens.velans* (from *suaveolens* × *lamarckiana* or reciprocal), P and R are completely linked. Since the 1·2 chromosome containing R is in this case incorporated into a circle of 4 (1·2 - 2·3 - 3·4 - 4·1) P, which cannot be in 1·2, must be in 3·4 (it is in a *velans* chromosome, not a *flavens* chromosome). In *flavens.stringens* (both containing 1·4 3·2), P (from *stringens*) is independent of *fr* (pollen sterility factor in *stringens*). The latter is in 1·4 since it segregates independently of the lethals in *albicans.stringens* which has only one pair (1·4). Therefore P cannot be in 4 and must be in 3.

(d) The relative positions of genes that belong to the same chromosome arm have been ascertained in the case of arm number 3, by determining the relative crossover percentages between each gene and the lethals in cases where these genes are included within large circles with balanced lethals (Renner, 1933; Oehlkers, 1933). The nearer a gene is to the end of a chromosome, the more chance there is that a crossover will occur between it and the rest of the complex. Thus the relative positions of the genes in the same arm can be obtained by determining the relative frequencies with which they become transferred to the opposing complex.

It can thus be seen that detailed genetic analysis in *Oenothera* is beset with special problems and must utilize special techniques. In spite of complicating factors, however, Table 2 shows that some success has attended attempts to define and locate gene loci. The genes listed in this table are probably not all of equal validity, however. The location of some is tentative, and some of the genes defined by Harte (1948) may prove to be identical to or at least allelic with some of those determined by Renner. In addition to the genes listed, Harte and others have recognized a goodly number of genes for which evidence regarding location has not been obtained.

X. Crossing-Over

One of the striking features of *Oenothera* genetics is the relative paucity of detectable crossing-over. Gates (1922, 1925) originally considered crossing-over to be impossible in *Oenothera*, on the ground that telosynapsis does not permit pairing until the chromosomes have become condensed. Shull (1923a,b), however, described cases of apparent crossing-over in *lamarckiana* or its derivatives. Thus, he obtained the following "crossover percentages" within his linkage group 1: 9% between the *rubricalyx* (r^h, = p^r of other authors) and *nanella* (n) loci; less than 1% to 3.08% between *funifolia* (f) and *rubricalyx*; 4.7–9.2% between *sulfurea* (s) and *nanella*. Within his linkage group 2, he obtained the following: 0.057–2.83% between *vetaurea* (v, = *vet* of other authors) and *supplena* (sp) (Shull, 1925); 0.87% between *vetaurea* and *bullata* (bu) (Shull, 1928).

In this work, Shull was inclined to deny a relationship between chromosome catenation and genetic linkage. To him, all linkage in *Oenothera* is conventional linkage and all crossing-over is conventional crossing-over. As a matter of fact, it did turn out that some of the cases that he analyzed were examples of conventional crossing-over. The genes v, sp, and bu are all resident in chromosome 1·2. This forms the pair in *lamarckiana*, so that linkage and crossing-over between these genes have the same cytological basis as linkage and crossing-over in other organisms. Furthermore, two of the genes which Shull ascribed to his group 1 (r^h and s) reside in a single chromosome segment (segment 3), and any separation of these is a result of ordinary crossing-over.

Other cases that Shull analyzed, however, involve loci that are not linked in the conventional manner in *lamarckiana* since they lie in chromosomes that are only partially homologous in opposing complexes. For instance, r^h (*rubricalyx*) belongs to end 3 of the *velans* chromosome 3·4; n (*nanella*) belongs to end 4 of the *gaudens* chromosome 4·9. In the race itself, therefore, these loci are not linked except as they both belong to the single circle of 12. However, n can transfer to the 3·4 chromosome and thus become linked with r^h in *velans*. It is not, on the other hand, possible for r^h, by crossing-over, to become linked with n in *gaudens*, since it

TABLE 2
List of Loci in *Euoenothera* Which have been Assigned to Particular Chromosomes or Arms of Chromosomes

Chromosome or Chromosome arm	Symbol	Phenotypic effect	Reference
1·2	R^{rub}	red midrib, homozygous lethal	Renner, 1942
	R^{exc}	red midrib, not homozygous lethal	Renner, 1942
	v	*vetaurea*, old gold flower color	Shull, 1921; Emerson, 1932
	bu	*bullata*	Shull, 1928
	s_p	*supplena*, double flowers	Shull, 1925
2 (probably)	Su	large flowers (suppresses Co)	Harte, 1948
1·4	fr†	sterile anthers in *stringens*	Emerson and Sturtevant, 1931
1,2,3, or 4	lat	broad bracts (*franciscana*)	Harte, 1948
	ang	narrow leaves (*franciscana*)	Harte, 1948
	lon	long style (*velans*)	Harte, 1948
	de	flattened tips of spikes	Harte, 1948
	bco	short bracts (*hookeri*)	Harte, 1948
	ra	red stem tips (*hookeri* and *stringens*)	Harte, 1948
	lin	narrow leaves (*hookeri* and *franciscana*)	Harte, 1948
	den	toothed leaves (*hookeri* and *franciscana*)	Harte, 1948
	ser	calyx hairs soft-silky	Harte, 1948
3	s	sulfur petals	Renner, 1942
	P-series	papilla and cone color	Renner, 1942
	M	red leaf margins, short sepal tips (allele of P-series?)	Renner, 1942
	str	striped calyx (close to P)	Harte, 1948
3 or 2	na	dwarf habit (= n of Renner?)	Harte, 1948
3·4	A^{cro}	pointed sepal tips (*flavens* and *stringens*)	Harte, 1948
	tub	red hypanthium (= *rubricalyx*)	Harte, 1948
	rub	red stems (*franciscana* and *hookeri*)	Harte, 1948
	vit	red-striped capsule (*franciscana* and *hookeri*)	Harte, 1948
4	$ster$	pollen sterility (*velans brevistylis*)	Harte, 1948
	ov	ovate leaves (*velans* and *velans brevistylis*)	Harte, 1948
	$serp$	tortuous stem (*velans brevistylis*)	Harte, 1948
	Fl	bent stem tips, spreading sepal tips, homozygous semilethal	Renner, 1942
	n	*nanella* (dwarf) in *gaudens*	Renner, 1942
	pus	dwarf in *punctulans* (allele of n?)	Renner, 1942
	Ce	bent stem tips, probable allele of Fl (*argillicola*)	Renner, 1942

TABLE 2 (Continued)

Chromosome or Chromosome arm	Symbol	Phentoypic effect	References
	Cu	bent stem tips, homozygous sublethal	Renner, 1942
5·6	let^{fl}	lethal in *flavens*	Renner, 1942
	Sp	acute leaves	Renner, 1942
5·8	let^{vel}	*velans* lethal	Baerecke, 1944
7·8	B	broad leaves	Renner, 1942
	kr	crippled habit in *stringens*	Harte, 1948
	ery	red-striped capsules in *flavens*	Harte, 1948
7 or 8	Cer	dark red hypanthium (*velans brevistylis*)	Harte, 1948
8	pil	homozygous lethal, brownish hairs	Renner, 1942
9·10 or 11·12	acu	long, thin sepal tips (*flavens* and *stringens*)	Harte, 1942
	car	red-striped capsules (*stringens*)	Harte, 1942
	ru	red stem tips (*stringens*)	Harte, 1942
	hir	long hairs on capsule (*flavens*)	Harte, 1942
11·12	def	defective sepals (*flavens*)	Renner, 1942
	br‡	short, defective styles	Catcheside, 1954
11·13*	lor^2	semilethal (narrow, thread-shaped leaves)	Renner, 1942
12·14*	lor^1	semilethal (narrow, thread-shaped leaves)	Renner, 1942
13·14	ano	crippled stem-tip (*flavens*)	Harte, 1948
	un	crinkled bracts and leaves (*flavens*)	Harte, 1948
	co‡	large flowers	Harte, 1948
	uni	solid calyx color (*hookeri, franciscana*)	Harte, 1948
	vir	green stem color (*franciscana*)	Harte, 1948
	tec	long bracts (*flavens*)	Harte, 1948
	lob	narrow bracts (*franciscana*)	Harte, 1948
	tri	long hairs on buds (*franciscana*)	Harte, 1948
	his	strong hairs (*franciscana*)	Harte, 1948
	A^{de}	glandular hairs on buds (*franciscana*)	Harte, 1948
	dil	pale anthocyanin (*franciscana*)	Harte, 1948

* 11·13 = 12·13, and 12·14 = 11·14 on Cleland system.
† Harte (1948) places *fr* in ends 2 or 3.
‡ Emerson and Sturtevant (1932) found that *Co* and *br* are linked with 15% crossing-over. Catcheside has found *br* to be in 11·12. Presumably, *Co* should therefore be in 11·12, not in 13·14 as stated by Harte.

would then lie in 3·12 of *gaudens*, not in 4·9. The 9% crossing-over said to occur between *rubricalyx* and *nanella* presumably refers to a situation where a 3·4 chromosome was present in duplicate. In *lamarckiana* itself (Renner, 1942), n is only rarely transferred from the 4·9 of *gaudens* to the 3·4 of *velans* (frequency of 1% or less). Alleles of the r^h locus are even less frequently crossed over to the 3·12 of *gaudens*.

Darlington in 1931 drew attention to the following facts: (1) since circle-bearing Oenotheras are highly heterozygous, and the two complexes differ as a rule in many characters, one would expect to find a considerable amount of detectable crossing-over, but little or no crossing-over occurs between genes involved in determining the differences between complexes; (2) on the other hand, the chromosomes within a circle maintain their end-to-end attachment with a high degree of regularity. If *Oenothera* chromosomes are held together by terminalized chiasmata, resulting from crossing-over, these chromosomes must in most instances have achieved an average of at least one crossover per pairing end, in order for the circle to be maintained intact [Japha (1939) claims as many as five per bivalent].

These facts suggest, as Darlington pointed out (1931), that crossing-over is confined to the end segments, and that the various complexes are for the most part homozygous with respect to these end segments, so that crossing-over has for the most part no genetically detectable result. We may assume, therefore, as a working hypothesis that crossing-over is largely confined to homozygous pairing segments, a situation that makes possible the union of chromosomes to form chains, and at the same time makes it possible for the various Renner complexes to remain intact, unmodified by exchange of genes that differentiate between associated complexes. These genes are confined to the central segments in which crossing-over ordinarily does not occur.

While it is probably the normal situation in a circle chromosome for crossing-over to be confined to the pairing ends, where the genes are mostly homozygous, exceptions have been described as occurring in both directions, i.e., (1) heterozygous genes may occasionally be found in the pairing segments, thus making possible genetically detectable crossing-over, and (2) on the other hand, crossing-over is said to take place occasionally or rarely within the proximal segments.

(1) There is considerable evidence, some of which has already been presented, that detectable crossing-over does occur occasionally in the pairing segments. The evidence in this regard has been summarized by Renner (1942, 1948). While he agrees with Darlington that the genes which distinguish one complex from another lie, for the most part, in the middle region of the chromosome where chiasmata are not ordinarily produced, and that chiasmata are confined mostly to the pairing ends

where the genes are mostly homozygous, he is nevertheless able to list some thirteen loci which have been known to show crossing-over, at least under certain circumstances.

Crossing-over involving such genes may be detected and measured in one of two ways: (a) If two heterozygous genes reside in the same segment, crossing-over between them may be detectable and its frequency determined. This is true of P and s, which both lie in end 3 and which may be separated by crossing-over. (b) Even where a heterozygous gene pair is the only pair known in a given end segment, it is possible to determine the frequency with which crossing-over occurs between it and the rest of the complex. This is owing to the fact that such a gene, if it crosses over, is transferred to the opposite complex, and becomes associated with a new set of lethals. A recessive gene carried by the alpha complex of a given race may be transferred to the beta complex which normally carries the dominant allele. The new beta complex may combine with a normal alpha complex and thus produce a plant homozygous for this recessive, thus exposing the fact of a prior crossover. It is important to realize that the frequency with which a gene in a pairing end is crossed over to the opposite complex is a function of its distance beyond the innermost point at which crossing-over ceases to be possible. If a gene lies far enough out on the chromosome to permit an average of one crossover proximally to it, this gene will be crossed over 50% of the time—in other words, it will behave as though independent of the complexes. As mentioned above, two genes of this sort are known—br (*brevistylis*, producing short styles), and Co (flower size). Catcheside (1954) has located one of these linked genes (br) in end 12 (11 of Cleland series). While no other genes are known which begin to approach these two in crossover frequency, the pairing ends of all chromosomes must approximate, as Darlington has suggested, 50 crossover units in length, since chiasma frequency approaches very closely to a minimum of one per end segment in most cases.

A few genes cross over from one complex to the other in the wild races in which they are normally found. For instance (as summarized by Renner, 1942), n has shown 0.1%–0.2% crossing-over in *biennis* de Vries (1915) and 1.0% in *lamarckiana* (de Vries, 1913); s has shown as high as 0.3% in *biennis* and 0.1% in *suaveolens* (de Vries, 1918); P has been found to cross over in one or two cases in *lamarckiana*. Shull's mut. *pervirens* is probably a case of *p-velans.p-gaudens*, and Renner has found *p-velans* in one or two trisomics from *lamarckiana* but not in normal *lamarckiana* itself. *P-gaudens* has not been found by Renner or others in ordinary *lamarckiana*. The *flavens* lethal (*let*) crosses over to the associated *albicans* with a frequency of about 2%, resulting in *flavens.flavens* offspring (*lutescens*). The same probably occurs in the *grandiflora* of de Vries, which throws a small percentage of *acuens.acuens* plants. When one considers the large number of *Oenothera* races that have been critically studied and recalls that most of them are highly heterozygous, the amount of crossing-over that has been detected is small indeed.

Renner considers that the genes that show crossing-over in the races lie in the transitional zone between the regularly pairing end segments and the proximal segments in which the genes that differentiate the complexes lie.

A second fact of interest was brought out by Renner (1942). He showed that the frequency of crossing-over is sometimes markedly greater in F_1 hybrids than in the parental races but tends to dwindle in succeeding generations of hybrid progeny. Thus, *n* (*nanella*) appeared in as many as 10–15% of the plants in early generations of certain hybrid complex-combinations although in the parental forms and in subsequent generations of the hybrids the percentage was reduced, often to zero. Renner ascribes the sudden increase in frequency to the presence of longer pairing segments in these hybrid complex-combinations and suggests that the falling off in later generations might be due to the development of factors interfering with crossing-over or to the elimination of such crossovers as are produced. To the reviewer it would seem that the increased frequency in early hybrid generations might well be due to the fact that the complexes newly brought together have in certain cases end segments which are capable of synapsing over a longer distance than was the case in the parent races; hence, a given gene finds itself in the hybrid farther removed from the inner end of the pairing segments. The dropping off in percentage of crossing-over in later generations might well be the result of unconscious selection. In the case of a character such as *n*, for instance, plants homozygous for the dominant (NN) might easily be selected to carry on the hybrid line since they are indistinguishable from the heterozygote (Nn). Renner found in most cases cited that the high frequency of crossing-over continued for several generations, dropping suddenly to zero.

The conclusion that a given pairing end is capable of synapsing for different distances back from the end with the pairing ends of different complexes is supported by evidence adduced by various authors that a given gene crosses-over more frequently in certain complex-combinations than in others. For instance, as between the *P* and *s* loci, Shull (1923a) found 6.1–8.7% crossing-over in hybrids between "*franciscana sulfurea*" and *rubricalyx;* Emerson (1931a) found 5–8% in *franciscana sulfurea* pollinated by a derivative of *lamarckiana* containing *rubricalyx;* and Emerson and Sturtevant (1932) found 12.5% in a backcross of this hybrid to the double recessive. On the other hand, Oehlkers (1933, Table 32; 1940, Table 8) found in the pollen mother cells of certain complex-combinations the crossover percentages between the P and s loci indicated in the tabulation.

Complex-combination	Percentage of crossing-over
flavens.stringens	0.25–0.28
flavens.ʰhookeri	0.00–1.82
flavens.velans	1.13
albicans.stringens	1.09
albicans.ʰhookeri	0.00–0.57
albicans.velans	0.00
curtans.flavens	0.03
deprimens.flavens	0.20
gaudens.ʰhookeri	2.44
excellens.ʰhookeri	0.36

Discordant results such as these can be understood if the pairing ends of arm 3 in Oehlkers' material are capable of synapsing for a shorter distance than these same arms in Shull's and Emerson's material. As shown in Fig. 9, if the pairing ends do not synapse as far in from the distal end

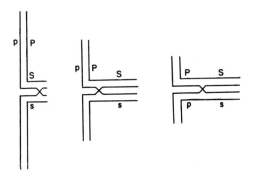

Fig. 9. Percentage of crossing-over between two loci situated in a pairing end will depend in each complex-combination on the distance over which synapsis can occur in that combination. See text, p. 178.

as the S locus, there will be no crossing-over between P and S; if synapsis extends slightly beyond the S locus but does not reach the P locus, a small amount of crossing-over can occur; if both the S and P loci are included in the synapsed segment, the full amount of crossing-over corresponding to the map distance between the two loci can take place.

Part of this difference, however, may be due in some cases to differing abilities of crossover and non-crossover gametophytes to compete in megaspore formation or pollen tube growth. This is evidenced by the different crossover percentages obtained through sperm and egg in the case of certain complex-combinations. Thus, Oehlkers (1940) obtained 0.0–0.57% crossing-over between the P and s loci when the hybrid albicans.ʰhookeri was used as male, and 4.24% when it was used as female; 0.0–1.82% in flavens.ʰhookeri used as male, 0.61–3.06% when used as female.

With respect to the extent and frequency of crossing-over in paired chromosomes vs. circle chromosomes, Emerson (1932) and Renner (1942) have both shown that no appreciable difference is to be found between chromosomes in these two categories. It might be expected that chromosomes that are completely homologous would be able to synapse throughout their entire length; and, hence, one might expect genes which are incapable of exchange in circle chromosomes to show crossing-over in paired chromosomes, because of their central position. This does not seem to be the case. Actually there is little evidence that chiasmata are produced in the proximal segments even of paired chromosomes; they seem to be confined largely to the pairing ends of both paired and circle chromosomes.

While crossing-over, therefore, is normally confined to the pairing ends, where the genes are mostly homozygous, heterozygous genes may exceptionally be found in pairing segments, thus making possible genetically detectable crossing-over.

(2) Exceptions to the rule that crossing-over is confined to homozygous pairing ends have also been claimed in the opposite direction by Darlington, who suggests that crossing-over may occasionally involve loci within the proximal segments. It has already been pointed out (p. 163) that Darlington considers it likely that breaks resulting in interchange may occur at other points than the centromere. If this were to take place, homologous segments would occasionally find themselves situated in the proximal portions of chromosomes that are otherwise non-homologous. Synapsis might occasionally occur between these homologous segments and crossing-over could follow, thus involving the central regions of one or both chromosomes. Darlington (1931) claims to have observed chiasmata involving the proximal regions of the chromosomes.

Darlington (1931) called these homologous segments in otherwise non-homologous or partially non-homologous chromosomes interstitial or differential segments. Later he made a distinction between the two (Darlington, 1936): interstitial segments are flanked on one end by homologous segments, at the other end by non-homologous segments; differential segments are flanked on both ends by non-homologous segments. Crossing-over between corresponding interstitial segments results in sterile crossover chromatids; between homologous differential segments it may, with proper orientation at the point of crossing-over, result in viable gametes carrying an altered segmental arrangement. While such a result is theoretically possible, the reviewer has never observed an interstitial chiasma in *Oenothera* and on this basis is skeptical of the reality of crossing-over in differential segments. Changes in segmental arrangement probably occur solely as a result of reciprocal translocations.

If the terms interstitial and differential segments had not been given such specific meanings by Darlington, they would be very useful terms. The cytological chromosome shows a rather clear distinction between what are probably the pairing ends and the proximal region which could very logically be designated the interstitial region. This same region in the genetic chromosome is the one which carries the genes which differentiate one Renner complex from another—it might well be called, therefore, the differential segment of the chromosome.

XI. Position Effect

Until the work of Catcheside, the phenomenon of position effect was definitely known only in *Drosophila*. One might expect to find in *Oenothera* considerable evidence of position effect inasmuch as the evolutionary history of the genus has involved the transfer of segments from one position to another in rather wholesale fashion. If, on the other hand, the breaks and reunions involved in an exchange occur in heterochromatin and not in the immediate vicinity of active genes, or if the exchange is a result of ordinary crossing-over within variously displaced homologous segments, one would not expect a position effect to result from such a transfer of segments.

The reviewer has had one experience in which a plant suffered interchange but showed no evidence, as a result, of position effect. This was the case of the *grandiflora* of de Vries in which an interchange occurred in 1936 leading to the appearance of a plant in 1937 with an altered configuration (⊙ 12, 1 pair, instead of ⊙ 14). This plant has given rise to a line of *grandiflora* which, in place of *acuens*, has an altered complex (*neo-acuens*), with the chromosomes 1·13 4·14 instead of 1·4 13·14. The altered *grandiflora* is phenotypically indistinguishable, however, from the unaltered race—and, as a result, it was not realized that the change had occurred until it was discovered that, in crosses to and from the altered 1937 plant, plants which were supposed to carry *acuens* did not have the configurations predicted for them. A check on the chromosome configuration of the 1937 *grandiflora* parent showed that it had ⊙12. Critical examination of the selfed progeny of this plant (also carrying ⊙ 12) showed that they differed in no respect phenotypically from their ancestors with ⊙ 14. In other words, no position effect was detectable.

Catcheside (1939, 1947a,b), however, has uncovered what seems to be a clear case of position effect in *O. blandina*, a derivative of *lamarckiana*. The complex ^{h}*blandina* (haplo-*blandina*) carries in end 3 the gene P^s (broad red stripes on flower cone, red papillae on stem). This is the second in a series of alleles, ranging from P^r (*rubricalyx*, = intense reddening of cone and hypanthium, red papillae) through P^s, through P (red papillae, green buds) to p (green papillae, green buds). Catcheside obtained via X-ray a complex *blandina-A*, with an altered segmental arrangement—3·11 4·12 (3·12 4·11 according to Cleland's system) instead of 3·4 11·12. When this complex was combined with normal ^{h}*blandina*, the combination (*blandina-A*.^{h}*blandina* with ⊙ 4) showed narrower leaves, and considerable reduction in the amount of pigment in the flower buds—instead of the broad red stripes normally produced by P^s, the stripes tended to be narrow with irregular green stripes separating or running through them, producing a mottled effect. This could be due theoretically to a mutation of P^s simultaneous with the interchange from 3·4 11·12 to 3·11 4·12, or it might be a position effect. That it was the latter was indicated when P^s *blandina-A*.P^s ^{h}*blandina* was pollinated with P^r ^{h}*blandina* and the resultant P^s *blandina-A*.P^r ^{h}*blandina* backcrossed to P^s ^{h}*blandina*.P^s ^{h}*blandina*. This should yield equal numbers of P^s *blandina-A*.P^s ^{h}*blandina* (showing mottling) and P^r ^{h}*blandina*.P^s ^{h}*blandina* (*rubricalyx*). This was the result except that one P^s ^{h}*blandina*.P^s ^{h}*blandina* plant (normal, unmottled) appeared. The appearance of the normal P^s plant would have to be explained either by postulating a mutation from P^r to P^s, a phenomenon never before observed, or by assuming the exchange by crossing-over of the P^s of *blandina-A* and the P^r of ^{h}*blandina*, thus producing a normal P^s ^{h}*blandina* genome—in other words, a normal P^s gene whose effect had been modified when transferred from a 3·4 to a 3·11 chromosome, became normal in its effect when returned to a 3·4 chromosome. Catcheside considered the latter alternative to be the more likely.

This assumption of position effect was based, of course, on only one plant. Consequently, Catcheside carried out further experiments, published in 1947. He used both P^s and P^r and found that both gave a mottled effect when present in *blandina-A* combined with ^{h}*blandina*. When the above-mentioned backcross was repeated (P^s *blandina-A*.P^r ^{h}*blandina* × P^s ^{h}*blandina*.P^s ^{h}*blandina*), 7 crossovers were obtained out of 699 plants [both crossover classes being obtained, i.e., normal P^s ^{h}*blandina*.P^s ^{h}*blandina* and mottled *rubricalyx* (P^r *blandina-A*.P^s ^{h}*blandina*)]. Crossover percentage between the P locus and the interchange point was, therefore, about 1%.

P lies in the same chromosome end with *s* (sulfur petal color), the two genes being about 8 units of map distance apart, with P lying closer to the centromere. Catcheside obtained various combinations of P alleles with S or s and tested the combinations for position effect. Amazingly, the S locus also showed position effect (large sulfur patches on yellow petals), although it lies 8 units away from P. A typical cross among the eight performed is as follows:

P^s S blandina-$A.P^r$ s ^{h}blandina (P^s S -$A.P^r$ s)

$\times$ P^s s ^{h}blandina.P^s s ^{h}blandina(P^s $s.P^s$ s)

This gave:

75 P^r $s.P^s$ s (normal rubricalyx sulfur),
112 P^s S -$A.P^s$ s (mottled red and green cones, mottled yellow and sulfur petals),
4 P^s $S.P^s$ s (normal red, yellow),
19 P^r $S.P^s$ s (normal rubricalyx, yellow),
6 P^s s -$A.P^s$ s (mottled red and green cones, sulfur),
1 P^r S -$A.P^s$ s (mottled rubricalyx and green cones, mottled yellow and sulfur),
1 P^r s -$A.P^s$ s (mottled rubricalyx and green cones, sulfur).

Analysis of this cross will show that P^s, P^r, and S all produce a mottling effect when present in blandina-A and that they all give their normal effects when returned via crossing-over to ^{h}blandina, showing that they have not been altered by their temporary transfer to blandina-A. The evidence from all eight backcrosses is extensive and consistent and leads to the same conclusions. Out of a total of 1112 progeny, there were 1001 non-crossovers, 109 single crossovers, and 2 double crossovers, all giving the expected phenotypes.

In his 1939 paper, Catcheside refers to the fact that, in every case examined, blandina-$A.^h$blandina combinations (i.e., those showing mottling) have been found to have $\odot$ 4, 5 pairs (vs. ^{h}blandina.^{h}blandina with 7 pairs). The reviewer has grown certain of Catcheside's crosses and has tested the configurations of the various combinations. The results confirm those of Catcheside fully—all plants which show mottling and therefore have the combination blandina-$A.^h$blandina, have shown $\odot$ 4; those homozygous for ^{h}blandina have shown 7 pairs. It was found that blandina-A.blandina-A plants are normally crippled and fail to mature, reduced viability being one of the effects of shift in position. Catcheside reports one case, however, of a family of P^r blandina-A homozygotes which came to flower. These plants showed the mottling.

This is an interesting case for at least three reasons: (1) It is the first clear-cut case of position effect in the flowering plants; (2) it involves an extraordinary length of the genetic map, more than 8 units (this may, of course, correspond to a very short distance on the cytological chromosome); (3) the mottling is in the form of large patches, suggesting that the abnormality has occurred in one cell and has been perpetuated through a long series of cell divisions. It is of interest that the patches tend to be larger and more striking early in the season than later.

XII. Gene Conversion

The genetic trait *cruciate petals* is a condition in which the petals are crippled, reduced in width until in the extreme form they are linear and sepaloid. This character was first studied by de Vries (1903b, p. 593) who found that certain material derived from the northeastern United States (*O. cruciata Nutt.*) bred true for the cruciate condition, which behaved in outcrosses as a recessive. In other material, however (*O. cruciata varia*, of uncertain origin), the condition was inconstant in expression, some plants showing the trait more strongly than others, some flowers on a single plant being cruciate, others normal, still others showing varying degrees of intermediacy between the two extremes.

The first intensive study of this character was that of Oehlkers (1930a,b, 1935, 1938), who used a constant race from the botanical garden in Leiden (*O. biennis cruciata apetala*) and two inconstant races, *O. biennis cruciata* (*sulfurea A_2a*) derived from Klebahn, and *O. lamarckiana cruciata* from the botanical garden in Tübingen. As in the case of de Vries' material, the character appeared to be a simple recessive in the constant race, which bred true for the cruciate condition; its behavior, however, in the inconstant races was highly irregular. Within a single individual, marked fluctuation in the degree of expression of the trait was found; and different plants, and succeeding generations, differed greatly among themselves. In the case of A_2a, selection was found to have a positive effect. For instance, he self-pollinated a plant that had a branch on which the petals were nearly perfect and another on which the flowers were strongly cruciate. Selfed seed from the former produced a line in which the flowers were nearly perfect ("*A_2a weakly cruciate*"); the latter gave rise to a line of "*A_2a strongly cruciate*" plants. In the case of *lamarckiana cruciata*, early experiments showed little effect of selection, but experiments carried out some years later showed that by that time the race had become labile, and selection was effective. A character, therefore, that seemed to be a clear recessive in one race, behaved in other races as though subject to fluctuating dominance, even within a single individual.

As a result of an analysis of a number of crosses between cruciate and normal races, Oehlkers found that the cruciate character expresses itself in varying degree in different complex-combinations. Seven levels of expression were recognized, ranging from normal (Class I) to completely cruciate petals (Class VII). The varied levels of expression in different complex-combinations Oehlkers ascribed to the fact that different complexes carry genes of differing strength in the production or expression of the cruciate condition. Thus, he classified the complexes with which he dealt as shown in the tabulation (Oehlkers, 1930b).

Descending degrees of dominance

Cr_1 ^{h}*hookeri*
Cr_2 *albicans* from *suaveolens*
     flavens from *suaveolens*
Cr_3 *albicans* from *biennis* Hannover
Cr_4 *rubens* from *biennis* Hannover

Ascending degrees of expression of cruciata condition

cr_1 *albicans* from *biennis cruciata* A_2a
cr_2 *rubens* from *biennis cruciata* A_2a
cr_3 *gaudens* from *lamarckiana cruciata*
cr_4 *velans* from *lamarckiana cruciata*
cr_5 *albicans* from *biennis cruciata apetala*
     rubens from *biennis cruciata apetala*

Cr_1 is the strongest normal; cr_5, the strongest cruciate factor.

This situation has arisen, according to Oehlkers, because the *cr* locus is a highly mutable one, as a result of which multiple alleles have developed in the population, which differ from one another in their degrees of dominance or recessiveness. All members of the allelic series are considered to be mutable. Apparently stable strains, such as *hookeri* on the one hand, or *biennis cruciata apetala* on the other, remain stable only because mutations do not extend far enough along the series to reverse the phenotypic expression. Complex-combinations closer to the center of the series, however, show an increased tendency to transgress the phenotypic boundary; combinations that are ordinarily normal may exhibit flowers with defective petals, while ordinarily cruciate or sub-cruciate lines may show an occasional normal flower.

One frequently gets unexpected results in particular crosses involving *cr* and these Oehlkers ascribed either to crossing-over between associated complexes or to change in dominance resulting from mutation.

As an example of the first, Oehlkers obtained the following results (1930a): *biennis cruciata apetala* × [*suaveolens sulfurea* × *apetala*] *flavens.-rubens* (i.e., *cr-albicans.cr-rubens* × *Cr-flavens.cr-rubens*) gave

albicans.flavens   41 Class I–II, 75 Class V–VI
albicans.rubens   6 Class I–II, 22 Class VII

The *albicans.flavens* would be expected to be *crCr* and all plants should be normal or nearly so, but the majority were sub-cruciate. The *albicans.- rubens* should be *crcr*, and, therefore, all cruciate; but 6 plants were normal to subnormal. Oehlkers attributed the unexpected results to the transfer of *cr* from *rubens* to *flavens* in the first case, and the transfer of *Cr* from *flavens* to *rubens* in the second.

That crossing-over of the *Cr* locus from one complex to the other can occur was indicated by results of crosses between *biennis cruciata* A_2a and *lamarckiana brevistylis* (Oehlkers, 1938). The *brevistylis* character segregates independently of the lethals in forms with ⊙ 14 as a result of 50% crossing-over. In the progeny of (A_2a strong *cr* × *brevistylis*) *cr-Br- albicans.Cr-br-velans* (⊙ 14) × self, the following was obtained: 37 *Br Cr*, 10 *Br cr*, 14 *br Cr*, 4 *br cr*. This close approach to 9:3:3:1 showed that both loci were segregating independently of the lethals and of each other in this hybrid with ⊙ 14; both were apparently showing 50% crossing-over. On the other hand (Oehlkers, 1938), crossing-over of *Cr* and *cr* occurs very seldom in *albicans.hhookeri* (⊙ 14) and in *albicans.gaudens* (⊙ 6, ⊙ 8).

A case in which a change of dominance is postulated involves *lamarckiana cruciata*, a race in which the cruciate character is not extreme in its expression (Class V). Although somewhat inconstant in its expression, it was at first not sufficiently labile to respond positively to selection. In later years, however, it became sufficiently labile to respond to such treatment. As a result, two lines were obtained, one strongly cruciate, the other weakly so. The cross *lamarckiana* de Vries × *lamarckiana cruciata strong* gave in F_1 both normal and cruciate plants. The latter yielded, upon selfing, normals as well as cruciates. One of the normals was selfed, and bred true, producing nothing but normals. Three of the cruciate plants were also selfed. One of them bred true, producing only cruciates. The other two produced a few normals as well as cruciates. These results indicate that *cr* had become dominant in this strain. *Cr* plants bred true because they were homozygous for what was now the recessive condition. Some of the cruciates bred true and thus behaved as homozygotes. Other cruciates produced both cruciates and normals, indicating that they were heterozygous, with *cr* as dominant.

Oehlkers' conclusion, therefore, was that the *cr* locus is a highly mutable one. As a result, multiple alleles have developed in the population which account for the varying degrees with which the character is expressed in different complex-combinations. Some alleles strongly favor the cruciate condition; some strongly dominate over this condition; some are less vigorous in their effect. Cases where different levels of expression are seen in single plants are the result of intermediate balance possibly supplemented by the occurrence of new somatic mutations producing changes of dominance. According to Oehlkers, such mutations can occur in homozygotes as well as heterozygotes.

Renner, in a series of papers extending from 1937 to 1959, has confirmed and extended Oehlkers' findings, but has arrived at a different explanation regarding the physical basis of the phenomena observed. His conclusion is that mutations affecting the *cr* locus occur only in heterozygotes, and he attributes such genic alterations to a process of "somatic conversion," adopting as a working hypothesis a modification of Winkler's idea (1930, 1932) that in a heterozygote one allele may cause the other to mutate to its own condition. Winkler applied his concept to meiosis and presented it as an alternative to the concept of crossing-over. Renner suggested that gene conversion in the case of the *cr* locus occurs in the soma, and expressed doubt as to whether it also occurs in meiosis. Conversion can take place in either direction: *cr* can convert *Cr* to *cr*, or the reverse can happen. He considered that conversion leading to *CrCr* homozygotes is commoner than the reverse. In either case, heterozygotes are converted to homozygotes.

In this respect, Oehlkers' and Renner's ideas differ materially. According to Oehlkers, *cr* and *Cr* are mutable genes which may mutate in either the homozygous or heterozygous condition, and the result may be either a heterozygote or a homozygote. The presence of large circles and lethals makes it difficult to determine in every case in *Oenothera* whether a plant is heterozygous or homozygous for a given locus—the lethals prevent Mendelian segregation after selfing; and one must resort to outcrossing, and the comparison of the behavior of the two complexes in crosses to the same tester complex, to determine whether alleles are identical or different in their effect upon the phenotype. The probability of the presence of different modifiers in opposing complexes makes such comparisons of doubtful validity. It is not easy, therefore, to determine whether mutations are confined to heterozygotes for the *cr* locus or whether they may occur in homozygotes.

Renner recognized the existence of differences in strength or "prepotency" among genes of different races. Some genes are more dominant than others at the phenotypic level, some have a greater potency at the genic level in the matter of being able to convert or to resist conversion. Among the three races which he analyzed in greatest detail, the *cr* of *biennis cruciata* and that of *atrovirens* seemed to show essentially equal strength; and he considered it possible that they are identical alleles. The *cr* of *lamarckiana cruciata*, on the other hand, is weaker and is probably a different allele.

Renner came to the conclusion that the stability of a *cr* allele is influenced by the genotype as a whole. He found, for instance, that *lamarackiana cruciata* which carries *Co* (small flower size derived from *biennis*) in double dose is a stable cruciate. Individuals that are heterozygous for *Co* are mostly cruciate, but they can be normal. Large flowered plants (*coco*) are mostly normal, with occasional slight defects in the petals. Even the rare large flowered cruciate produces mostly normal (*Cr*) offspring. The stabilization of the *cr* gene (which has arisen by conversion from *Cr*) is thus strengthened by the *Co* gene from *biennis*. This influence of *Co* is of interest because *Cr* and *Co* are probably situated in different chromosomes. Renner places *Cr* in 5·6, while *Co* is either in 13·14 (Harte) or 11.12 (Catcheside).

Another case of this sort involves the gene *R* for red midribs, situated in 1·2: *crcr-hookeri* is stable when small flowered; but, when it is large flowered, it remains stable only if *R* is present in heterozygous condition (*RR* is lethal). If the plant is *rr*, *cr* tends to mutate to *Cr*.

Mosaics in which both *Cr* and *cr* appear are explicable on the basis of conversion, and the pattern of mosaicism will be determined by the stage of development at which conversion occurs. Since converted nuclei are homozygous for this locus, it is not surprising that *Cr* branches tend to produce *Cr* progeny and *cr* branches produce *cr*.

Renner offers no suggestions as to the mechanism by which conversion takes place but considers that the empirical findings support the reality of this process. Goldschmidt (1958), in a review of this situation in *Oenothera*, has suggested a possible mechanism in terms of position effect resulting from unequal crossing-over.

The cruciate character, therefore, is one which expresses itself in varying degree in different races; it appears to exhibit differing levels of dominance and recessiveness. Hybrids commonly show a mosaicism in which the flowers on a single plant may show every transition from normal to cruciate. In some cases, the progeny of *cr* flowers tend to be cruciate; those of normal flowers, even from the same plant, tend to be normal. We are confronted with two hypotheses to account for the behavior of this locus: either the locus is a highly mutable one, at which either allele may mutate at any time, irrespective of whether the individual is heterozygous or homozygous for this locus; or a mutation may take the form of gene conversion, which can only occur in a heterozygote, and in which one gene converts the other to its own state, thus producing a homozygous condition, which in some cases appears to be stable, in others, labile. Decisive experiments have not been devised to distinguish between the concepts of Oehlkers and Renner—such experiments are difficult to accomplish because of the peculiar cytological behavior which characterizes most Oenotheras.

XIII. The Nature of de Vries' Mutants

Had it not been for the propensity of *Oenothera* to throw "mutants," de Vries' attention would not have been drawn to the genus. Genetic analysis of the few aberrant individuals which he found in a large population of what he took to be *O. lamarckiana* Ser. convinced him that they were new phylogenetic entities arising suddenly from the parent form. These he termed "elementary species," and considered that some of them might have sufficient survival value to compete successfully and thus become new species, while others might be unfit to survive.

Since de Vries was concerned primarily with the origin of species, it is natural that his first classification of mutants should be based upon his conception of their mode of origin. He classified them, therefore (de Vries, 1903), in terms of his theory of Intracellular Pangenesis (de Vries, 1889).

According to this theory, living protoplasm consists of pangenes, one kind for each character. All kinds of pangenes in an organism are present in the nucleus of every cell, where they multiply. Some of the products of multiplication of each pangene remain in the nucleus and serve as the vehicles of heredity; others may migrate to the cytoplasm where they multiply further and determine the characteristics of the cell. Pangenes may exist in several states. Those which migrate to the cytoplasm are in the active state. Those that remain in the nucleus are in an inactive state. Pangenes whose activities are not called for in particular cells will remain wholly within the nucleus and hence inactive. Other pangenes may be genetically latent or inactive and these will remain within the nucleus in all cells of the body. Pangenes may also exist in a semi-active or a semi-latent state. Furthermore, they may, at certain times, enter a labile state, in which condition they can change from any state to any other. Finally, new pangenes may arise, usually in the latent condition, from which they may change to the active condition. The change of pangenes from one state to another, or the creation of new pangenes, is considered to be a mutational change.

On the basis of this theory, de Vries classified his mutants as follows:

(1) Progressive mutants are those which arise as a result of the creation of one or more new pangenes (e.g., *gigas*).

(2) Retrogressive mutants arise when one or more active pangenes change to the inactive or latent condition, resulting in the inability of certain characters to manifest themselves (e.g., *brevistylis*, in which ability to produce long styles is lost; *rubrinervis*, in which vascular bundles have lost their highly lignified character, making the stems brittle).

(3) Degressive mutants arise from other changes of state (latent to

active or semi-latent, active to semi-active, or vice versa). *Lata, scintillans*, and *albida* were considered examples. Progressive mutations were considered (de Vries, 1903b) to give rise to new elementary species, retrogressive and degressive mutations to new varieties. According to de Vries' interpretation, therefore, all hereditary modifications are the result of a change of state of one or more pangenes. De Vries was impressed with the fact that different organisms seem to show very different propensities to mutate, and concluded that species pass through periods of mutability, when at least some of their pangenes are in the labile condition. He considered that *O. lamarckiana* is at present in a mutable condition, producing a continuing burst of mutations (de Vries, 1905).

De Vries' attempts to explain the various aberrants in terms of mutating pangenes brought forth a varying response from different workers. Heribert-Nilsson (1912) insisted that "mutations" in the Swedish *lamarckiana* were new combinations of previously present factors, and not the putative beginnings of new species or varieties. Davis (1911, 1916) claimed that de Vries' "mutations" were merely segregants resulting from the fact that *lamarckiana* is of hybrid origin (which de Vries denied) and consequently highly heterozygous.

Gates (1911a; de Vries and Gates, 1928) and Bartlett (1915), on the other hand, tended to accept the aberrants from *lamarckiana* as bona fide mutants carrying new hereditary characteristics, rather than as mere segregants or new combinations of old genes.

Whatever their attitudes in this regard, however, workers outside of de Vries' own school were unable to accept a classification of mutants based upon the theory of Intracellular Pangenesis; and this classification soon fell into disuse. In his later years, de Vries himself referred to this classification with decreasing frequency, although he was never able to divorce himself completely from this concept.

His own, as well as Renner's breeding experiments, however, led him in time to another classification of mutants based upon breeding behavior (de Vries, 1929). Those that were able to transmit their mutant characteristics through both sperm and egg he termed "isogamous" mutants. Those which could transmit mutant characteristics only through the egg he termed "heterogamous." The inability of the latter class to transmit its mutant characters through the pollen he ascribed to the presence of "androlethal factors" (de Vries, 1925b). The heterogamous mutants he further subdivided into "dimorphic" and "sesquiplex" mutants. Dimorphic mutants produce segregating progenies; they throw in each generation a certain proportion of *lamarckiana* in addition to plants like themselves. Sesquiplex mutants breed true. They possess two kinds of egg and one kind of sperm. Half of the eggs carry the mutant genome; the other half carry one of the (lethal-bearing) complexes of *lamarckiana;* the sperms carry the same lethal-bearing *lamarckiana* genome. These mutants will therefore produce 50% bad seeds and will give rise to progeny only like themselves. The distinction between these two kinds of heterogamous mutants will be clearer when the cytological basis for their origin is explained below.

A fourth type of mutant based on breeding behavior was described in 1917—the so-called "half-mutant." This carries in both sperm and egg a modified alethal genome plus an unmodified *lamarckiana* genome. It throws in each generation, therefore, a large class of "full mutants," containing the modified alethal genome in double dose.

In the meantime it had been shown (Lutz, 1907, 1908) that *Oenothera* mut. *gigas* had 28 ($4n$) chromosomes, and that certain other mutants from *lamarckiana* had 15 ($2n + 1$) chromosomes. Although de Vries considered that the departure from the $2n$ chromosome number was a result of mutation rather than a cause of the aberrancy, he could not fail to take note of the relationship between chromosome number and the type of deviation from normality displayed. Beginning in 1923 he attempted, with Boedijn, to classify mutants on the basis of the chromosomes mainly responsible for them. Assuming the presence of 7 pairs of chromosomes in *lamarckiana*, and assuming also that each mutant depends primarily upon genes in a single chromosome, these authors attempted to group the known mutants into 7 groups, one for each chromosome. The first of these they called the "central group," which included the group of genes differentiating the genomes of *lamarckiana* ("*laeta*" and "*velutina*"), the lethals, and all of the 14-chromosome mutants plus one 15-chromosome form (*pulla*). This group, much larger than any of the others, they considered to correspond to the group of genes assigned by Shull to his Chromosome I. In addition, 6 other groups were recognized, all the mutants in these groups containing 15 chromosomes ($2n + 1$). In each of these groups, the one which appeared most frequently was thought of as a "primary" mutant, after which the group was named. The others were termed "secondary" mutations. The primary trisomic mutations and some of their secondaries, as classified by de Vries and Boedijn (1923) were as shown in the tabulation.

Three of these groups of mutants were larger than the other four. De Vries and Boedijn claimed to find in Cleland's 1922 paper on *franciscana* evidence for the existence in this form of 3 larger and 4 smaller chromosomes, although the latter author did not think that size differences were consistent or uniform. Later, Boedijn (1925) claimed to have found in *lamarckiana* 4 large and 3 small pairs and assigned the *cana* group to one of the larger chromosomes.

The paper in which this classification of mutants into 7 groups was made (de Vries and Boedijn, 1923) was written before any description of large circles had been published. The first description of large circles (Cleland, 1923) appeared in the same year as de Vries and Boedijn's paper, and the first announcement of the presence of a circle of 12 and 1 pair in *lamarckiana* came 2 years later (Cleland, 1925). These authors, therefore, did not have the benefit of knowledge regarding chromosome behavior in *lamarckiana*. When the chromosome configuration in *lamarckiana* was announced, however, they refused to accept this finding, and in 1928 Boedijn confirmed his earlier account of the cytology of *lamarckiana* (Boedijn, 1925) in which he claimed the presence of 7 pairs.

	Primary trisomic mutants	Secondary trisomic mutants
Central group	*pulla*	—
Lateral groups	*lata*	*albida, semi-lata, flava,* etc.
	scintillans	*oblonga, aurita, auricula, nitens,* etc.
	cana	*candicans*
	pallescens	*lactuca*
	spathulata	—
	liquida	—

* Other secondaries were added in later papers of de Vries and Boedijn.

The discovery of a circle of 12 chromosomes and 1 pair in *lamarckiana*, and its explanation in terms of reciprocal translocation, placed all of these matters in an entirely new light. As Hoeppener and Renner (1929) pointed out, *lamarckiana* does not have 7 kinds of chromosomes, but in reality, 13 different kinds. Each chromosome in the circle of 12 is structurally unlike every other chromosome. There are, therefore, 12 kinds of chromosomes in the circle, the chromosomes of the pair constituting the 13th kind.

Obviously, therefore, the chromosomal situation calls, not for 7 groups, but for 13 groups of mutants, if mutations are based primarily on single chromosomes. While the newer discoveries, therefore, completely negated de Vries' attempt to classify mutants in terms of the chromosomes responsible, his attempt to make this correlation was a forward step.

Since this time, much further light has been shed on the nature of de Vries' and similar so-called mutants. Catcheside (1933, 1936), Emerson (1935, 1936), and Renner (1943a, 1949) have contributed most significantly to an understanding of their origin and nature; and Håkansson (1926,

1930b, 1931) has analyzed in considerable detail the chromosome configurations found in various forms with extra chromosomes.

Catcheside (1936) and Emerson (1936) undertook independently an analysis of the types of mutant possible in the progeny of *lamarckiana* and the basis of their origin. While some of the aberrants found by de Vries, Shull, etc., were undoubtedly the result of true point mutation (*brevistylis, nanella, sulfurea, vetaurea, bullata, rubricalyx,* etc.) most of the early mutants of de Vries were more complex in nature, having come about as the result of alteration in chromosome number. Catcheside and Emerson independently devoted special attention to the mode of origin of the trisomics which constitute the bulk of the mutants. Both showed how various types of irregularity in the zigzag arrangement within the circle of 12 could produce the different types of mutants found by de Vries. This may be briefly summarized as follows (see Table 3 for terminologies used):

TABLE 3
Terminologies Used in Classification of Trisomic Mutants*

Reference			Characterization
de Vries (1929)	Catcheside (1936)	Renner (1949)	
dimorphic	dimorphic	isotrisomic	Extra chromosome is one of the pair (one chromosome therefore present three times)
dimorphic	dimorphic	anisotrisomic additive	Extra chromosome is one of circle chromosomes (one chromosome present twice)
sesquiplex	monomorphic	anisotrisomic compensated	Extra ends derived from two different chromosomes (no chromosome present more than once)

* Devised for *lamarckiana* mutants, but applicable to those from other sources.

(a) The dimorphic mutants based on circle chromosomes result from an irregularity in which a certain chromosome goes to the same pole as both its neighbors—3 chromosomes in a row going, therefore, to the same pole (Fig. 10). This type of configuration has been frequently seen. This will give an 8-chromosome gamete containing one entire genome plus one chromosome from the other complex. If a *velans* genome, for instance, receives in this way an added *gaudens* chromosome, and the resultant 8-chromosome gamete combines with normal *gaudens*, a trisomic mutant will be formed which will owe its phenotypic peculiarities to the presence of a *gaudens* chromosome in double dose. Since there are 12 chromosomes

CYTOGENETICS OF OENOTHERA 193

FIG. 10. Irregularity in the zigzag arrangement that will produce an 8-chromosome gamete capable of giving rise to a dimorphic trisomic.

in the circle in *lamarckiana*, there may be theoretically 12 different trisomics of this type, half of them with an extra *gaudens* chromosome, half with an extra *velans* chromosome. Such a trisomic will produce the following types of eggs: *velans*, *gaudens*, *velans* + extra, *gaudens* + extra. The sperms will ordinarily not function if the extra is present, so that only *velans* and *gaudens* sperm will be produced. On selfing, such a mutant will then produce both *lamarckiana* (*velans.gaudens*) and the mutant (*gaudens* + *velans* + extra). Such a mutant will, therefore, be dimorphic. To these 12 should be added the trisomic based on non-disjunction of the paired chromosomes, which will show the same breeding behavior (isotrisomic mutant of Renner).

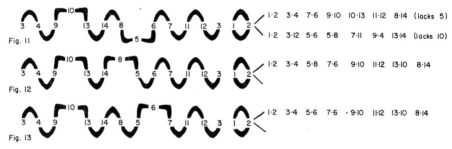

FIGS. 11–13. The irregularity in the zigzag arrangement shown in Fig. 11 will produce only inviable gametes. Those in Figs. 12 and 13 will produce 8-chromosome gametes capable of giving rise to sesquiplex trisomics.

(b) Sesquiplex (monomorphic) mutants arise occasionally when 2 adjacent chromosomes pass to the same pole and are compensated by 2 other adjacent chromosomes going to a single pole. If 2 adjacent chromosomes go to one pole and the compensating 2 go to the opposite pole, the products will be nonviable, since they will be deficient for certain segments (Fig. 11).[3] If, however, the 2 compensating chromosomes go to the same pole as the first 2, a viable 8-chromosome genome will be produced, which, by combining with a normal haploid genome will produce a sesquiplex mutant. In a circle of 12, the 2 sets of 2 chromosomes going to the same pole may be separated by 1 chromosome in one direction around the circle and 7 chromosomes in the other direction; or they may be separated by 3 chromosomes in one direction and 5 in the other (Figs. 12 and 13). Altogether, after eliminating duplications, 24 different kinds

[3] The segmental arrangements which have been worked out for *velans* and *gaudens* are as follows:

velans   1·2   3·4   5·8   7·6   9·10   11·12   13·14
gaudens 1·2   3·12 5·6   7·11 9·4   10·13   8·14

(According to Catcheside and Renner, *gaudens* has 3·11 and 7·12.)

of 8-chromosome gametes can be formed in this way in *lamarckiana* with its circle of 12.

Eight-chromosome genomes so formed will have mixtures of *velans* and *gaudens* ends, but they will have all the ends necessary to complete the segmental arrangement of one complex or the other, plus 2 ends as extras which are not normally associated in any single chromosome of the parent plant. Thus, in Fig. 12, the 8-chromosome genome has all the ends necessary to make the *velans* segmental arrangement (ends 13 and 14, however, from *gaudens*), plus ends 8 and 10 (also from *gaudens*).

Genomes so constituted, if combined with the appropriate normal complex (*velans* or *gaudens*), will produce trisomics which will be unable to segregate *lamarckiana*. They will breed true—in other words, they will be

Fig. 14. Chromosome configuration in a sesquiplex trisomic. Only two kinds of gamete are formed, one of which is a normal *lamarckiana* gamete, the other an 8-chromosome gamete carrying, in this case, ends 8 and 10 in duplicate. It is impossible for this trisomic to segregate *lamarckiana*, as do the dimorphics; instead, it will breed true.

sesquiplex mutants. To illustrate, suppose that the 8-chromosome genome obtained in Fig. 12 unites with normal *gaudens* to form a trisomic plant. In meiosis in this plant, the chromosome configuration shown in Fig. 14 will be present if all pairing ends unite. (This may be modified by failure of one or more ends to unite, thus forming one pair attached to a chain of 11, plus a free pair, or a chain of 13 plus a pair.)

In any event, but two kinds of gamete will be produced, barring further irregularities. There will be two kinds of egg but only *gaudens* sperm, since extra-chromosome genomes usually fail to survive in the male gametophyte. Since *gaudens.gaudens* is lethal, only the reunion of the 8-chromosome egg with *gaudens* sperm will produce viable offspring. The mutant therefore breeds true.

While Catcheside and Emerson agreed in regard to the above facts, they differed in their estimate of the total number of trisomic mutants that could be produced directly from *lamarckiana*. This difference of opinion arose from the fact that Emerson reasoned on the basis of the presence of a single lethal factor in each of the complexes, whereas Catcheside produced evidence suggesting that no such specific lethal exists in *velans*. As a consequence, Catcheside claimed that 36 possible 8-chromosome gametes could be formed based on non-disjunction in the circle of 12 and that each of these could combine with either *velans* or *gaudens*, making a total of 72 possible trisomics, to which must be added the trisomic for the chromosome of the free pair. Emerson, on the other hand, showed that, if *velans* and *gaudens* both have specific lethals at specific loci, many combinations of 8-chromosome gametes with *velans* and *gaudens* will be impossible. He found, assuming one lethal in each complex, the maximum number of trisomics derivable directly from *lamarckiana* to be 42 (based on non-disjunction in the circle) plus one based on the pair, a total of 43.

The difference of opinion is of slight significance; the chief fact, upon which both agree, is that many trisomics are theoretically obtainable from a plant with a large circle as the result of non-disjunction. It is not surprising, therefore, that the great majority of mutants arising from *lamarckiana* and other races are trisomics.

In recent years, the extra chromosome or chromosome ends in several trisomics have been identified. Catcheside was the first to make such a determination (1937b) in the case of *lata*, one of de Vries' mutants. Herzog (1940) and Renner (1940b, 1943a, 1949) have added a few others, some of them de Vriesian mutants, from *lamarckiana* or its derivatives, some derived from other sources (Table 4). In certain cases, these determinations have been based primarily on cytological evidence, in others on a combination of cytological and genetic data. It is worth noting that one of these mutants, *tripus* (Renner, 1943a), presents the only known case where a trisomic involves the paired chromosomes of *lamarckiana*— where a single chromosome is present three times. *Tripus* is therefore a "primary trisomic" in the sense in which this term was applied to *Datura* by Blakeslee—the only such trisomic known in *Oenothera*. It did not appear in *lamarckiana* itself but in a cross between *lamarckiana subcruciata* and *biennis*.

As a result of the studies outlined above, the nature of most of de Vries' mutants is now well understood. They are not what he thought they were, and they do not have the profound evolutionary significance which he ascribed to them. They are nevertheless of real interest for the light that they throw on the genetic and phylogenetic mechanisms in *Oenothera*. *Nanella* and *brevistylis* are point mutations, recessive to wild type. *Gigas* is a tetraploid; *semi-gigas*, *perennis*, and *quadrata* are triploids. The great bulk of de Vriesian mutants are trisomics and owe their peculiarities to the extra chromosome or chromosome segments present. *Rubrinervis* and *erythrina* are half-mutants with a circle of 6 and 4 pairs.

Of the various types of mutant, the half-mutants are the most puzzling. The mechanisms by which other categories of mutant are formed are not difficult to visualize, but it is not easy to postulate the method by which these 14-chromosome aberrants have arisen. The half-mutants that have been analyzed segmentally, although they have arisen inde-

TABLE 4
Trisomics for Which the Extra Chromosome or Chromosome Ends Have Been Determined

Name of trisomic	Origin	Classification	Extra chromosome or chromosome ends	Author
tripus	*O.(lamarckiana subcruciata* × *biennis) velans.rubens**	dimorphic (isotrisomic of Renner)	1·2	Renner, 1943a, 1949
lata	de Vriesian mutant	dimorphic	5·6 (of *gaudens*)	Catcheside, 1937b
dependens	(= *pallescens* of de Vries) from *lamarckiana* and *M-lamarckiana*	dimorphic	3·11† (of *gaudens*)	Renner, 1943a
scintillans	de Vriesian mutant	dimorphic	13·10 (of *gaudens*)	Renner, 1943a, 1949
cana	de Vriesian mutant	dimorphic	3·4 (of *velans*)	Herzog, 1940
incana	*biennis* crosses	dimorphic	4·9 (of *rubens*)	Renner, 1949
macilenta	from *M-lamarckiana* of Renner via *dependens* and *cana*	dimorphic	3·2 (of *M-gaudens*)‡	Renner, 1943a, 1949
mira (*MM dependens*)	from *M-lamarckiana* and *M-lam. dependens*	sesquiplex (monomorphic)	3 and 11 (of *gaudens*)§	Renner, 1949
candicans (*mm cana*)	from *M-lamarckiana cana*	sesquiplex (monomorphic)	1 and 4 (of *velans*)	Herzog, 1940 Renner, 1949
glossa	from *albicans.velans* and other sources	sesquiplex (monomorphic)	2 and 3 (of *velans*)	Renner, 1943a, 1949
lonche	from (*biennis* × *hookeri*) *albicans.hhookeri*	sesquiplex (monomorphic)	3 and 13 (of *hookeri*)	Renner, 1949

* Since *rubens* has the same segmental arrangement as *gaudens*, *velans.rubens* = *lamarckiana* segmentally (⊙12, 1 pair).

† Renner has followed Catcheside's numbering system. This is chromosome 3·12 of Cleland.

‡ *M-gaudens* has 1·11 3·2 (= 1·12 3·2 of Cleland) instead of 1·2 3·11 (3·12). The 3·2 chromosome was derived from *flectens* (of *atrovirens*). *M-lamarckiana* has ⊙14, instead of the usual ⊙12, 1 pair.

§ The 11-end = 12 of Cleland. The 3-end is derived ultimately from *flectens*.

pendently in different strains of *lamarckiana*, have in several cases shown the same chromosome configuration (circle of 6) and identical segmental arrangements. *Rubrinervis* and *erythrina*, as well as the *rubrisepala* of Heribert-Nilsson, are identical in segmental arrangement. They all possess a segmentally unaltered *velans* and a lethal-free complex derived by translocation, which includes both *velans* (3·4 9·10 11·12) and *gaudens* (5·6) chromosomes plus two interchange chromosomes (7·14 13·8). The 1·2 chromosome which is common to *velans* and *gaudens* is also present. The interchange chromosomes could not have come about by a single interchange from *lamarckiana*. Two interchanges involving either three or four chromosomes would be necessary. It seems surprising that mutants requiring such a complicated process for their origin should appear on as many different occasions as they have, identical in chromosome configuration and segmental arrangement, as well as in hereditary behavior. What was considered by de Vries to be *rubrinervis* arose 66 times between 1890 and 1900 from *lamarckiana* (de Vries, 1919a). It has also appeared in the progeny of at least one trisomic mutant (*oblonga*). *Erythrina* arose at least 9 times in de Vries' cultures in five different families (de Vries, 1919b) descended from a different strain of *lamarckiana* from the one that produced *rubrinervis*. *Rubrisepala* arose several times in Heribert-Nilsson's Swedish *lamarckiana* (Håkansson, 1930a, p. 393). Thus, these mutants have arisen many times independently; and it is difficult to avoid the assumption that there is some structural basis that makes this particular structural alteration easier of accomplishment, in spite of its apparent complexity, than other possible alterations.

Darlington (1931) suggested that half-mutants may arise through crossovers between homologous segments in otherwise non-homologous chromosomes. Such a process would give two altered chromosomes but it would not give a complex capable of forming four pairs with *velans* and only two pairs with *gaudens*. Emerson (1935, 1936) adopted Darlington's explanation as did Catcheside (1936) who called attention, however (Catcheside, 1940), to the problem arising from the fact that the alethal complexes of half-mutants could not have arisen by a single interchange from the complexes of *lamarckiana*. Renner (1943c) suggested as a possible mechanism the frequent interlocking that takes place between chromosome pairs, or between circles and pairs. It is conceivable that, even within a single large circle, a chromosome could lie prior to synapsis in such a way that it would become interlocked with one or more of the other chromosomes after synapsis. If a chromosome became interlocked with two other chromosomes, shortening of the chromosomes might bring about stresses at the point where all three were interlocked; and, as a result, breakage and reunion of all three might occur simultaneously, giving rise to two chromo-

TABLE 5
Partial List of Mutants from *Oenothera lamarckiana*

Class of mutant	Name of mutant	Chromosome number	Chromosome configuration	Reference
Gene mutations	*nanella*	14	⊙ 12	de Vries, 1901b
	brevistylis	14	⊙ 12	Pohl, 1895; de Vries, 1901b
	funifolia	14	⊙ 12	Shull, 1921
	vetaurea	14	⊙ 12	Shull, 1921
	supplena	14	⊙ 12	Shull, 1925
	bullata	14	⊙ 12	Shull, 1928
	pollicata	14	⊙ 12	Shull, 1934
	flavescens	14	⊙ 12	Håkansson, 1926
	planifolia	14	⊙ 12	Håkansson, 1926
	angustissima	14	—	de Vries, 1929
	pustulata	14	—	de Vries, 1929
	stenophylla	14	—	de Vries, 1929
Half-mutants	*rubrinervis*	14	⊙ 6, 4 pairs	de Vries, 1901b, 1919a
	erythrina	14	⊙ 6, 4 pairs	de Vries, 1919b
	rubricalyx	14	⊙ 6, 4 pairs	Gates, 1911b, 1915
	rubricalyx, "Afterglow"	14	⊙ 8, 3 pairs	Cleland, 1925, 1931
	proxima	14	—	de Vries, 1929
Segregants from half-mutants	*blandina* from *pro-blandina*	14	7 pairs	de Vries, 1917
	deserens from *rubrinervis*	14	7 pairs	de Vries, 1917
	decipiens from *erythrina*	14	7 pairs	de Vries, 1917
	latifrons from *rubricalyx* "Afterglow"	14	7 pairs	Shull, 1923a
	retardata from *proxima*	14	—	de Vries, 1929
Tetraploids	*gigas*	28	variable	de Vries, 1900c, 1901b
Triploids	*semigigas*	21	variable	Stomps, 1912; de Vries and Boedijn, 1924
	perennis	21	variable	de Vries, 1923
	quadrata	21	variable	de Vries, 1929
	excelsa	21	variable	Håkansson, 1926

TABLE 5 (Continued)
Partial List of Mutants from Oenothera lamarckiana

Class of mutant	Name of mutant	Chromosome number	Chromosome configuration	Reference
Trisomics, dimorphic	cana	15	(see footnote†)	de Vries, 1916
	cucumis	15		de Vries, 1929
	lactuca	15*		de Vries, 1916
	lata	15		de Vries, 1900d, 1901b
	lingua	16		de Vries, 1929
	liquida	15		de Vries, 1916
	pallescens	15		de Vries, 1916
	pulla	15		de Vries, and Boedijn, 1924
	scintillans	15		de Vries, 1900d, 1091b
	spathulata	15		de Vries and Boedijn, 1923
Sesquiplex	albida	15		de Vries, 1900d, 1091b; de Vries, and Boedijn, 1923
	aneria	15*		de Vries, 1929
	auricula	15*		de Vries, and Boedijn, 1923
	aurita	15*		de Vries, and Boedijn, 1923
	candicans	15		de Vries, and Boedijn, 1923
	compacta	15*		de Vries, and Boedijn, 1923
	crinita	15*		de Vries, 1929
	delata	15*		de Vries and Boedijn, 1923
	diluta	15*		de Vries and Boedijn, 1923
	distans	15*		de Vries and Boedijn, 1923
	elongata	15*		de Vries, and Boedijn, 1923
	flava	15*		de Vries and Boedijn, 1923
	hamata	15		de Vries and Boedijn, 1924
	militaris	15*		de Vries and Boedijn, 1923

TABLE 5 (*Continued*)

Class of mutant	Name of mutant	Chromosome number	Chromosome configuration	Reference
	nitens	15*		de Vries and Boedijn, 1923
	oblonga	15		de Vries, 1900d de Vries and Boedijn, 1923
	opaca	15		de Vries, 1929
	persicaria	15*		de Vries, 1929
	planifolia	15*		de Vries, 1929
	semilata	15*		de Vries and Boedijn, 1923
	tardescens	15*		de Vries, 1924, 1925a
	venusta	15*		de Vries and Boedijn, 1923

* Chromosome number not reported. Classification based on genetic behavior.
† Chromosome configurations of trisomics variable where studied. See Håkansson (1926) and Emerson (1936).

somes with completely new associations. Such an event could produce the segmental arrangement found in the alethal complex of a half-mutant.

The question may be raised, however, as to why the interlocking should involve the same chromosomes in most cases. The fact that most half-mutants have a circle of 6 and 4 pairs (a few have ⊙ 8, 3 pairs, or ⊙ 4, 5 pairs) may be the result of limitations which reduce drastically the number of exchanges that will produce viable gametes. Many exchanges of the type just described are theoretically possible, but most of them may result in inviable gametes. The location of lethals is no doubt an important factor. An alteration that causes a lethal-bearing chromosome or segment to be present in double dose will result in inviable offspring. It may well be, as Renner has pointed out, that the large number of empty pollen grains found in all races that possess large circles may result from the many deviations from normality, including such exchanges as here envisaged. Only a few such exchanges may be capable of producing functional gametes, and these may give rise to the so-called half-mutants with their unaltered *velans* or *gaudens* complex combined with an alethal, segmentally altered complex.

De Vries' so-called mutants, therefore, have proved to be, in a few cases point mutations, in most cases plants with altered chromosome numbers, and in some cases the result of exchange of segments among the chromosomes of the ring of 12. In no case can they be regarded as the incipient species that de Vries thought them to be.

Table 5 lists some of the mutants of *lamarckiana* and indicates the category to which each belongs.

XIV. Induced Mutations

The foregoing discussion has involved only spontaneous alterations in the genes or in chromosome structure or number. In recent years, however, successful attempts have been made to induce gene or chromosome mutations in *Oenothera*. Most of this work has been done by Oehlkers and his students with a view to elucidating various aspects of the physiology of meiosis.

The first attempt to induce mutations in *Oenothera* was that of Michaelis (1930) who subjected *hookeri* plants to cold treatment and used pollen from cells undergoing meiosis at the time of treatment to pollinate untreated plants. He obtained $2n - 1$, $2n + 1$, and $2n + 2$ plants. The first to attempt irradiation was Brittingham (1931), who used unfiltered radon tubes on inflorescences of *franciscana* and *lamarckiana* and obtained a great variety of abnormalities in F_1 rosettes, which subsequently succumbed except in one case where selfing was accomplished and abnormal F_2 plants were obtained. This work was not followed up. Catcheside (1935, 1937a, 1939) later experimented with X-rays, irradiating *blandina* pollen and placing this on untreated *blandina* stigmas. In F_1 he obtained, in addition to normal fertile plants, some aberrants with 50% bad pollen (three of them with ⊙ 4), and some normal looking plants with 50% sterile pollen. Among the latter were seven plants with changed chromosome configurations (five plants with ⊙ 4, one with ⊙ 6, one with ⊙ 4, ⊙ 4, as opposed to the 7 pairs normally found in *blandina*). In five of these seven plants, the interchanges were essentially equal, but in two plants they were unequal. Only non-interchange gametes were transmitted to F_2 by most of the interchange hybrids, suggesting that the interchange complements were defective, probably as a result of deletions. F_2 cultures from normal F_1 plants yielded several morphological aberrants. One of the translocations resulted in the position effect discussed above.

Rudloff and Stubbe (1935) irradiated pollen of *hookeri* and used it on untreated *hookeri* plants. They obtained many aberrants in F_1 and F_2. Detailed descriptions of twenty-four F_2 variations, considered to be gene mutations, were given, these being the first cases of mutation observed in *hookeri*, a strictly homozygous race. No cytological studies were made, but the presence of normal pollen, in normal amounts, suggested that chromosome abnormalities were not involved. Marquardt (1948) irradiated pollen of *O. hookeri* (7 pairs) with 4500–5500 r of X-rays and obtained two plants in each of which there were 2 circles of 4. Analysis of the behavior of the circles showed more tendency toward failure of chiasmata in rings than in pairs, and the existence of a small percentage of what were interpreted as interstitial chiasmata between ring chromosomes. Both plants as a result had greatly reduced seed fertility.

That other agents can be effective in producing structural alterations in the chromosomes was shown by Oehlkers and his students. In 1935, Oehlkers showed that in a hybrid between *suaveolens* and *hookeri* with ⊙ 4, 5 pairs, exposure to both low and high temperature during meiosis caused a tendency for the chromosomes in the circle to fall apart. Together with his students (Oehlkers, 1936; Zürn, 1937a,b; Haselwarter, 1937; Kisch, 1937) he showed that chiasma frequency, and hence the ability of the circles to remain intact, was influenced, not only by temperature, but also by osmotic relations and by the proportion of chlorophyll and carotinoids in the plastids. In 1943, Oehlkers announced that various chemicals are able to bring about breakage, resulting in translocations and consequent alteration in chromosome configuration. Using *suaveolens* × *hookeri* (*flavens.^{h}hookeri*) or its reciprocal (*^{h}hookeri.flavens*) with ⊙ 4, 5 pairs, he found that certain inorganic salts gave significant results, KCl giving 2.5% and KNO_3 4.5% of cells with altered configurations. Certain organic substances were also effective, but the most striking results were obtained with certain combinations of inorganic and organic compounds, especially with ethyl urethane (1/20 M) and KCl (1/200 M). This combination gave 38% of cells with altered configurations in a total of 100 cells. The following new chromosome configurations were found:

Configurations representing a single translocation: ⊙ 6; ⊙ 4, ⊙ 4.

Configurations representing two translocations: chain of 8; ⊙ 6, ⊙ 4; ⊙ 4, ⊙ 4, chain of 4 (= ⊙ 4, ⊙ 4, ⊙ 4, 1 pair).

Configurations representing three or four translocations: chain of 10 (= ⊙ 10, 2 pairs); chains of 8 and 4 (= ⊙ 12, 1 pair, or ⊙ 8, ⊙ 4, 1 pair).

These results compare qualitatively and quantitatively with the results obtained by Marquardt, using X-rays. As in the case of X-rays, the results are random and unspecific, and the magnitude of the effect is comparable with the effect of X-rays. However, Oehlkers and Linnert (1951) showed that the cytoplasm may also exercise an influence on the magnitude of the effect in the case of urethane + KCl, which gives twice as powerful an effect when *flavens* and *^{h}hookeri* are present in *suaveolens* plasma as when they are in *hookeri* plasma (23% of cells with altered configurations vs. 10%).

While some of the chromosome configurations figured in these papers may be open to other interpretations than those given, there can be no doubt that the effect of certain chemicals on chromosome breakage is strong and that it differs greatly with different substances.

In later experiments (Oehlkers, 1949), other substances were tested. Certain narcotics, especially acetophenone, glycol, and acetanilide, yielded high percentages of translocation (28–36% of cells). Certain alkaloids, including morphine (14–30%) and colchicine (21%) were also very effective. Urethane is the only substance, however, which shows a cytoplasmic influence, giving results of a different order of magnitude in reciprocal hybrids. In all cases, the reactions have appeared to be non-specific. The effect has been relatively independent of concentration and length of exposure and seems to be the indirect result of general cellular disturbances. The most susceptible stage appears to be the metabolic stage, possibly including early prophase. On the whole, these chemicals give results comparable with those obtained with X-rays.

The amount of work done on the induction of mutations in *Oenothera* is as yet small, and the attempts which have been made have been more effective in bringing about translocations than any other type of change. They have been of greatest interest in demonstrating the effectiveness of certain chemicals, especially mitotic poisons, such as urethane, in producing structural alterations in the chromosomes.

XV. Incompatibility

The first case of self-incompatibility of the *Nicotiana* type in the genus *Oenothera* was found by Emerson (1938) in *O. organensis*, an outlying member of the subgenus *Euoenothera*. According to East (1940) Emerson also found self-incompatibility in *O.* (*Anogra*) *californica*, *O.* (*Pachylophis*) *caespitosa* var. *marginata* and *O.* (*Lavauxia*) *acaulis*. Gates in 1939 reported this condition in *O.* (*Megapterium*) *missouriensis*. In 1944, Hecht found self-sterility in *O.* (*Anogra*) *latifolia*, *pallida*, *runcinata*, and *trichocalyx*, as well as in *O.* (*Raimannia*) *heterophylla* and *rhombipetala*. Hagen (1950) confirmed this condition in *O. trichocalyx*, *missouriensis*, and *caespitosa* var. *marginata* (also var. *montana*). He failed to find it, however, in the strain of *O. acaulis* which he studied. Crowe (1955) did not find self-incompatibility in either *O.* (*Lavauxia*) *acaulis* or *O.* (*Anogra*) *trichocalyx*. Hagen added to the list of self-incompatible forms *O.* (*Hartmannia*) *speciosa*, *O.* (*Salpingia*) *Greggii* and *Hartwegi*, and *O.* (*Sphaerostigma*) *bistorta* var. *Veitchiana*. Self-incompatibility, therefore, has been found in nine of the subgenera of *Oenothera*.

In these groups, self-incompatibility is always associated with the presence of small circles or none. Furthermore, wherever they have been investigated, it has been shown that the numbers of S-factors in a population is very large (Emerson, 1940; Cleland, 1960). When many alleles are present, a very efficient system is set up for preventing inbreeding and maintaining hybrid vigor. If the numbers of S-factors in the population were not large, the chances of successful outcrossing would be so low that the necessity of having to depend upon pollen from other plants would constitute a major hazard. This system contrasts strongly with the system of balanced lethals in large circles, so characteristic of *Euoenothera*, which also maintains maximum hybrid vigor, but without the uncertainty arising from cross-pollination, although at the cost of partial sterility resulting from the lethals, which, however, is more than compensated for by the self-pollinating habit.

Within the subgenus *Euoenothera*, S-factors have been known until recently only in *O. organensis*. This species is an atypical member of the North American Euoenotheras. Phenotypically it is closer to *Raimannia* and the South American Euoenotheras than it is to the North American branch of this subgenus. It does not cross readily with other Euoenotheras, or with *Raimannia*. Emerson (1938) found it cross-sterile with five species of *Raimannia*. Hecht found it cross-sterile with *O. (Raimannia) rhombipetala*, but obtained viable seeds with a strain of *longiflora* (also a *Raimannia*), which produced yellow seedlings (Hecht, 1950). Crosses with the other North American Euoenotheras rarely succeed, and hybrids when obtained have been completely sterile. Since the other North American Euoenotheras are readily crossable among themselves, it is evident that *organensis* is not a typical member of the subgenus, being classified as a *Euoenothera* only because of its seed morphology. It is unlike the other Euoenotheras also in having balanced S-factors.

The *organensis* population when first studied consisted of fewer than 1000 plants, growing in the Organ Mountains of southern New Mexico. The population has since dwindled to a very few individuals. The plants are perennial. Emerson (1940) found about 45 alleles of the S-gene. The reaction of the pollen is gametophytically controlled, i.e., is controlled by the S-allele in the pollen grain itself. The other S-factor has no effect. Incompatibility is brought about by reaction between pollen tube and stylar tissue. Some S-factors permit the tubes to grow farther down the style than others. Incompatible tubes are normal in morphology, but differ from compatible tubes only in their inability to grow in a given style.

Emerson (1941) found associated with S_{20} a pollen lethal that prevented tubes from growing in any style. The two loci were closely linked with a recombination value of 0.3–0.5%.

Lewis studied the mutability of the S locus in *organensis*, using as evidence of mutation the appearance of compatible tubes in the style, or seed set in capsules following self- or cross-incompatible pollination. The rate of spontaneous mutation (Lewis, 1948, 1949) was very low—an average of 1.87 per million pollen grains for mutations occurring at post-pachyphase stages, 0.60 per million for those taking place prior to pachyphase. The frequency of induced mutations, although higher, was still low. From data published in 1951 it appears that some alleles have shown no mutability whatever, and the figures for others have been low. Thus (Lewis, 1951, p. 407, Table 6) S_2 and S_3 showed no mutability, either spontaneous or following 500–700 r of X-rays: the mutation rates for S_6 were (induced) 9.5, and (spontaneous) 0.20 per million genes; for S_4 the corresponding figures were 7.3 and 0.9.

Mutations occurring at post-pachyphase stages resulted in single pollen tubes in the style or single seeds in the ovary. Those occurring 5 days before metaphase I gave rise to pairs of tubes or to two seeds (more if mutations were produced in pre-meiotic cell cycles). Apparently, replication of the chromosomes takes place about 5 days prior to metaphase I, after which a mutation will affect only one chromatid and will result in a single mutated pollen grain. A mutation occurring prior to replication will be found in two chromatids, and because pollen grains from the same and neighboring sporocytes are held together by viscid strands, the two mutant pollen grains will probably find themselves close together on the stigma and their tubes will grow in close association.

Capsules with single seeds ordinarily fail to develop in *Oenothera*, but spraying ovaries with 160 ppm β-naphthoxyacetic acid 24 hours after pollination stimulated the ovaries to grow even when they contained only a single seed.

Certain additional facts brought out by Lewis should be mentioned:

(1) No mutants have been found in which the mutated gene has retained an active and specific incompatibility reaction in the pollen. Mutated *S*-alleles cease to function as incompatibility factors, so far as the pollen is concerned. They are, however, wholly unchanged so far as their operation in the style is concerned, retaining their original specificity of reaction. This suggests that the *S*-gene consists of at least two parts, one operating in the pollen, the other in the style. Attempts to detect crossing-over between the two have failed. None of the mutant alleles has affected deleteriously the vigor or fertility of the plant, even when in homozygous condition, which condition becomes possible when the mutant gene loses its power to function in the pollen.

(2) Mutations proved to be of two kinds, permanent and revertible (Lewis, 1951). Among induced mutants, about half were revertible and most of the spontaneous mutants were of this type. A revertible mutation

is one which converts a stable allele into an unstable one, which soon reverts to the normal.

The rarity of mutations at the S locus, coupled with the fact that no cases have been found where new functional incompatibility factors have been produced as a result of mutation, poses a problem in regard to the origin of the many alleles found in nature. If mutations occur with extreme rarity it is difficult to account for the extraordinarily large number of alleles that exist in the population. Possibly factors as yet unrecognized exist that can bring about mutations to new functional sterility alleles with relative ease. Without the presence of numerous alleles in the population an S-factor situation would be highly deleterious. It will be of interest to try to discover an agent capable of bringing about, not merely the inactivation of an allele in the pollen, but an actual transformation of one allele into another which is equally functional and equally specific in its reactivity.

Although balanced S-factors are unknown in the rest of the Euoenotheras, the work of Steiner (1956, 1961) suggests that what were originally S-factors may still exist in the population in an unbalanced condition and serve as one form of pollen lethal. Steiner has found that when the alpha (egg) complexes of two races belonging to *"biennis I"* (the midwestern group of races possessing the so-called *biennis* characters), are brought together, the resulting alpha.alpha plants are self-sterile, although in combination with beta (sperm) complexes they produce fertile plants. Steiner postulates that the present *biennis I* races are descended from hybridizations that occurred between individuals of two originally isolated populations, one of which possessed a balanced S-factor situation, the other being free of incompatibility factors. Because different histories of interchange had been experienced in the two populations, some of the hybrids between them had ⊙ 14. Such a hybrid would have received from one of its parents, and only one, an S-factor. Since the styles of this hybrid would contain the same S-factor, the half of the pollen grains receiving this factor would be unable to develop pollen tubes; hence, the S-factor would become a male gametophytic lethal. If the complex possessing the S-factor was able to win out in the competition for the production of embryo sacs, and thus to suppress the other complex, a balanced lethal situation such as characterizes complex-heterozygotes would be brought into being at once. Steiner, therefore, visualizes that the *biennis I* group of complex-heterozygotes became established as the result of crosses between a *biennis*-like population carrying an S-factor and a *strigosa*-like population, some individuals of which, at least, did not possess incompatibility factors.

Later work (Steiner, 1961) points out that the situation is somewhat complicated by the presence in some races of pollen lethals of a different sort—lethals that prevent a complex from functioning in the male gametophyte, no matter what kind of style receives the pollen. Such pollen lethals will mask the presence of an associated S-factor. In alpha. alpha combinations that have 7 pairs of chromosomes, however, it should be possible, unless they are linked, to separate them and demonstrate the S-factor if it is present.

Steiner has demonstrated that S-factors are present in at least two races of the *strigosa* group, and one *biennis II* race, as well as in *biennis I*. In many other races, the probability is that S-factors are present but masked by pollen lethals.

XVI. Plastid Behavior

The role of the plastids in the determination and transmission of hereditary traits has long been a center of interest in the case of *Oenothera*. De Vries noted as early as 1913 (p. 131) that reciprocal crosses between *lamarckiana* and *hookeri* differ in respect to plastid behavior. With *hookeri* as mother, both classes of progeny (*hookeri-velutina* = ^{h}*hookeri.velans* of Renner and *hookeri-laeta* = ^{h}*hookeri.gaudens*) are green; with *lamarckiana* as mother, the *laetae* are green, the *velutinae* yellowish and weak. Later generations of the cross *hookeri* × *lamarckiana* are wholly green, but the *laeta* from *lamarckiana* × *hookeri* splits off yellowish "*velutinae.*" In terms of Renner's complex concept the situation is summarized in the tabulation.

Cross	F_1	F_2
hookeri × *lamarckiana*	^{h}*hookeri.gaudens* (green)	^{h}*hookeri.gaudens* (green)
		^{h}*hookeri*.^{h}*hookeri* (green)
	^{h}*hookeri.velans* (green)	^{h}*hookeri.velans* (green)
		^{h}*hookeri*.^{h}*hookeri* (green)
lamarckiana × *hookeri*	*gaudens*.^{h}*hookeri* (green)	*gaudens*.^{h}*hookeri* (green)
		^{h}*hookeri*.^{h}*hookeri* (yellow)
	velans.^{h}*hookeri* (yellow)	*velans*.^{h}*hookeri* (yellow)
		^{h}*hookeri*.^{h}*hookeri* (yellow)

In 1924, Renner showed that about 15% of the *velans.^{h}hookeri* hybrids derived (as shown above) from *lamarckiana* × *hookeri* have green flecks or sectors. If a flower from a green sector is selfed, the progeny are green; if one from a yellow area is selfed, the progeny are yellow. He postulated that plastids may occasionally be brought into the zygote by the pollen tube. Plastids brought in from *hookeri* as the male parent will form green sectors, since *hookeri* plastids are able to function in association with *velans* and *ʰhookeri* genomes whereas *lamarckiana* plastids are not. He also found that yellow spots are often present in the reciprocal cross, *hookeri* × *lamarckiana* (*ʰhookeri.velans*). These represent islands of *lamarckiana* plastids brought into the hybrid through the sperm.

Renner found the same behavior in crosses between other races. From this he reached the general conclusion that the plastids of one race may differ in quality from those of another race—that different classes of plastids exist in the genus as a whole. Thus the plastids of *lamarckiana* and *biennis* differ, inasmuch as *biennis* plastids will function in the genomic combination *ʰhookeri.ʰhookeri*, whereas *lamarckiana* plastids are yellow; *biennis* and *muricata* plastids differ since the former are colorless in the presence of *rubens.curvans*, while the latter are green. *Biennis* and *suaveolens*, however, have the same kind of plastids since they behave alike in all complex-combinations.

The differences observed in the various races must have come about originally through mutation, but the changes in behavior observed during the course of such experiments as these are not the result of mutation; for instance, the *lamarckiana* plastids which become yellow in *velans.ʰhookeri* have not been altered genetically by being brought into association with this particular genome combination. Renner emphasized this point in numerous papers, basing his conclusion on the fact that if plastids that are deleteriously affected by their genetic milieu are transferred to compatible genic environments, they again function normally. For instance, (Renner, 1937a) *cruciata* (*pingens.flectens*) × *biennis* (*albicans.rubens*) gives *pingens.rubens* with green *cruciata* plastids, but containing yellow sectors that have *biennis* plastids. If a flower on a yellow sector is pollinated by *albicans.velans*, the progeny will be partly *pingens.velans*, partly *rubens.velans*, both with predominantly *biennis* plastids. These are, however, green in both combinations. *Biennis* plastids that were unable to function in *pingens.rubens* are now able once more to turn green.

That the change in behavior of the plastids is the result of differences inherent in the plastids themselves, and not ascribable to the cytoplasm, was shown by Renner (1936) in a variety of crosses. For example, as seen in Table 6, *suaveolens*, *lamarckiana*, *hookeri*, and *biennis* plastids are yellow in both of the reciprocal crosses with *muricata*, whether they are in their own or foreign cytoplasm. *Muricata* plastids are green in all of these crosses, whether they are in their own or foreign cytoplasm. It is not the cytoplasm that determines whether plastids will be yellow or green; it is the nature of the plastids themselves that determines whether they can function in association with particular combinations of genomes.

Most cases of failure of plastids to turn green in *Oenothera* are the result of incompatibility between a given class of plastids and a specific combination of gene complexes. In the wild, such incompatibilities can only arise through outcrossing between races, which rarely occurs; but in artificially produced hybrids between races, such incompatibilities are frequently encountered.

TABLE 6
Plastid Behavior in Certain *Oenothera* Crosses

Cross	Plastids		Cytoplasm
	Source	Color	
suaveolens × *muricata* (*flavens.curvans*)	*suaveolens*	yellow	*suaveolens*
	muricata	green	
muricata × *suaveolens* (*curvans.flavens*)	*muricata*	green	*muricata*
	suaveolens	yellow	
lamarckiana × *muricata* (*gaudens.curvans*)	*lamarckiana*	yellow	*lamarckiana*
	muricata	green	
muricata × *lamarckiana* (*curvans.gaudens*)	*muricata*	green	*muricata*
	lamarckiana	yellow	
hookeri × *muricata* (*ʰhookeri.curvans*)	*hookeri*	yellow	*hookeri*
	muricata	green	
muricata × *hookeri* (*curvans.ʰhookeri*)	*muricata*	green	*muricata*
	hookeri	yellow	
biennis × *muricata* (*rubens.curvans*)	*biennis*	yellow	*biennis*
	muricata	green	
muricata × *biennis* (*curvans.rubens*)	*muricata*	green	*muricata*
	biennis	yellow	

There are a few cases where specific gene loci are known to be responsible for failure or partial failure of plastid function. For example, Renner (1943b) reported the case of the gene *lor*, belonging to the complex *flectens*, which, when associated in homozygous condition with *lamarckiana* plastids, prevents the latter from becoming green. If *lor lor* is associated with *Fl Fl* in the presence of foreign plastids, seed germination is prevented. Another case is reported by Stubbe (1955) in *suaveolens*, where a recessive gene causes the bases of the leaves in the young rosette to become blanched. This character, known as "Weissherz," is the only case reported where a single locus in *Oenothera* is known to produce within a race a chlorophyll defect. The chlorosis in this case disappears with age, and Weissherz plants are indistinguishable at maturity from normal plants.

In addition to these cases, several instances are known where plastids have mutated spontaneously to conditions where they are incapable of developing chlorophyll in any gene environment to which they have been subjected. The first six cases were reported by Renner in 1936. That these were true plastid mutations rather than the results of gene mutation was shown by the fact that each was cytoplasmically inherited: green × white yielded green (at the most, with white flecks derived from the pollen), white × green yielded white (with occasional green flecks). Furthermore, these mutant plastids proved incapable of becoming green, no matter what genes were associated with them. Renner reported that plastid mutations occur in a frequency of about 5 in 10,000 plants.

In 1954 Schötz published a study involving an additional plastid mutant. This study is interesting because it was possible to use this mutant as a tool with which to analyze the relative speed of multiplication of the plastids in different races. This particular mutant arose in a line of *hookeri* descended from the cross *biennis* × *hookeri*, in which the plastids were derived from *biennis*. A single individual in this strain in 1946 was mottled with white, and crosses made to other races using flowers on white sectors of this plant showed the white areas to contain a mutant plastid incapable in every complex-combination tested of becoming green. These crosses also revealed that the plastids of different races differed from each other in their ability or lack of ability to surpass the mutant plastids in rate of multiplication. The mutant plastids thus became a yardstick against which the speeds of reproduction of various plastids could be measured and compared. That the differing speeds of multiplication were primarily the result of innate qualities of the plastids themselves and were not due to the genomes present could be shown in specific cases. For instance, *albicans.^{h}hookeri* derived from *suaveolens* × the mutant *hookeri* could be compared with the *albicans.^{h}hookeri* obtained from *biennis* × the mutant; the plastids were derived from *suaveolens* in the first cross, from *biennis* in the second, and it was found that the two kinds of plastids differed from each other in their ability to compete with the mutant plastids. Since the two hybrids were identical in their genomes, the difference must have been due largely to the plastids themselves. There was evidence, however, that genic influence is not entirely lacking in determining speed of multiplication, for in certain crosses where twin hybrids were produced, slight differences were recognized in the rate of plastid multiplication, as for instance in crosses involving *lamarckiana* and the mutant *hookeri*, where *velans.^{h}hookeri* and *gaudens.^{h}hookeri* differed somewhat; or crosses involving *rubricaulis*, where *rubens.^{h}hookeri* and *pingens.^{h}hookeri* showed minor differences.

Schötz placed the various races into categories, according to the percentage of white surface that they showed when crossed as female with *hookeri* containing the mutated plastid. A hybrid in which the mutated plastids multiplied rapidly in comparison with their green associates would show a relatively large area of white surface, and vice versa. The races studied fell into the following sequence when the mutant *hookeri* was used as male parent: *blandina* and *hookeri* (0% white surface)–*Bauri–lamarckiana–rubricaulis–suaveolens–biennis–syrticola–parviflora–rubricuspis–ammophila–atrovirens–pingens.rubens–pingens.rubens cruciata*. *Hookeri* and *blandina* have the most rapidly proliferating plastids, completely overgrowing the mutant plastids brought in with the sperm. *Pingens.rubens cruciata* plastids, at the other extreme, grew so slowly that approximately 20% of the surface of the hybrid was white.

When the mutant *hookeri* was used as female parent, the progeny was composed of plants that were wholly or predominantly white, but green plastids from some of the races were able to proliferate in the seedlings. The order was somewhat different from that obtained in the reciprocal crosses. *Lamarckiana* plastids were the most successful, coming to occupy about 20% of the area of the seedlings. These were followed in descending order by *blandina, rubricaulis, hookeri, Bauri, biennis, suaveolens*, and a group in which green plastids never appeared; namely, *pingens.rubens, parviflora, syrticola, ammophila*, and *atrovirens*. Various lines of evidence ruled out the possibility that these differences are the result of differences in the number of proplastids introduced by sperm from the various races, or differing abilities of eggs to resist the intrusion of male cytoplasm.

The behavior of plastids in the seedling and later stages of development has been studied especially by Stubbe (1955, 1957, 1959) and Schötz (1954, 1958a,b). These authors have shown that cells containing mixtures of plastids derived from the two parents occur only in early stages of development. At a relatively early stage the two classes of plastid become segregated, giving rise to various types of chimeras in the developing plant. Cells with mixtures of plastids are rarely found in mature plants. A cell does not divide until it has a full complement of plastids (Schötz, 1954), consequently cells receiving the more slowly proliferating class of plastids will divide more slowly than those receiving the other class. It is not always the green plastids that outgrow the yellow or colorless ones, however. In certain cases (*atrovirens* × *biennis*) *pingens. rubens*, for instance, the yellow *biennis* plastids multiply faster than the green *atrovirens* plastids.

Schötz (1958a,b) has analyzed the manner in which plastid defects appear. Whether the defects are the result of plastid mutation or the presence of incompatible genomes, they do not consist in the loss of ability to manufacture chlorophyll, but rather result from excessive photo-oxidation of pigments. Plastids that show loss of chlorophyll tend to pass through three stages, first a greening stage, then a bleaching stage, and later a second greening stage. The plastids of different races

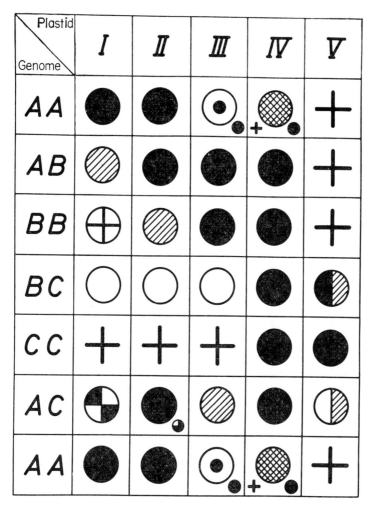

FIG. 15. Chart from Stubbe (1959). He recognizes five classes of plastid, listed at the top, from the standpoint of their reaction to various genome combinations. He also recognizes three categories of genome based upon their effect on plastids. The various combinations of these are listed at the left. Behavior of the different classes of plastid in the presence of the various combinations of genome categories is shown graphically. Reprinted from *Z. Vererbungslehre* by permission.

differ in regard to the time of onset and duration of these stages. In some, bleaching does not begin until a week or two after germination, in others it may begin before or at germination. The bleaching period may last from a few days to several weeks, only those plants surviving that have received enough green plastids through the sperm to maintain life. The second greening period may be suppressed in some combinations; in others, the plants recover completely and flower normally. In many cases, hybrids that become bleached and die in full sunlight can be brought to maturity if grown in dim light. Schötz recognized six "bleaching types" among the many hybrid combinations studied.

The question as to whether the differences in structure and behavior displayed by the various classes of plastids are due to the influence of the cytoplasm in which they lie, or whether they are innate, has been further discussed by Stubbe (1957). He studied early stages in the development of hybrid plants at a time when mixtures of the different kinds of plastids still exist, and found that easily recognizable and distinct types of plastids could be distinguished under the microscope. In one case, he was able to obtain a hybrid with three different types of plastid, all identifiable within the same cell. The fact that all three could exist side by side in the same cytoplasm without losing their distinctive characteristics indicates that these characteristics are not determined by the cytoplasm but are inherent in the plastids themselves. This supports the earlier conclusions of Renner (1936) who showed that whether a given type of plastid is in its normal cytoplasm or in foreign cytoplasm does not influence its ability or inability to become green. This is determined by the genome combination with which it happens to be associated.

The work of Renner which showed the existence of several genetically

TABLE 7
Classes of Plastid and Genome in the North American Euoenotheras

Group or superspecies	Genome type	Plastid type
hookeri	A_2A_2	I
elata	AA ?	?
strigosa	A_1A_1	I
biennis I	BA	III
biennis II	AB	II
biennis III	BB	III
grandiflora	BB	III ?
parviflora I	BC	IV
parviflora II	AC	IV
argillicola	CC	V

distinct classes of plastid has been followed up especially by Stubbe (1959) who, on the basis of a study of more than 400 genome combinations involving fourteen distinct races, has recognized in the North American Euoenotheras and their European derivatives five distinct classes of plastid. Each of the groups of races or "superspecies," to use Stubbe's terminology, which the writer has found in North America, is characterized by the presence of plastids belonging to one of these classes. He also finds that the genomes in this assemblage fall into three classes with respect to the effect that combinations of them have on the five classes of plastids. These findings are summarized in Table 7, and Fig. 15, taken from his 1959 paper.

XVII. Evolutionary Considerations

The North American Euoenotheras have been a very troublesome group to the taxonomist. On the one hand, they show a great variety of phenotypic expression. On the other hand, these variations often grade into one another so insensibly that the taxonomist finds it very difficult to discover clear-cut boundary lines separating one taxon from another. The reason for this is to be found in the peculiarities of *Oenothera* evolution, which has produced throughout much of its range a unique type of population structure. It will be of interest to discover what these unique features of population structure are, how they evolved, and what the possibilities may be for future evolutionary development in the group.

Because the plants found in northern Mexico and California have mostly paired chromosomes, lack lethals, and have large open-pollinated flowers, there is nothing very unusual about their population structure.

As one moves away from this geographical area, however, one finds that the situation begins to change. In Arizona, Nevada, Utah, New Mexico, etc., the Euoenotheras, while still retaining large open-pollinated flowers and an absence of lethals, have developed greater heterogeneity in segmental arrangement, so that several chromosome configurations will be found in a local population, mostly intermediate between large circles and the all-pairing condition. Plants which possess one or more circles do not ordinarily breed true for these circles, since they lack, as a rule, lethal factors. Thus, a plant with ⊙ 4, ⊙ 6, 2 pairs, may segregate plants with ⊙ 4, 5 pairs or ⊙ 6, 4 pairs, or 7 pairs. Since outcrossing is the rule, however, crosses between plants whose gametes have diverse segmental arrangements are constantly producing plants with circles, so that a balance tends to be maintained between plants with and without circles.

Farther east, these races are supplanted by forms with ⊙ 14, balanced lethals, and small self-pollinating flowers. From the Rocky Mountains to the Atlantic, practically all plants show the concurrent presence of these three characteristics. As a result, this portion of the range has acquired a very peculiar population structure. The large circle and balanced lethals mean that each plant will breed true when self-pollinated. The self-pollinating habit reduces outcrossings to very small proportions. As a result of these conditions, innumerable true-breeding lines are set up, within any one of which all individuals of all generations tend to be genetically identical; reproductive barriers to the flow of genes from one such line to another are also set up. Mutations occurring in a given line tend to be imprisoned within this line. If recessive, they get no chance to express themselves in the permanently heterozygous strain. If dominant, they are still confined within a single line of descent.

Occasional outcrosses may occur, however, when for some reason a flower fails to develop pollen. Thus, hybrids will occasionally, though rarely, be produced. Some of these will have configurations other than ⊙ 14 and will tend to be weeded out because of their failure to breed true and because maximum heterosis will not be maintained. Other hybrids, however, will have a circle of 14, balanced lethals (one from each parent), and self-pollination, in which case a new line will be established with a new combination of genomes. By such a process, the number of more or less isolated lines has undoubtedly increased in the past, and is still increasing, each with its own more or less distinct phenotypic and genotypic characteristics. This process has led to a curiously anomalous situation with regard to variability. No genus shows a greater tendency than *Oenothera* toward absolute uniformity within a line, yet few genera show as great a variability within the population as a whole. The variability is due to the great number of lines rather than to any tendency toward segregation within a line.

The reviewer and students have attempted to determine the nature and degree of the relationships existing between these various lines, and to elucidate the process by which the present situation has evolved. The key to the solution of this problem came from the work of Hoeppener and Renner (1928), who presented the results of a genetic analysis of a number of complexes derived from various races of *Euoenothera*. Some of these complexes were found to be closely related to each other genetically, while others showed little or no evidence of close relationship. On comparing their findings with the results obtained by examination of the chromosome configurations in hybrids combining these various complexes, the present author showed that those complexes which Hoeppener and Renner found to be most closely related genetically have very similar segmental arrangements and give mostly pairs or small circles when combined; those which are more distantly related are more diverse in segmental arrangement. This, together with other lines of evidence, seemed to suggest that similarity in segmental arrangement between complexes can be taken in most cases as evidence of recent common ancestry and therefore of close phylogenetic relationship. Certain exceptional cases were pointed out where this correlation did not hold strictly (Cleland, 1942), but on the whole the correlation was so strong that it seemed to constitute a valid basis for such a hypothesis. Using this assumption as a working hypothesis, therefore, an extensive analysis of the North American Euoenotheras has been undertaken from the standpoint of the segmental arrangement and genotypic composition of the complexes of the various races or lines. The findings from such a study, considered in relation to geographical distribution, should give us an understanding of the extent to which complexes resemble or differ from one another, and the degree to which they are found to fall into cytogenetic and geographical groups. It may thus be possible to determine whether there exist in nature larger groupings of the isolated lines which make up the population, groupings to which the designation "species" may be validly assigned. If such groupings are found to exist, then the question may be asked as to how they came into existence.

Approximately 360 complexes from North American material have been analyzed to date in the sense that their segmental arrangements have been completely determined, and their chief genetic characteristics ascertained. To describe the methods by which segmental arrangements are analyzed and genetic characteristics determined would take more space than can be allowed in this review. The reader is referred to Cleland and Hammond (1950) for a detailed account of methods. The races of which these complexes are a part have been found to fall into certain definite groupings, each with its distinctive cytogenetic and geographic limitations, some easily distinguished phenotypically from other groups, others not so easily distinguished. These groups are as follows:

1. *The hookeris* are found in northern Mexico, and the southwestern part of the United States. In Mexico and California they lack circles, lethals, and self pollination; but in the eastern portion of the range they have developed circles of intermediate size, and in a few cases lethals are present. The commonest arrangement of end segments in California (the so-called "Johansen" arrangement) is considered the ancestral arrangement in the North American Euoenotheras.

2. *The elatas* are found from Mexico southward. They show some resemblance to the hookeris in phenotype and agree with them in having open-pollinated flowers and paired, alethal chromosomes. One of them (*"Guatemala"*) has formed robust hybrids with some members of the subgenus *Raimannia*. Segmentally they are two to three interchanges

away from the hookeris and their arrangements bear little relationship to the arrangements found in the rest of North America. It is clear that the hookeris are much closer to the ancestral line of the North American Euoenotheras than are the elatas, but the latter may be closer to the common line from which both South and North American Euoenotheras have sprung.

3. *The strigosas* range from the Rocky Mountains to the eastern edge of the great plains. In the northern part of their range they extend westward to the Pacific. They have ⊙ 14, balanced lethals and small self-pollinated flowers. The *alpha* (see Section II) complex is transmitted almost wholly through the egg, the *beta* complex through the sperm. The alpha complexes of the various races differ markedly in segmental arrangement from the betas, so much so that with very few exceptions any alpha complex so far analyzed will give ⊙ 14 with any beta complex. The betas resemble one another, however, very closely in segmental arrangement; and the alphas, though very different from the betas, also show much similarity among themselves. For instance, 1·2 and 3·4 have exchanged segments in the evolution of the betas to form (in apparently all cases) 1·4 3·2; in the alphas, however, there is no evidence of this exchange. In the development of the betas, 7·10 has exchanged with 5·6, and 9·8 with 13·14; or 7·10 and 9·8 have exchanged, as well as 5·6 and 13·14. In the origin of the alphas, however, 7·10 is often found to have exchanged with 1·2, and 5·6 with 9·8. Both alpha and beta complexes differ from the original by an average of three interchanges. Phenotypically, the alphas and betas give very similar effects—producing thick, rather narrow leaves with appressed hairs, and woody stems.

4. *The biennis I group* is in general midwestern. It occupies the area east of the great plains to the Appalachians, breaking through these mountains in the latitude of Virginia and southward and reaching the Atlantic coast. The plants have ⊙ 14, balanced lethals, and rather small self-pollinated flowers. The alpha complexes produce the *"biennis"* phenotype—broad, thin, crinkly leaves with relatively sparse pubescence, often brittle stems. The beta complexes produce a phenotype with *strigosa* characteristics. Some *biennis I* races are strictly heterogamous, others are isogamous (i.e., both complexes are transmitted through both sperm and egg). Other races show transitions between these two extremes. Segmentally, the alpha complexes are close to the original, most of them being only one interchange removed. The alphas are highly homogeneous in segmental arrangement, the great bulk of them having the same arrangement. The beta complexes are less homogeneous, though there are two arrangements, two interchanges apart, which are most frequently found. In phenotypic effect and segmental arrangement they resemble

rather closely the beta strigosas; for instance, 1·4 3·2 have been formed during their evolution, and 5·6 is often found to have exchanged with 7·10, 9·8 with 13·14. Further interchanges have often tended to obscure these earlier exchanges.

The *biennis I* races which are found in the extreme southeastern part of the country have developed narrower leaves, and a *grandiflora*-like habit (the so-called *grandiflora* of de Vries is actually an outlying number of the *biennis I* alliance). Their *strigosa*-like complexes have departed markedly from the original, segmentally and phenotypically. The *strigosa*-like complex in the *grandiflora* of de Vries is called *truncans*, and the *truncans* segmental arrangement is rather widespread among the southeastern *biennis I* races.

5. *The biennis II group* is almost indistinguishable phenotypically from *biennis I* as it grows in nature. It differs markedly from the latter, however, in geographical position, and in certain cytogenetic features. *Biennis II* occupies the northeastern portion of the United States and eastern Canada, extending westward in the northern part of the range to western New York and Ontario. Further south, it is confined to the region of the Alleghenies and eastward, and extends southward as far as Virginia, and possibly North Carolina. Its races have with one or two exceptions ⊙ 14, also balanced lethals, and small self-pollinated flowers. In *biennis II*, the gametophytic lethals have been reversed with respect to *biennis I*. The alpha complexes carry genes for the *strigosa* characters and the beta complexes tend to produce the *biennis* phenotype. The races tend to be rather strictly heterogamous, though not uncommonly alpha complexes are transmitted through the sperm, and/or beta complexes through the egg, producing "metacline" hybrids. Segmentally, the commonest arrangement among the alpha complexes is close to the beta *strigosa* arrangements. Following the exchange to 1·4 3·2, and that to 5·7 6·10, a further exchange between 1·4 and 6·10 has apparently occurred to produce 1·6 4·10. The beta complexes show no evidence of the exchange 1·2 3·4 → 1·4 3·2. A series of exchanges between 3·4, 9·8, 11·12, and 13·14 has produced the commonest arrangement found among these complexes.

6. *The biennis III group* has been found in Virginia, North Carolina, eastern Tennessee, and in western Pennsylvania, in the general region where *biennis I* and *biennis II* overlap. Its races have ⊙ 14, balanced lethals, and small, self-pollinated flowers. They tend to be rather strongly heterogamous, though metacline hybrids are produced in varying numbers in the different races. The alpha complexes produce a *biennis* phenotype. Segmentally, these are very close to the alpha *biennis I* complexes. The beta complexes also produce a *biennis* phenotype—in other words, complexes of the *strigosa* type do not exist in *biennis III*. Segmentally, the beta

biennis III complexes are indistinguishable from the beta *biennis II* complexes—the commonest arrangement in both groups of complexes is the same. It is quite probable that the *biennis III* races have originated by hybridization between *biennis I* and *biennis II*, where these have overlapped, giving rise to hybrids with alpha *biennis I* and beta *biennis II* complexes, both of which produce the *biennis* phenotype.

7. *The parvifloras* occupy essentially the same range as *biennis II*. They have with one exception (⊙ 4, ⊙ 10) shown ⊙ 14. They have balanced lethals and small, self-pollinated flowers. They tend to be strongly heterogamous. The alpha complex never comes through the sperm, though rarely in some races the beta complexes are transmitted through the egg. From the standpoint of phenotypic expression some races have alpha complexes of the *biennis* type *(parviflora I)*; in others the alpha is of the *strigosa* type *(parviflora II)*. Segmentally, these complexes show a considerable heterogeneity, few cases of identical arrangement having been found. The beta complexes give to the parvifloras their distinctive phenotype. In addition to very narrow, often almost glabrous foliage, they produce a very peculiar type of bending of the young stem tips, downward behind the tip, upward at the tip, and also cause the sepal tips to be inserted subterminally. These two characters have never been separated genetically and are either the products of closely linked genes or of a pleiotropic gene. They are not characteristic of any of the other Euoenotheras except the argillicolas, but they are found not uncommonly in the subgenus *Raimannia*, which suggests that the beta *parviflora* complexes may be closely related to the latter. In general, *Euoenothera* and *Raimannia* cannot be crossed, but it is significant that successful combinations have been made by the reviewer of alpha *parviflora* and *Raimannia* genomes (unpublished).

8. *The argillicolas* occupy dry shale barrens in the eastern Appalachian regions, ranging from Pennsylvania southward into Virginia. They are a very distinctive group, both phenotypically and in their cytogenetic characters. They resemble the hookeris in having large, open-pollinated flowers, small circles or none, and an absence of lethals. In these respects they are almost unique among eastern forms. Segmentally, their complexes are close to the original but show some heterogeneity ranging from the original arrangement to arrangements which are two interchanges removed from the original. In all these respects the argillicolas seem to be primitive and close to the hookeris. On the other hand, they resemble the beta parvifloras and many of the Raimannias in having bent stem tips and subterminal sepal tips, and they have the linear, hairless leaves which also characterize the beta parvifloras. According to Stinson (1953) they may be ancestral to the beta parvifloras.

9. *The grandifloras* are a small assemblage found near Mobile, Alabama. The plants have large, open-pollinated flowers, 7 pairs, the original segmental arrangement, and no lethals. They seem to be primitive in respect to all these characters.

The above summarizes the groupings into which the many races of North American Euoenotheras fall. Phenotypically *biennis I* and *biennis II* cannot be told apart with certainty. *Biennis III* can be distinguished from the other *biennis* groups only after a little practice. The other groups are rather distinctive in appearance.

Although there is little gene flow within most of these groupings, because of the way in which the self-pollinating habit has compartmentalized them into more or less isolated lines, each group has sufficient geographic and cytogenetic distinctness, and most of them have sufficient phenotypic individuality, to warrant their being considered species. As will be apparent from what follows, each group represents the end point of a separate and distinct line of evolutionary development.

We may now consider the process by which these species have come into existence and the relationships which exist between them.

Cleland and colleagues (1950) have assumed that the center of origin of the Euoenotheras was somewhere in Central America, perhaps in Mexico. This is suggested by the following facts: (a) The Euoenotheras are found throughout both North and South America. (b) The alethal, all-pairing condition, which no doubt is primitive, is characteristic of Central America and contiguous areas. The farther one goes into either North or South America, the more preponderant do large circles and balanced lethals become. This is especially striking in North America. (c) There is evidence that other organisms have spread from tropical or subtropical America into North America, evolving as they have spread. The initial assumption, therefore, is that the North American Euoenotheras have originated south of the United States.

From this region, successive migrations are conceived to have occurred, giving rise to successive populations. Perhaps the first for which we have evidence was a population (population 1) which was still close to the Raimannias, from which it had only recently become distinct. It was probably a population in which the leaves were narrow, relatively free of hairs, and in which subterminal sepal tips and bent stem tips were shared with the nearly related Raimannias. This population in time became swamped over much of the continent but it has survived over a restricted range in the small assemblage of the argillicolas, and its descendants are also probably represented by the beta *parviflora* complexes. The descendants of this population 1 are now found far from the original center of distribution. Those which have survived as the argillicolas have deviated only slightly from the original in segmental arrangement, but those which have crossed with later invaders have become quite diverse segmentally, a process which may have occurred in part after these crossings had taken place, since many interchanges producing viable gametes in a complex-heterozygote may result not only in the formation of modified chromosomes, but also in the transfer of one or more chromosomes from one complex to the other and hence produce marked changes in the segmental arrangement of the affected complexes.

The second distinctive population (population 2) to emerge from the center of origin probably developed at a time when the southwestern part of the continent was moist, since this population developed distinctively mesophytic characters, including thin leaves with sparse erect pubescence.

This population, which was ancestral to the complexes carrying *biennis* characters, spread over the continent, including the eastern portion. Later, when conditions in the west became drier, it dissappeared in this area except in a few isolated spots where relics may still be considered to exist. In the moister east, however, it survived. Over a goodly portion of its range its complexes retained the original segmental arrangement or departed but slightly from it. It probably retained, for the most part at least, large, open-pollinated flowers as well as paired chromosomes or small circles. In the northeastern portion of its range, however, it seems to have deviated more noticeably in segmental arrangement. In the northeast, some plants with more or less divergent segmental arrangements apparently crossed with the earlier *Raimannia*-like population mentioned above and, since the histories of interchange of the two populations had been different, the resulting hybrids had, in some cases at least, a circle of 14. These became the progenitors of the present *parviflora I* assemblage. The balanced lethals found in these plants probably arose by the process mentioned above (p. 207), as postulated by Steiner (1956, 1961). At first, these hybrids were open-pollinated, but the appearance at a later date of self pollination, as a result of the shortening of the style relative to the anthers, brought increased survival value because it increased the chance of pollination and thus helped to offset the sterilizing effect of the lethals, and also because it resulted in true-breeding.

A third migration from the center of origin (population 3) came along at a time when conditions were drier in the southwest. The plants developing under these conditions were more xerophytic and were able to conquer the mountainous areas and the great plains which were too dry for the earlier developed *biennis*-like plants. They probably possessed large open-pollinated flowers and small circles or none but in other respects resembled the strigosas of the present day. These plants not only conquered the drier central and western areas—they also spread eastward into the territory already occupied by the *biennis*-like population mentioned above. It is probable that populations 2 and 3 still possessed open-pollinated flowers and so were able to cross; but because they had suffered different histories of interchange, some of the hybrids thus produced had a circle of 14. Thus there arose here and there plants with ⊙ 14, one of whose complexes had genes producing *biennis* characters, the other genes for a *strigosa* phenotype. The phenotype of such hybrids might be expected to be intermediate between that of *strigosa* and *biennis* plants. In actuality, however, the *biennis* phenotype overshadows the *strigosa* phenotype to such an extent that taxonomists examining the descendants of these plants have been unconscious of the presence of a *strigosa* genome in them. In this way, by hybridization occurring at various points in the region of overlap, the present *biennis* population arose. The balanced lethals found in these plants probably arose by the process mentioned above.

Population 3 also found members of population 1 that had not crossed with population 2 and still had large open-pollinated flowers. It crossed with these and as a result the progenitors of *parviflora II* were formed. The fact that *rigens* of *muricata* (*parviflora II*), as shown by Steiner, carries an *S*-factor, is in line with the fact that the complexes of *parviflora II* that have descended from population 3 are egg complexes, those descended from population 1 are sperm complexes.

The situation as regards gametophytic lethals is variable among the *biennis* forms. On the whole, in *biennis I* (the midwestern group), the complex that carries the *biennis* potentialities tends to be transmitted only or predominantly through the egg in outcrosses. The *biennis II* population shows the reverse situation. Similarly, the *strigosa*-like complex tends to compete unsuccessfully in the female gametophyte in *biennis I*, in the male gametophyte in *biennis II*. However, the difference between *biennis I* and *biennis II* in this respect is not a hard and fast one. Many races in both *biennis I* and *biennis II* approach a condition of isogamy, and it is probable that when hybrids were first formed between the ancestral populations 2 and 3 there were some cases where the *biennis* complex from population 2 tended to win out in competition with the *strigosa* complex from population 3 in megaspore formation, but to lose out in pollen grain production, and the reverse was true in other cases. In still other instances, the competing complexes found themselves of essentially equal potency in megaspore or pollen formation, or both. On the whole, however, the complexes of the northeastern members of the ancestral population 2, those which had gotten farther away from the original segmentally, tended to compete more successfully in the male gametophyte and less successfully in the female when brought into competition with *strigosa* complexes, the reverse being true in the remainder of the population. While both *biennis I* and *biennis II* groups include essentially isogamous forms, and forms approaching this condition, there are few *biennis I* forms in which the typical *biennis II* lethal situation prevails, and the reverse is true for *biennis II*. It is not clear why the two groups have become so distinct with regard to the nature of the so-called gametophytic lethals.

The fact that *biennis II* and the parvifloras occupy essentially the same geographical range predisposes one toward a hypothesis that couples the origin of *biennis II* with the parvifloras. The fact also that the beta *biennis II* complexes in general have chromosome arrangements not more than two or three interchanges removed from those of the alpha parvifloras also suggests common origin. However, these resemblances seem to be as strong in the case of the alpha *parviflora I* as of the alpha *parviflora II* complexes, which is not easy to explain. We have not yet devised a satisfactory explanation of the origin of *biennis II*.

Biennis III has obviously come about by crosses between *biennis I* as female and *biennis II* as male. Chromosome end arrangements, as well as phenotypic characters and geographical distribution, all point to this as the probable origin of *biennis III*.

In one small portion of its range, the ancestral *biennis* population failed to become contaminated with *strigosa* and has remained to the present day essentially unmodified. This is the small collection of alethal, large-flowered, open-pollinated plants known as the grandifloras and found near Mobile, Alabama.

A fourth migration (population 4) appears to have come out of subtropical America; this group developed a phenotype essentially like that of the preceding *strigosa*-like population, but experienced a different series of reciprocal translocations. It was well adapted to moderately dry situations and spread over the western part of the United States. In this region it encountered the previously established *strigosa*-like population 3 and, both populations being no doubt still open-pollinated, they crossed with each other. Since the segmental interchanges which had occurred in the two populations were different, hybrids with ⊙ 14 resulted here and there. These developed lethals and so the circles became fixed. In some cases, Steiner indicates that S-factors were involved. Self-pollination later became established. In this way, the present-day strigosas came into existence. The second and later *strigosa*-like population gave rise to the alpha *strigosa* complexes, the earlier population to the beta complexes.

The ancestors of the alpha *strigosa* complexes seem, therefore, to have contaminated the ancestors of the beta strigosas, just as the latter, at an earlier date, had contaminated the ancestors of the *biennis* type of complex and the beta parvifloras. The alpha *strigosa* progenitors, however, seem not to have combined with plants having the *biennis* phenotype, probably because these had already combined with *parviflora*-like (population 1) plants, or with the ancestors of the beta strigosas (population 3), and thus had developed into self-pollinating and therefore isolated races, not easily outcrossed.[4]

The fact that the present-day complex-heterozygous populations contain such a large number of isolated lines, differing in details of phenotype and in segmental arrangement from each other, suggests that these groups did not arise as a result of a single, or a very few crosses, producing true-breeding hybrids which subsequently swamped the rest of the Oenotheras. It is more likely that the invading populations penetrated far and wide into the areas occupied by earlier populations and that crossing occurred frequently, and in many localities, giving rise gradually to a welter of hybrids with ⊙ 14. Subsequent, more rarely occurring, outcrosses between these hybrid lines would result in increased numbers of hybrid races bearing ⊙ 14.

Wherever it has been possible to test for their presence, complex-heterozygous races have been found to have zygotic lethals and these have probably arisen as deficiencies that are carried along indefinitely because of the enforced heterozygosity of the race. In most cases, however, gametophytic lethals are present and their presence makes it difficult to test for the presence of zygotic lethals.

In the case of alpha complexes, we have seen that the male gametophytic lethal may be in many cases an S-factor, and the complex possessing it is therefore unable to pass through the sperm in selfed line. In outcrosses, however, such a complex may function through the sperm. On the other hand, a male gametophytic lethal that is not an S-factor may partially or wholly inactivate all gametes which receive it, thus preventing or limiting their ability to function in outcrosses as well as in selfings. In either case, zygotes homozygous for the alpha complex will not be formed in selfed line; and, consequently, the presence of zygotic lethals cannot be detected.

The beta complexes, on the other hand, do not have S-factors. They are transmitted through the sperm without difficulty, but their behavior with respect to the egg is determined by their ability or lack of ability to compete with the associated alpha complex in megaspore formation.

[4] Since this was written, one race has been found near Elgin, Illinois, which has a *strigosa* complex apparently of the alpha type associated with a *biennis I* type of complex. It remains to be discovered how widespread this situation is.

If a plant is outcrossed as the female parent and its beta appears in a certain percentage of the progeny, this is indication that a proportion of the eggs tend to carry the beta complex. If such a plant fails to produce beta.betas when selfed, it is probable that a zygotic lethal is present in the beta complex which prevents the appearance of beta.betas.

Finally, we have the hookeris which now occupy northern Mexico, California, and the states immediately to the north and east of California. As stated above, these plants have large, open-pollinated flowers, no lethals (one or two cases of lethals are known at the extreme eastern edge of the range), and their commonest segmental arrangement (the so-called ^{h}*Johansen* arrangement) is apparently the one which is ancestral to the North American Euoenotheras as a group. This is shown by the fact that chromosomes belonging to the ^{h}*Johansen* segmental arrangement are the most widely scattered and commonest chromosomes throughout the entire continent, where they no doubt exist as relict chromosomes, i.e., as chromosomes which, in the course of evolution, have escaped translocation. As shown above (p. 160) there are 91 possible kinds of chromosomes from the standpoint of ends present. All have been found in our studies, most of them only a relatively few times, but the 7 chromosomes making up the ^{h}*Johansen* formula have each been found many times (Fig. 8). Of the nine associations of end segments that have been found in 100 or more of the complexes whose segmental arrangements have been determined by the reviewer and his students, seven represent the associations considered to be the original ones. The other two (1·4 and 3·2) owe their high frequency to the fact that the exchange 1·2 3·4 → 1·4 3·2 occurred very early in the evolution of population 3, which population has given rise not only to the beta strigosas but also to the beta *biennis I*, alpha *biennis II*, and alpha *parviflora II* complexes, almost all of which still carry either 1·4 or 3·2 or both.

It is an open question as to whether the hookeris are to be considered truly ancestral to the North American Euoenotheras or merely a recent migration which has retained old primitive characters. In any event, this group may be taken to represent the condition which was ancestral at least to the North American Euoenotheras since it has retained the primitive characters which the ancestor must have had, even to the original arrangement of segments. It may not, however, be ancestral to the South American Euoenotheras. It is possible that a segmental arrangement present in Central America may be truly ancestral to all of the euoenotheras, having been altered through interchange in the northern part of its range to the "*Johansen*" condition from which the North American Euoenotheras have in turn been derived.

Summing up this account of the evolutionary history of the North

American Euoenotheras, it appears (1) that the parvifloras arose as a result of overlapping and consequent hybridization between the most ancient population of *Euoenothera* of which we have evidence in North America (population 1) and subsequent invaders, in part *biennis* in character (producing *parviflora I*), in part *strigosa* in character (producing *parviflora II*); (2) that the argillicolas may represent a relic of population 1; (3) that *biennis I* and *biennis II* represent the result of hybridization between population 2 and a later *strigosa*-like population 3; (4) that *biennis III* resulted from the overlapping and hybridization of *biennis I* and *biennis II*; (5) that *grandiflora* represents a relict of population 2; (6) that the strigosas are the result of the overlapping and hybridization of population 3 (producing the beta *strigosa* complexes) with a later population 4 (producing the alpha *strigosa* complexes); (7) that the hookeris either represent the original population of North American Euoenotheras or they are a later offshoot which has retained in whole (in Mexico and California) or in part (in Utah, Colorado, and New Mexico) the original characters of the North American Euoenotheras.

According to this scheme, the following factors have been important in the evolution of the North American Euoenotheras (see Cleland, 1957):

(a) An original set of chromosomes all members of which were identical in size, with median centromeres and probably so constructed as to favor equal translocations (possibly because of heterochromatin contiguous to the centromere). As a result, translocation-heterozygotes have chromosomes with unaltered morphology and have usually been able to produce regular zigzag arrangements in metaphase, and so have suffered little in the way of non-disjunction and consequent sterility, so that translocations have tended to survive.

(b) Overlapping and hybridization of populations whose histories of interchange have been sufficiently diverse to make it possible for the resultant hybrids in many cases to have ⊙ 14.

(c) The presence of ⊙ 14 has favored retention of lethals, since only one pair of balanced lethals will preserve the heterozygosity of the whole complement, and thus the benefits of maximum hybrid vigor can be attained with minimal loss of fertility; furthermore, balanced lethals have transformed segregating hybrids into true-breeding strains.

(d) Development of self pollination has largely overcome the deleterious sterilizing effect of the lethals by insuring rich pollination.

Thus, three different factors (translocations, lethals, self pollination) which by themselves might have been expected to have harmful effects, constitute a mechanism which combines maximum hybrid vigor and true-breeding with rich pollination and seed-set—all of which makes for maximum survival value so far as the individual is concerned.

This peculiar mechanism, which enhances the possibilities of individual survival, suggests some interesting implications for the future evolution of the group. (1) It has had the effect of setting up, every time a new combination with ⊙ 14 is established by crossing, a true-breeding, relatively isolated line. Any mutation occurring in this line will be more or less confined to that line since outcrossing will be rare. Consequently, innumerable barriers are set up to the free flow of genes in the population. This would seem to hamper the further evolutionary development of the Oenotheras.

(2) On the other hand, this is an admirable mechanism for the preservation, though not for the immediate exposure, of recessive mutations. Because homozygotes are ordinarily not produced, recessive genes will tend to remain under cover and not be subject to elimination through natural selection. They may, therefore, accumulate indefinitely. The genetic material may, as a result, suffer major alteration with little effect on phenotype apart from possible overdominance, until such time as translocation, crossing-over, or rare outcrossing may permit certain nonlethal-bearing chromosomes or chromosome segments to exist in homozygous condition.

Even when partial homozygosity has been achieved by such means, however, it is likely to be short-lived. One would expect that random crossing would produce plants with about as many configurations other than ⊙ 14 as with this configuration. The almost total absence of forms with other than ⊙ 14 throughout most of the range suggests that plants with intermediate chromosome configurations have less survival value than those having ⊙ 14. When recessive genes, many of them possibly lethal or semilethal in character, are brought to light by some process making it possible for a chromosome segment to exist in homozygous condition, the plants carrying these genes are weeded out by natural selection.

Only if such genes become included by further crossing in plants with ⊙ 14 will they continue to be preserved, in which case they will probably find themselves again suppressed by dominant alleles. They may have been introduced, however, by this process into new biotypes and so may become more widely distributed than would have been the case had no outcrossing occurred.

It would seem, therefore, that any future evolutionary development in the subgenus is likely to take the form of the formation through rare outcrossing of new combinations of complexes having ⊙ 14. Some of these may have superior survival value. In some of them, mutant genes may find themselves only partially or not at all dominated by their new alleles. In this way, the population may come to have an ever increasing number of isolated lines or races and hence become increasingly complicated in structure; and the fact that all are permanent, true-breeding heterozygotes, with maximum heterosis maintained with little or no loss of reproductive capacity, will no doubt continue to make this a very successful genus. The barriers thus set up, however, to gene flow and to the exposure of recessive mutations would seem to offer little promise that the genus will undergo major evolutionary development, at least at a normal or average rate of speed.

In conclusion it might be asked whether there is not also the possibility that genetic changes taking place in the center of origin in Central America might result in new migrations, carrying modified segmental arrangements, into regions already occupied, thus bringing about overlapping and hybridization, resulting in entirely new populations with large circles, as has occurred repeatedly in the past.

If a new open-pollinated population invaded a region where existing races were still open-pollinated, as in California and adjacent areas, hybrids between the two populations might give rise to races with ⊙ 14, if the overlapping populations differed sufficiently in segmental arrangement. There would be little chance, however, that the same would happen if the new population invaded a territory already occupied by self-pollinating races. The self-pollinating habit would place such a barrier in the way of hybridization between the old races and the new that crossing would occur very rarely between them, and hybrids combining complexes from the invader and the invaded would rarely be produced. In all cases where overlapping and hybridization have occurred in the past, the populations involved have no doubt still retained the original open pollination at the time of crossing; self pollination has not developed until later in the circle-bearing hybrids thus produced.

It is probable, therefore, that the door has very nearly been closed to the possibility of new populations arising from the overlapping of populations that have experienced different histories of interchange—at least this would seem to be true with regard to the areas east of the Continental Divide, where existing populations are already almost all self-pollinating.

References

Alpinus, P. 1627. De plantis exoticis. I. Guerilium, Venice.

Baerecke, M. 1944. Zur Genetik und Cytologie von *Oenothera ammophila* Focke, *Bauri* Boedijn, *Beckeri* Renner, *parviflora* L., *rubricaulis* Klebahn, *silesiaca* Renner. *Flora (Jena)* **138**, 57–92.

Bartlett, H. H. 1915. The experimental study of genetic relationships. *Am. J. Botany* **2**, 132–155.

Bartlett, H. H. 1916. The status of the mutation theory with especial reference to *Oenothera*. *Am. Naturalist* **50**, 513–529.
Belling, J. 1927. The attachments of chromosomes at the reduction division in flowering plants. *J. Genet.* **18**, 177–205.
Belling, J., and Blakeslee, A. F. 1926. On the attachment of non-homologous chromosomes at the reduction division in certain 25-chromosome Daturas. *Proc. Natl. Acad. Sci. U.S.* **12**, 7–11.
Bhaduri, P. N. 1940. Cytological studies in *Oenothera* with special reference to the relation of chromosomes and nucleoli. *Proc. Roy. Soc.* **B128**, 353–378.
Boedijn, K. 1925. Der Zusammenhang zwischen den Chromosomen und Mutationen bei *Oenothera lamarckiana*. *Rec. trav. botan. néerl.* **22**, 173–261.
Boedijn, K. 1928. Chromosomen und Pollen der Oenotheren. *Rec. trav. botan. néerl.* **25A**, 25–35.
Brittingham, W. H. 1931. Variation in the evening primrose induced by radium. *Science* **74**, 463–464.
Catcheside, D. G. 1931. Critical evidence of parasynapsis in *Oenothera*. *Proc. Roy. Soc.* **B109**, 165–184.
Catcheside, D. G. 1932. The chromosomes of a new haploid *Oenothera*. *Cytologia (Tokyo)* **4**, 68–113.
Catcheside, D. G. 1933. Chromosome configurations in trisomic oenotheras. *Genetica* **15**, 177–201.
Catcheside, D. G. 1935. X-ray treatment of *Oenothera* chromosomes. *Genetica* **17** 313–341.
Catcheside, D. G. 1936. Origin, nature and breeding behaviour of *Oenothera lamarckiana* trisomics. *J. Genet.* **33**, 1–23.
Catcheside, D. G. 1937a. Recessive mutations induced in *Oenothera blandina* by x-rays. *Genetica* **18**, 134–142.
Catcheside, D. G. 1937b. The extra chromosome of *Oenothera lamarckiana lata*. *Genetics* **22**, 564–576.
Catcheside, D. G. 1939. A position effect in *Oenothera*. *J. Genet.* **38**, 345–352.
Catcheside, D. G. 1940. Structural analysis of *Oenothera* complexes. *Proc. Roy. Soc* **B128**, 509–535.
Catcheside, D. G. 1947a. The *P*-locus position effect in *Oenothera*. *J. Genet.* **48**, 31–42.
Catcheside, D. G. 1947b. A duplication and deficiency in *Oenothera*. *J. Genet.* **48**, 99–110.
Catcheside, D. G. 1954. The genetics of *brevistylis* in *Oenothera*. *Heredity* **8**, 125–137.
Cleland, R. E. 1922. The reduction divisions in the pollen mother cells of *Oenothera franciscana*. *Am. J. Botany* **9**, 391–413.
Cleland, R. E. 1923. Chromosome arrangements during meiosis in certain *Oenotheras*. *Am. Naturalist* **57**, 562–566.
Cleland, R. E. 1924. Meiosis in pollen mother cells of *Oenothera franciscana sulfurea*. *Botan. Gaz.* **77**, 149–170.
Cleland, R. E. 1925. Chromosome behavior during meiosis in the pollen mother cells of certain oenotheras. *Am. Naturalist* **59**, 475–479.
Cleland, R. E. 1926a. Meiosis in the pollen mother cells of *Oenothera biennis* and *Oenothera biennis sulfurea*. *Genetics* **11**, 127–162.
Cleland, R. E. 1926b. Cytological study of meiosis in anthers of *Oenothera muricata*. *Botan. Gaz.* **82**, 55–70.
Cleland, R. E. 1928. The genetics of *Oenothera* in relation to chromosome behavior with special reference to certain hybrids. *Z. Vererbungslehre Suppl.* **I** pp. 554–567.

Cleland, R. E. 1931. The probable origin of *Oenothera rubricalyx* "Afterglow" on the basis of the segmental interchange theory. *Proc. Natl. Acad. Sci. U.S.* **17**, 437–440.

Cleland, R. E. 1942. The origin of *ʰdecipiens* from the complexes of *Oenothera lamarckiana* and its bearing upon the phylogenetic significance of similarities in segmental arrangement. *Genetics* **27**, 55–83.

Cleland, R. E. 1950. Studies in *Oenothera* cytogenetics and phylogeny, Introduction and general summary. *Indiana Univ. Publ. Sci. Ser.* **16**, 5–9.

Cleland, R. E. 1954. Evolution of the North American euoenotheras: the strigosas. *Proc. Am. Phil. Soc.* **98**, 189–203.

Cleland, R. E. 1957. Chromosome structure in *Oenothera* and its effect on the evolution of the genus. *Cytologia (Proc. Intern. Genet. Symposia) 1956*, 5–19.

Cleland, R. E. 1958. The evolution of the North American oenotheras of the *"biennis"* group. *Planta* **51**, 378–398.

Cleland, R. E. 1960. The S-factor situation in a small sample of an *Oenothera (Raimannia) heterophylla* population. *Z. Vererbungslehre* **91**, 303–311.

Cleland, R. E., and Blakeslee, A. F. 1931. Segmental interchange, the basis of chromosomal attachments in *Oenothera*. *Cytologia (Tokyo)* **2**, 175–233.

Cleland, R. E., and Brittingham, W. H. 1934. A contribution to an understanding of crossing over within chromosome rings in *Oenothera*. *Genetics* **19**, 62–72.

Cleland, R. E., and Hammond, B. L. 1950. Analysis of segmental arrangements in certain races of *Oenothera*. *Indiana Univ. Publ. Sci. Ser.* **16**, 10–72.

Cleland, R. E., and Oehlkers, F. 1930. Erblichkeit und Zytologie verschiedener Oenotheren und ihrer Kreuzungen. *Jahrb. wiss. Botan.* **73**, 1–124.

Cleland, R. E., Preer, L. B., and Geckler, L. H. 1950. The nature and relationships of taxonomic entities in the North American euoenotheras. *Indiana Univ. Publ. Sci. Ser.* **16**, 218–254.

Correns, C. 1917. Ein Fall experimenteller Verschiebung des Geschlechtsverhältnisses. *Sitzber. preuss. Akad. Wiss.* **1917**, 685–717.

Crowe, L. K. 1955. The evolution of incompatibility in species of *Oenothera*. *Heredity* **9**, 293–322.

Darlington, C. D. 1929. Ring-formation in *Oenothera* and other genera. *J. Genet.* **20**, 345–369.

Darlington, C. D. 1931. The cytological theory of inheritance in *Oenothera*. *J. Genet.* **24**, 405–474.

Darlington, C. D. 1933. The behaviour of interchange heterozygotes in *Oenothera*. *Proc. Natl. Acad. Sci. U.S.* **19**, 101–103.

Darlington, C. D. 1936. The limitation of crossing over in *Oenothera*. *J. Genet.* **32**, 343–352.

Davis, B. M. 1909. Cytological studies on *Oenothera* I. Pollen development of *Oenothera grandiflora*. *Ann. Botany (London)* **23**, 551–571.

Davis, B. M. 1910. Cytological studies on *Oenothera* II. Reduction divisions of *Oenothera biennis*. *Ann. Botany (London)* **24**, 631–651.

Davis, B. M. 1911. Genetical studies on *Oenothera* II. Some hybrids of *Oenothera biennis* and *O. grandiflora* that resemble *Oenothera lamarckiana*. *Am. Naturalist* **45**, 193–233.

Davis, B. M. 1916. *Oenothera neo-lamarckiana*, hybrid of *O. franciscana* Bartlett $\times$ *O. biennis* Linnaeus. *Am. Naturalist* **50**, 688–696.

de Vries, H. 1889. "Intracellulare Pangenesis." G. Fischer, Jena.

de Vries, H. 1900a. Sur la loi de disjonction des hybrides. *Compt. rend. acad. sci.* **130**, 845–847.

de Vries, H. 1900b. Das Spaltungsgesetz der Bastarde. *Ber. deut. botan. Ges.* **18,** 83–90.
de Vries, H. 1900c. Sur l'origine expérimentale d'une nouvelle espèce végétale. *Compt. rend. acad. sci.* **131,** 124–126.
de Vries, H. 1900d. Sur la mutabilité de l'*Oenothera lamarckiana*. *Compt. rénd. acad. sci.* **131,** 561–563.
de Vries, H. 1901. "Die Mutationstheorie," Vol. 1. Veit & Co., Leipzig.
de Vries, H. 1903a. "Die Mutationstheorie," Vol. 2. Veit & Co., Leipzig.
de Vries, H. 1903b. Anwendung der Mutationslehre auf die Bastardierungsgesetze. *Ber. deut. botan. Ges.* **21,** 45–52.
de Vries, H. 1905. Ueber die Dauer der Mutationsperiode bei *Oenothera lamarckiana*. *Ber. deut. botan. Ges.* **23,** 382–387.
de Vries, H. 1913. "Gruppenweise Artbildung, unter specieller Berücksichtigung der Gattung *Oenothera*," 365 pp. Borntraeger, Berlin.
de Vries, H. 1915. The coefficient of mutation in *Oenothera biennis* I. *Botan. Gaz.* **59,** 169–196.
de Vries, H. 1916. New dimorphic mutants of the oenotheras. *Botan. Gaz.* **62,** 249–280.
de Vries, H. 1917. Halbmutanten und Zwillingsbastarde. *Ber. deut. botan. Ges.* **35,** 128–135.
de Vries, H. 1918. Mutations of *Oenothera suaveolens* Desf. *Genetics* **3,** 1–26.
de Vries, H. 1919a. *Oenothera rubrinervis*, a half mutant. *Botan. Gaz.* **67,** 1–26.
de Vries, H. 1919b. *Oenothera erythrina*, eine neue Halbmutante. *Z. Vererbungslehre* **21,** 91–118.
de Vries, H. 1923. *Oenothera lamarckiana* mut. *perennis*. *Flora (Jena)* **116,** 336–345.
de Vries, H. 1924. Die Mutabilität von *Oenothera gigas*. *Z. Vererbungslehre* **35,** 197–237.
de Vries, H. 1925a. Sekundäre Mutationen von *Oenothera lamarckiana*. *Z. Botan.* **17,** 193–211.
de Vries, H. 1925b. Androlethal factors in *Oenothera*. *J. Gen. Physiol.* **8,** 109–113.
de Vries, H. 1929. Ueber das Auftreten von Mutanten aus *Oenothera lamarckiana*. *Z. Vererbungslehre* **52,** 121–190.
de Vries, H., and Boedijn, K. 1923. On the distribution of mutant characters among the chromosomes of *Oenothera lamarckiana*. *Genetics* **8,** 233–238.
de Vries, H., and Boedijn, K. 1924. Doubled chromosomes of *Oenothera semigigas*. *Botan. Gaz.* **78,** 249–270.
de Vries, H., and Gates, R. R. 1928. A survey of the cultures of *Oenothera lamarckiana* at Lunteren. *Z. Vererbungslehre* **47,** 275–286.
Digby, L. 1912. The cytology of *Primula kewensis* and of other related *Primula* hybrids. *Ann. Botany (London)* **26,** 358–388.
East, E. M. 1940. The distribution of self-sterility in the flowering plants. *Proc. Am. Phil. Soc.* **82,** 449–518.
Emerson, S. H. 1931a. Genetic and cytological studies on *Oenothera* II. Certain crosses involving *Oe. rubricalyx* and *Oe. "franciscana sulfurea."* *Z. Vererbungslehre* **59,** 381–394.
Emerson, S. H. 1931b. The inheritance of certain characters in *Oenothera* hybrids of different chromosome configurations. *Genetics* **16,** 325–348.
Emerson, S. H. 1932. Chromosome rings in *Oenothera*, *Drosophila* and maize. *Proc. Natl. Acad. Sci. U.S.* **18,** 630–632.
Emerson, S. H. 1935. The genetic nature of de Vries' mutations in *Oenothera lamarckiana*. *Am. Naturalist* **69,** 545–559.
Emerson, S. H. 1936. Trisomic derivatives of *Oenothera lamarckiana*. *Genetics* **21,** 200–224.

Emerson, S. H. 1938. The genetics of self-incompatibility in *Oenothera organensis*. *Genetics* **23**, 190–202.
Emerson, S. H. 1940. Growth of incompatible pollen tubes in *Oenothera organensis*. *Botan. Gaz.* **101**, 890–911.
Emerson, S. H. 1941. Linkage relationships of two gametophytic characters in *Oenothera organensis*. *Genetics* **26**, 469–473.
Emerson, S. H., and Sturtevant, A. H. 1931. Genetical and cytological studies in *Oenothera* III. The translocation hypothesis. *Z. Vererbungslehre* **59**, 395–419.
Emerson, S. H., and Sturtevant, A. H. 1932. The linkage relations of certain genes in *Oenothera*. *Genetics* **17**, 393–412.
Gardella, C. 1953. Studies in the chromosome structure of *Oenothera* (abstract). *Dissertation Abstr.* **13**, 957.
Gates, R. R. 1907a. Pollen development in hybrids of *Oenothera lata* × *O. lamarckiana*, and its relation to mutation. *Botan. Gaz.* **43**, 81–115.
Gates, R. R. 1907b. Hybridization and germ cells of *Oenothera* mutants. *Botan. Gaz.* **44**, 1–21.
Gates, R. R. 1911a. Mutation in *Oenothera*. *Am. Naturalist* **45**, 577–606.
Gates, R. R. 1911b. Studies on the variability and heritability of pigmentation in *Oenothera*. *Z. Vererbungslehre* **4**, 337–372.
Gates, R. R. 1915. On the origin and behaviour of *Oenothera rubricalyx*. *J. Genet.* **4**, 353–360.
Gates, R. R. 1922. Some points on the relation of cytology and genetics. *J. Heredity* **13**, 75–76.
Gates, R. R. 1925. Present problems of *Oenothera* research. Mem. Publ. in honor of 100th birthday of J. G. Mendel. *Czechslov. Eugenics Soc.* pp. 135–145.
Gates, R. R. 1939. Self-sterility in *Oenothera*. *Nature* **143**, 245.
Geerts, J. M. 1907. Ueber die Zahl der Chromosomen von *Oenothera lamarckiana*. *Ber. deut. botan. Ges.* **25**, 191–195.
Geerts, J. M. 1909. Beiträge zur Kenntnis der Cytologie und der partiellen Sterilität von *Oenothera lamarckiana*. *Rec. trav. botan. néerl.* **5**, 93–208.
Goldschmidt, R. B. 1958. Genic conversion in *Oenothera*? *Am. Naturalist* **92**, 93–104.
Hagen, C. W., Jr. 1950. A contribution to the cytogenetics of the genus *Oenothera*, with special reference to certain forms from South America. *Indiana Univ. Publ. Sci. Ser.* **16**, 305–348.
Håkansson, A. 1926. Ueber das Verhalten der Chromosomen bei der heterotypischen Teilung schwedischer *Oenothera lamarckiana* und einiger ihrer Mutanten und Bastarde. *Hereditas* **8**, 255–304.
Håkansson, A. 1930a. Die Chromosomenreduktion bei einigen Mutanten und Bastarden von *Oenothera lamarckiana*. *Jahrb. wiss. Botan.* **72**, 385–402.
Håkansson, A. 1930b. Zur Zytologie trisomatischer Mutanten aus *Oenothera lamarckiana*. *Hereditas* **14**, 1–32.
Håkansson, A. 1931. Beobachtungen über die Chromosomenbindungen bei einigen triploiden Oenotheren. *Botan. Notiser* **1931**, 339–342.
Harte, C. 1942. Meiosis und crossing-over. Weitere Beiträge zur Zytogenetik von *Oenothera*. *Z. Botan.* **38**, 65–137.
Harte, C. 1948. Zytologisch-genetische Untersuchungen an spaltenden Oenotherenbastarden. *Z. Vererbungslehre* **82**, 495–640.
Haselwarter, A. 1937. Untersuchungen zur Physiologie der Meiosis V. *Z. Botan.* **31**, 273–328.
Hecht, A. 1944. Induced tetraploids of a self sterile *Oenothera*. *Genetics* **29**, 69–74.

Hecht, A. 1950. Cytogenetic studies of *Oenothera*, subgenus *Raimannia*. *Indiana Univ. Publ. Sci. Ser.* **16**, 255–304.

Heribert-Nilsson, N. 1912. Die Variabilität der *Oenothera lamarckiana* und das Problem der Mutation. *Z. Vererbungslehre* **8**, 89–231.

Heribert-Nilsson, N. 1920. Zuwachsgeschwindigkeit der Pollenschläuche und gestörte Mendelzahlen bei *Oenothera lamarckiana*. *Hereditas* **1**, 41–67.

Herzog, G. 1940. Genetische und cytologische Untersuchungen über 15-chromosomige Mutanten von *Oenothera biennis* und *O. lamarckiana*. *Flora (Jena)* **134**, 377–432.

Hoeppener, E., and Renner, O. 1928. Genetische und zytologische Oenotherenstudien I. Zur Kenntnis der *Oenothera ammophila* Focke. *Z. Vererbungslehre* **19**, 1–25.

Hoeppener, E., and Renner, O. 1929. Genetische und zytologische Oenotherenstudien II. Zur Kenntnis von *Oe. rubrinervis, deserens, lamarckiana-gigas, biennis-gigas, franciscana, hookeri, suaveolens, lutescens*. *Botan. Abhandl.* **15**, 3–86.

Japha, B. 1939. Die Meiosis von *Oenothera* II. *Z. Botan.* **34**, 321–369.

Kisch, R. 1937. Die Bedeutung der Wasserversorgung für den Ablauf der Meiosis. (Untersuchungen zur Physiologie der Meiosis VI). *Jahrb. wiss. Botan.* **85**, 450–484.

Langendorf, J. 1930. Zur Kenntnis der Genetik und Entwicklungsgeschichte von *Oenothera fallax, rubirigida* und *Hookeri-albata*. *Botan. Arch.* **29**, 474–530.

Lewis, D. 1948. Structure of the incompatibility gene I. Spontaneous mutation rate. *Heredity* **2**, 219–236.

Lewis, D. 1949. Structure of the incompatibility gene II. Induced mutation rate. *Heredity* **3**, 339–355.

Lewis, D. 1951. Structure of the incompatibility gene III. Types of spontaneous and induced mutation. *Heredity* **5**, 399–414.

Lewitsky, G. A. 1931. The morphology of chromosomes. *Bull. Appl. Botany Genet. Plant Breeding (U.S.S.R.)* **27**, 19–174.

Lutz, A. M. 1907. Preliminary note on the chromosomes of *Oenothera lamarckiana* and one of its mutants, *O. gigas*. *Science* **26**, 151–152.

Lutz, A. M. 1908. Chromosomes of somatic cells of the oenotheras. *Science* **27**, 335.

Marquardt, H. 1937. Die Meiosis von *Oenothera* I. *Z. Zellforsch. u. mikroskop. Anat.* **27**, 159–210.

Marquardt, H. 1948. Das Verhalten röntgeninduzierter Viererringe mit grossen interstitiellen Segmenten bei *Oenothera hookeri*. *Z. Vererbungslehre* **82**, 415–429.

Michaelis, P. 1930. Ueber experimentell erzeugte, heteroploide Pflanzen von *Oenothera hookeri*. *Z. Botan.* **23**, 288–308.

Muller, H. J. 1917. An *Oenothera*-like case in *Drosophila*. *Proc. Natl. Acad. Sci. U.S.* **3**, 619–626.

Munz, P. A. 1928. Studies in Onagraceae I. A revision of the subgenus *Chylismia* of the genus *Oenothera*. *Am. J. Botany* **15**, 233–240 (first of a series of taxonomic revisions of the genus).

Munz, P. A. 1949. The *Oenothera hookeri* group. *El Aliso* **2**, 1–47.

Oehlkers, F. 1924. Sammelreferat über neuere experimentelle Oenotherenarbeiten. *Z. Vererbungslehre* **34**, 259–283.

Oehlkers, F. 1926a. Sammelreferat über neuere experimentelle Oenotherenarbeiten II. *Z. Vererbungslehre* **41**, 359–375.

Oehlkers, F. 1926b. Erblichkeit und Zytologie einiger Kreuzungen mit *Oenothera strigosa*. *Jahrb. wiss. botan.* **65**, 401–446.

Oehlkers, F. 1930a. Studien zum Problem der Polymerie und des multiplen Allelomorphismus I. *Z. Botan.* **22**, 473–537.

Oehlkers, F. 1930b. Studien zum Problem der Polymerie und des multiplen Allelomorphismus II. *Z. Botan.* **23**, 967–1003.

Oehlkers, F. 1933. Crossing over bei *Oenothera* (Vererbungsversuche an Oenotheren V). *Z. Botan.* **26**, 385–430.

Oehlkers, F. 1935. Untersuchungen zur Physiologie der Meiosis I. *Z. Botan.* **29**, 1–53.

Oehlkers, F. 1936. Untersuchungen zur Physiologie der Meiosis III. *Z. Botan.* **30**, 253–276.

Oehlkers, F. 1938. Über die Erblichkeit des *cruciata*-Merkmals bei den Oenotheren; eine Erwiderung. *Z. Vererbungslehre* **75**, 277–297.

Oehlkers, F. 1940. Meiosis und Crossing-over. Zytogenetische Untersuchungen an Oenotheren. *Z. Vererbungslehre* **78**, 157–186.

Oehlkers, F. 1943. Die Auslösung von Chromosomenmutationen in der Meiosis durch Einwirkung von Chemikalien. *Z. Vererbungslehre* **81**, 313–341.

Oehlkers, F. 1949. Mutationsauslösung durch Chemikalien. *Sitzber. Heidelberg. Akad. Wiss. Math. Naturw. Kl. Abhandl.* **1949**, 1–40.

Oehlkers, F., and Linnert, G. 1951. Weitere Untersuchungen über die Wirkungsweise von Chemikalien bei der Auslösung von Chromosomenmutationen. *Z. Vererbungslehre* **83**, 429–438.

Pohl, J. 1895. Ueber Variationsweite der *Oenothera lamarckiana*. *Österr. Botan. Z.* **45**, 166–169, 205–212.

Renner, O. 1917a. Die tauben Samen der Oenotheren. *Ber. deut. botan. Ges.* **34**, 858–869.

Renner, O. 1917b. Versuche über die genetische Konstitution der Oenotheren. *Z. Vererbungslehre* **18**, 121–294.

Renner, O. 1917c. Artbastarde und Bastardarten in der Gattung *Oenothera*. *Ber. deut. botan. Ges.* **35**, (21)–(26).

Renner, O. 1919. Zur Biologie und Morphologie der männlichen Haplonten einiger Oenotheren. *Z. Botan.* **11**, 305–380.

Renner, O. 1921a. Heterogamie im weiblichen Geschlecht und Embryosackentwicklung bei den Oenotheren. *Z. Botan.* **13**, 609–621.

Renner, O. 1921b. Das Rotnervenmerkmal der Oenotheren. *Ber. deut. botan. Ges.* **39**, 264–270.

Renner, O. 1924. Die Scheckung der Oenotherenbastarde. *Botan. Zentr.* **44**, 309–336.

Renner, O. 1925. Untersuchungen über die faktorielle Konstitution einiger komplexheterozygotischer Oenotheren. *Bibliotheca Genet.* **9**, 1–168.

Renner, O. 1928. Ueber Koppelungswechsel bei *Oenothera*. *Z. Vererbungslehre (Verhandl. 5th intern. Kongr. Vererbungslehre)* **2**, 1216–1220.

Renner, O. 1933. Zur Kenntnis der Letalfaktoren und des Koppelungswechsels der Oenotheren. *Flora (Jena)* **27**, 215–250.

Renner, O. 1936. Zur Kenntnis der nichtmendelnden Buntheit der Laubblätter. *Flora (Jena)* **30**, 218–290.

Renner, O. 1937a. Zur Kenntnis der Plastiden-und Plasmavererbung. *Cytologia (Fujii Jubil. Vol.)* pp. 644–653.

Renner, O. 1937b. Über *Oenothera atrovirens* Sh. et Bartl. und über somatische Konversion in Erbgang des cruciata-Merkmals der Oenotheren. *Z. Vererbungslehre* **74**, 91–124.

Renner, O. 1940a. Kurze Mitteilungen über Oenothera IV. Ueber die Beziehungen zwischen Heterogamie und Embryosackentwicklung und über diplarrhene Verbindungen. *Flora (Jena)* **34**, 145–158.

Renner, O. 1940b. Zur Kenntnis der 15-chromosomigen Mutanten von *Oenothera lamarckiana*. *Flora (Jena)* **134**, 257–310.

Renner, O. 1942. Ueber das Crossing-over bei *Oenothera*. *Flora (Jena)* **136**, 117–214.

Renner, O. 1943a. Kurze Mitteilungen über *Oenothera* VI. Ueber die 15-chromosomigen Mutanten *dependens, incana, scintillans, glossa, tripus*. *Flora (Jena)* **137**, 216–229.

Renner, O. 1943b. Zur Kenntnis des Pollenkomplexes *flectens* der *Oenothera atrovirens* Sh. et Bartl. *Z. Vererbungslehre* **81**, 391–483.

Renner, O. 1943c. Ueber die Entstehung homozygotischer Formen aus heterozygotischen Oenotheren. II. Die Translocationshomozygoten. *Z. Botan.* **39**, 49–105.

Renner, O. 1948. Die cytologischen Grundlagen des Crossing-over bei *Oenothera*. *Z. Naturforsch.* **3b**, 188–196.

Renner, O. 1949. Die 15-chromosomigen Mutanten der *Oenothera lamarckiana* und ihrer Verwandten. *Z. Vererbungslehre* **83**, 1–15.

Renner, O. 1959. Somatic conversion in the heredity of the *cruciata* character in *Oenothera*. *Heredity* **13**, 283–288.

Renner, O., and Cleland, R. E. 1933. Zur Genetik und Cytologie der *Oenothera chicaginensis* und ihre Abkömmlinge. *Z. Vererbungslehre* **66**, 275–318.

Rudloff, F. 1929. Zur Kenntnis der *Oenothera purpurata* Klebahn und *O. rubricaulis* Klebahn. *Z. Vererbungslehre* **52**, 191–235.

Rudloff, C. F. 1931. Zur Polarization in der Reduktionsteilung heterogamer Oenotheren I. Die Embryosackentwicklung und ihre Tendenzen. *Z. Vererbungslehre* **58**, 422–433.

Rudloff, C. F., and Stubbe, H. 1935. Mutationsversuche mit *Oenothera hookeri*. *Flora (Jena)* **128**, 347–362.

Schötz, F. 1954. Ueber Plastidenkonkurrenz bei *Oenothera*. *Planta* **47**, 182–240.

Schötz, F. 1958a. Beobachtungen zur Plastidenkonkurrenz bei *Oenothera* und Beiträge zum Problem der Plastidenvererbung. *Planta* **51**, 173–185.

Schötz, F. 1958b. Periodische Ausbleichungserscheinungen des Laubes bei *Oenothera*. *Planta* **52**, 351–392.

Seitz, F. W. 1935. Zytologische Untersuchungen an tetraploiden Oenotheren. *Z. Botan.* **28**, 481–542.

Shull, G. H. 1921. Three new mutations in *Oenothera lamarckiana*. *J. Heredity* **12**, 354–363.

Shull, G. H. 1923a. Further evidence of linkage with crossing over in *Oenothera*. *Genetics* **8**, 154–167.

Shull, G. H. 1923b. Linkage with lethal factors the solution of the *Oenothera* problem. Eugenics, Genetics and the Family. *Proc. 2nd Intern. Congr. Eugenics* Williams & Wilkins, Baltimore, Maryland. **1**, 86–99.

Shull, G. H. 1925. The third linkage group in *Oenothera*. *Proc. Natl. Acad. Sci. U.S.* **11**, 715–718.

Shull, G. H. 1928. A new gene mutation (mut. *bullata*) in *Oenothera lamarckiana* and its linkage relations. *Z. Vererbungslehre (Verhandl. 5th intern. Kongr. Vererbungslehre)* **2**, 1322–1342.

Shull, G. H. 1934. *Oenothera* mut. *pollicata*, an interesting new mutation. *Am. Naturalist* **68**, 481–490.

Steiner, E. E. 1956. New aspects of the balanced lethal mechanism in *Oenothera*. *Genetics* **41**, 486–500.

Steiner, E. E. 1961. Incompatibility studies in *Oenothera*. *Z. Vererbungslehre* **92**, 205–212.

Stinson, H. T. 1953. Cytogenetics and phylogeny of *Oenothera argillicola* Mackenz. *Genetics* **38**, 389–406.

Stomps, T. J. 1912. Die Entstehung von *Oenothera gigas* de Vries. *Ber. deut. botan. Ges.* **30**, 406–416.

Stubbe, W. 1955. Erbliche Chlorophylldefekte bei *Oenothera*. *Phot. u. Wiss.* **4**, 3–8.

Stubbe, W. 1957. Dreifarbenpanaschierung bei Oenotheren I. Entmischung von drei in der Zygote vereinigten Plastidomen. *Ber. deut. botan. Ges.* **70**, 221–226.

Stubbe, W. 1959. Genetische Analyse des Zusammenwirkens von Genom und Plastom bei *Oenothera*. *Z. Vererbungslehre* **90,** 288–298.

van Overeem, C. 1922. Ueber Formen mit abweichender Chromosomenzahl bei *Oenothera*. *Botan. Centr. Beih.* *(Abt. I)* **39,** 1–80.

Weier, T. E. 1930. A comparison of meiotic prophase in *Oenothera lamarckiana* and *O. hookeri*. *La Cellule* **39,** 271–305.

Winkler, H. 1930. "Die Konversion der Gene." G. Fischer, Jena.

Winkler, H. 1932. Konversions-Theorie und Austausch-Theorie. *Biol. Zentr.* **53,** 165–189.

Wisniewska, E. 1935. Zytologische Untersuchungen an *Oenothera hookeri* de Vries. (In Polish, with German summary.) *Acta. Soc. Botan. Polon.* **12,** 113–164.

Zürn, K. 1937a. Untersuchungen zur Physiologie der Meiosis IV. *Z. Botan.* **30,** 577–603.

Zürn, K. 1937b. Die Bedeutung der Plastiden für den Ablauf der Meiosis. (Untersuchungen zur Physiologie der Meiosis IX). *Jahrb. wiss. Botan.* **85,** 706–731.

GENETICS AND THE HUMAN SEX RATIO

A. W. F. Edwards*

Eugenics Society Darwin Research Fellow, Department of Genetics, University of Cambridge, Cambridge, England

	Page
I. Introduction	239
II. The Importance and Origin of Sexual Differentiation	240
III. The Determination of Sex	241
A. Early Theories	241
B. The Chromosome Theory	242
C. The Human Sex Chromosomes	243
IV. The Genetic Problem of the Human Sex Ratio	244
A. The Problem and the Direct Approach to It	244
B. An Alternative Approach Using Mortality Rates	245
V. Evidence on the Variability and Heritability of the Human Sex Ratio	246
A. Types of Data and the Methods of Analysis	246
B. Early Investigations	248
C. Data Which Conform to the Simple Binomial Distribution	249
D. Geissler's Data	249
E. The Data of Ounsted and Hewitt, Webb and Stewart	252
F. Data in Which the Order of Birth Is Recorded	252
G. The Effects of Birth Control	255
H. Miscellaneous Investigations	256
I. Conclusion	257
VI. Evidence on the Variability and Heritability of the Sex Ratio in Other Species	258
VII. Natural Selection and the Sex Ratio	259
A. Introduction	259
B. Fisher's Theory	259
C. The Analytic Extension of Fisher's Theory	260
D. Kalmus and Smith's Theory	261
E. Further Examination of the Concept of Parental Expenditure	262
F. Artificial Selection for the Sex Ratio	264
VIII. Conclusion	265
References	265

I. Introduction

Sexual differentiation may be looked at in two ways: as a widely occurring polymorphism whose origin, purpose, mechanism, physiology, and effects need investigating; and as a genetic segregation requiring

* Present address: Istituto di Genetica, Università di Pavia, Via Sant'Epifanio 14, Pavia, Italy.

statistical analysis and the application of mathematical population genetics.

Many authors have discussed sex from the first viewpoint, and some of the subjects are mentioned in the following two sections. Many others have treated of the purely statistical aspects of the second viewpoint, sometimes, one feels, merely as an exercise in statistical manipulation. It is the purpose of this review to examine the statistical work on the sex ratio, and the genetic implications of the sexual segregation, in man, and in other species in so far as that will throw light on the problem in man. Some of the results reviewed will, of course, be relevant to other species, but this is incidental.

There are some aspects of the statistical work with which this review is not primarily concerned: temporal variations in the sex ratio, including such phenomena as its increase in time of war; variations in the sex ratio with the age of the parents, with parity, etc.; the sex ratio in multiple births and stillbirths; the effects of inbreeding, wishes and occupations of the parents, and exposure to radiation; and associations of the sex ratio with blood groups, and length of gestation. Some of the more recent papers on these topics are included in the list of additional references at the end of this review.

II. The Importance and Origin of Sexual Differentiation

Genetic systems need explaining in evolutionary terms no less than the action of individual factors within those systems. Sexuality is a genetic system which promotes outbreeding—the incompatibility mechanism *par excellence*. Fisher (1930), in "The Genetical Theory of Natural Selection," summarizes the advantage of sexual reproduction over asexual as follows:

> A consequence of sexual reproduction which seems to be of fundamental importance to evolutionary theory is that advantageous changes in different structural elements of the germ plasm can be taken advantage of independently; whereas with asexual organisms either the genetic uniformity of the whole group must be such that evolutionary progress is greatly retarded, or if there is considerable genetic diversity, many beneficial changes will be lost through occurring in individuals destined to leave no ultimate descendants in the species. In consequence an organism sexually reproduced can respond so much more rapidly to whatever selection is in action, that if placed in competition on equal terms with an asexual organism similar in all other respects, the latter would certainly be replaced by the former.

In "Evolution of Genetic Systems" Darlington (1958) devotes a chapter to the evolution of sex:

> Its origins can be seen in the Protozoa and Algae where all degrees of differentiation, all stages in its evolution, occur. In its simplest and probably original form

this differentiation was a differentiation within the individual which therefore bore cells of both kinds. Most of the higher plants are still hermaphrodite, bearing both pollen and eggs. They have . . . various special devices which assure cross-fertilisation. This end is achieved in most of the higher animals by having the sexes separated in different individuals. Here and there in a number of different families of plants we can see the same mechanism of sex differentiation coming into existence. In animals it is long established and indispensable. In plants it is a sporadic and short-lived alternative to other systems.

Darlington goes on to consider the origin of sex in detail, and the interested reader is left in his hands.

III. The Determination of Sex

A. Early Theories

Unlike the subjects of the last section, the mechanism of the determination of sex has been a controversial point for many centuries.

The ancient Greeks speculated a great deal on the mechanism, and Aristotle (ca. 350 B.C.) wrote a review of the theories then current:

Anaxagoras (500–428 B.C.) thought that the male parent determined the sex of the embryo by producing "male" semen from the right testis and "female" from the left. Empedocles (494–434 B.C.) and Democritus (470–370 B.C.), on the other hand, thought that sex was determined in the womb, Empedocles supposing that the time in the menstrual cycle was important, and Democritus that the sex depended on the origins of the parents' contributions.

Aristotle criticized Empedocles' theory on several grounds, the most substantial being that it cannot explain the formation of unlike-sexed twins. Democritus' theory was also criticized because it was clear to Aristotle that semen is not drawn from all parts of the body. Furthermore, if a child is female because of the greater influence of its mother it should take after its mother in all respects, which is not so. Even Anaxagoras' theory was criticized on similar grounds, but Aristotle finally dismissed it with the evidence that men who have had one testis excised are still capable of producing children of both sexes.

Aristotle's own theory stated that females are formed due to the parents' lack of "heat": when there is a deficiency of "heat" the reproductive fluids degenerate and become female-producing. In support of this theory he quoted the "fact" that very young and very old parents tend to beget females because they are not yet in their prime, or are failing. Furthermore, the fact that males tend to be born if the wind was in the north at their conception is explained, since the body is well known to be "hotter" then! Similarly, the effect of the phases of the moon is explained.

Aristotle did not refer to Hipprocrates (ca. 420 B.C.), who supposed that both men and women secrete substances of varying degrees of maleness and potency according to their nature. When both parents secrete substances of the same "sex" normal individuals are produced, but when the secretions are equally opposed hermaphrodites result. This theory, claimed Hippocrates, also explains the different degrees of maleness and femaleness observed.

In the ensuing twenty-two centuries little advance on these theories was made. Thomson (1908) remarks that "the number of speculations as to the nature of sex has been well-nigh doubled since Drelincourt, in the eighteenth century, brought together two hundred and sixty-two 'groundless hypotheses,' and since Blumenbach caustically remarked [sic] that nothing was more certain than that Drelincourt's own theory formed the two hundred and sixty-third. Subsequent investigators have long ago added Blumenbach's theory . . . to the list."

Eventually, of course, the chromosomal theory was accepted, although as late as 1921 (Doncaster) there was still some doubt over its exact nature, due mostly to the fact that in some species the male is heterogametic, and in others the female.

B. The Chromosome Theory

The historical circumstances of the discovery of the chromosome mechanism of sex determination have been fully described by Crew (1954). Briefly, Mendel (1865) himself speculated that sex might be a genetic segregation,[1] and at the beginning of this century several workers found experimental evidence for this. After reports of different chromosome complements in males and females of the same species, McClung (1901, 1902) suggested that sex determination might be vested in the chromosomes. Several cytologists set out to verify this suggestion, and Sutton (1902), Stevens (1905), and Wilson (1905) were successful with various species. Some initial confusion as to the nature of the theory was due to the fact that in some of the species studied one sex has one more chromosome than the other, while in other species the sexual difference was found to correspond to the segregation of a pair of chromosomes of unequal length. However, by the end of the first decade of this century the situation had been almost completely resolved, and a useful review of the work is provided by a paper of Wilson's (1911), "The Sex Chromosomes."

---

[1] According to Crew. I can find no such speculation in the English translation of Mendel's 1865 paper published in Bateson's "Mendel's Principles of Heredity" (1913), but Bateson does mention that Mendel wrote of the matter in his correspondence with Nägeli. At the end of the letter dated September 27, 1870 (see *Genetics* **35**), the subject is indeed broached in connection with a sex ratio of one-quarter.

Shortly afterwards both Morgan (1914) and Doncaster (1914) devoted books to the determination and other genetic aspects of sex. In man, the existence of sex chromosomes was confirmed cytologically by Painter (1924).

Proof that the sex chromosome constitution determines, rather than is determined by, the sex of an individual, was supplied by Bridges' (1916) studies on non-disjunction of the sex chromosomes in *Drosophila*. He also concluded that the Y-chromosome is not functional in sex determination. In a later paper Bridges (1925) extended his work to show that the genes responsible for female development are on the X-chromosome, while those responsible for male development are autosomal. Thus the sexual development of an individual depends upon a balance between the X-chromosomes and the autosomes. In general, equal numbers of X-chromosomes and sets of autosomes in a polyploid individual lead to female development; but added X-chromosomes give rise to superfemales, while added sets of autosomes lead to males, or, in the case of severe unbalance, supermales. Some nearly balanced arrangements give rise to intersexes.

Goldschmidt's work on sex determination in *Lymantria*, started earlier than Bridges' work and summarized in 1934, also exposed a system of balance, although of a somewhat different kind.

On the other hand, in the mouse Welshons and Russell (1959) showed that loss of the Y-chromosome leads to female development, so that this chromosome appears to be male-determining. A similar situation is now known to operate in man: the Y-chromosome is necessary, although not sufficient, for male development (see Ford, 1961).

C. The Human Sex Chromosomes

The ability to identify the sex-chromosome constitution in man is recent, and stems from the observations of Barr and Bertram (1949) on sexual dimorphism in the nuclei of cells from the nervous system of the cat. In 1953 Moore and associates extended this method of differentiation to human skin, and determined the nuclear sex in man.

A more readily observable sex difference was described by Davidson and Smith (1954), who found that a proportion of the polymorphonuclear neutrophile leucocytes from female blood contained a small appendage to the nucleus, which they termed a "drumstick." No drumsticks were found in the blood of normal males, and Davidson and Smith conjectured that the drumstick was a manifestation of the two X-chromosomes of the female, thus implying that the presence or absence of drumsticks indicates the nuclear, but not necessarily the phenotypic, sex. This implication was verified when it became possible to examine the human chromosomes themselves. However, this is not the place to review the great advances that have recently been made in the fields of cell culture, diagnostic sex determination and chromosome identification in man: the reader is referred to the *Symposium on Cytology and Cell Culture Genetics of Man*, edited by G. Allen (1960), and "Sex Determination: Diagnostic Methods" by Davidson (1960). It is sufficient to state here that it is now possible to ascertain the sex-chromosome constitution of an individual accurately; that in the case of XX and XY individuals the nuclear sex as determined by the presence or absence of "sex-chromatin" is confirmed; and that abnormalities of the sex-chromosome constitution have been found and identified (see Ford, 1961).

IV. The Genetic Problem of the Human Sex Ratio

A. THE PROBLEM AND THE DIRECT APPROACH TO IT

The preceding chapters have briefly discussed how and why sexual reproduction has evolved, and the nature of sex determination. Since the chromosomal determination of sex implies a Mendelian segregation of 1:1 in the absence of differential viability, it is a natural sequel to the foregoing discussions to enquire whether this ratio, which is approximately that which is observed in mammals at least, is more advantageous than any other ratio, or whether it is merely a satisfactory by-product of a system of sexual differentiation which is sufficiently advantageous in itself to survive.

The relative abundance of the sexes is commonly measured in terms of the proportion of males, and this proportion is loosely referred to as the sex ratio, a misnomer whose use is too widespread to correct. The primary sex ratio is defined as that at conception. It is the proportion of ova that is fertilized by Y-chromosome-bearing spermatozoa, and may be influenced by the differential capacities of X- and Y-bearing spermatozoa in motility, fertility, and viability. The secondary sex ratio is defined as that at birth, and is the primary sex ratio modified by the differential mortality rate *in utero*. It is the ratio most often observed, and from a knowledge of it and the relevant mortality rates inferences may be made about the primary ratio. Stillbirths are sometimes excluded in calculating the secondary sex ratio, but whether this has been done in a particular case is usually stated. The tertiary sex ratio is ill defined in the literature: it is here defined as the sex ratio at the age at which the children become independent of their parents. The reason for this definition will become apparent later on.

In order that selection may act on the sex ratio at any stage it is necessary that the ratio at that stage be genetically variable. Now it is clear that there is no scope for variability in the initial segregation of X- and

Y-bearing spermatozoa unless some major change in the mechanism is at hand. However, in the ensuing stages the sex ratio can be modified by sex-differential mortality. Such mortality may be genetically controlled to some extent, with the result that the primary and subsequent sex ratios may be expected to be genetically variable.

Since the initial segregation of the two types of spermatozoa, presumably in equal numbers, is an established system, it would be surprising if an enquiry into the effect of natural selection on the sex ratio showed that it is disadvantageous. It is more likely that selection will be concerned with the sex ratio at some later stage, and provided the optimum ratio at that stage is not too different from one-half there should be enough variability in the mortality rates for the optimum to be achieved. In this case it would seem likely that the X-Y mechanism of sex determination survived because it was capable of producing the optimum sex ratio at some stage, while alternative mechanisms mostly disappeared because of their inability to do this. It would be interesting to hear from zoologists whether this hypothesis is tenable in the light of the alternative mechanisms which do exist in some species.

Logically, the first step in solving this problem in man is to establish that genetic variability in the sex ratio exists. It happens that the great majority of data on the human sex ratio refers to the secondary ratio, and, independently of whether this ratio transpires to be the important one in evolution, the most fruitful approach seems to be to examine it for variability. A review of attempts to demonstrate this variability constitutes the next section.

B. An Alternative Approach Using Mortality Rates

An alternative approach is to examine the evidence for variability in the mortality rates at various stages after conception. There is, however, little agreement on what these rates are, let alone on whether they are genetically variable. They are important because they represent the only method of getting information about the primary sex ratio at the moment, since an examination of the sex-differential properties of spermatozoa has not yet been accomplished [see Rothschild et al. (1960) who doubts whether the two types of spermatozoa have ever been separately identified in man, although Gordon (1957) has claimed some success with experimental separation in the rabbit]. Until recently it had been thought that the primary sex ratio was considerably greater than one-half. This estimate was obtained by sexing aborted fetuses and then calculating what primary sex ratio would be necessary to give rise to the known secondary sex ratio in view of the mortality rates. However, it is difficult to sex early abortions, as McKeown and Lowe (1951a) mention in their review of the evidence for a high primary sex ratio, in which they compare their own results with those of previous workers. They state that the existing data are inadequate to justify any assumption about its value; that "at least half of all abortions occur in the first three months" of pregnancy; and that the sex ratio of live fetuses is about one-half at the seventh month. But reliable information about the sex ratio of early abortions may soon be forthcoming through the method of nuclear sexing.

Columbo (1957) came to the same conclusion as McKeown and Lowe by dividing nearly ten thousand women into two classes with high and low abortion rates and finding that the secondary sex ratio did not differ significantly between the two classes. However, his argument is weak because it rests on the assumption that there is no association between the abortion rate and the primary sex ratio. Bunak (1934) advanced an interesting theory to explain the supposed excess of male abortions: he suggested that some XX fetuses develop male characteristics and are thus aborted, only to be classified as males. Dahlberg (1951), also believing the primary sex ratio to be high, suggested that "old" ova tend to be fertilized predominantly by Y-bearing spermatozoa, and also tend to develop into fetuses with a high relative probability of being aborted. Hart and Moody (1949) provided some evidence for this by showing that, in rats, a long interval between ovulation and semination increases the proportion of males born. The technique of nuclear sexing should finally settle whether such hypotheses are necessary.

V. Evidence on the Variability and Heritability of the Human Sex Ratio

A. Types of Data and the Methods of Analysis

A consideration of the variability of the human sex ratio is necessarily couched in terms of probability, and it would therefore seem appropriate to start this section with the definition of probability. But unfortunately there is little general agreement on how probability should be defined, and this review is not the place to conduct a search for a satisfactory definition.

Fisher (1958) has outlined his approach to the concept of probability in an article in which he uses the determination of sex as an example, and it is convenient to state the present statistical problem in Fisher's terms.

The majority of data on the human sex ratio gives the number of boys and girls in a family. Sometimes the actual sequence of the sexes is given, but this is rare on account of the space and labor involved in the presentation. The analysis appropriate to these cases will be described below. It must not be supposed that this type of data is particularly well suited to the present problem: a program designed to investigate the variability and heritability of the sex ratio would pay much more attention to pedigree data.

The object of statistical analysis of the available data is to try to demonstrate that subsets comprising families, or groups of families, exist, and, if possible, to identify them. For example, if certain data in the form of binomial distributions for the different family sizes exhibit variances consistently greater than those expected on the theory of the simple binomial distribution, the most elementary hypothesis would be that subsets of families exist in the data, although they cannot be identified with particular groups of families. Although this hypothesis is the simplest, it is by no means the only one, and the multiplicity of acceptable hypotheses renders the interpretation of "binomial" sex ratio data very difficult.

In the following description of ways in which the binomial variance may be altered, the concept of variations in probability will be used, it being understood that this is just another way of describing the existence of subsets.

Edwards (1960a) has described how variations in the probability of a birth being male between and within families, and a correlation between the sexes of successive births in a family, can affect the binomial variance. Briefly, variation in the probability between families (Lexian variation) increases the variance over that of the simple binomial distribution, while variation within families, such as a parity effect (Poisson variation), decreases it. A correlation between the sexes of successive children will increase the variance if it is positive but decrease it if it is negative.

Any of these variations or correlations, if they exist, are undoubtedly very small, with the result that the modified binomial distributions that arise by taking them into account are, for a given variance, almost identical whether that variance is due to variability of the probability, or a correlation, or a combination of both. Even if data fit the simple binomial distribution satisfactorily they may do so because, for example, the effects of a positive correlation and variability in the probability within families cancel out.

Thus it transpires that the only purpose of fitting modified binomial distributions to data on the human sex ratio is to have the satisfaction of knowing that there *exist* hypotheses that can explain them, without knowing precisely which hypotheses. Of course, there is the possibility that *no* modified binomial distribution will fit satisfactorily, in which case a further hypothesis will be necessary to account for the remaining anomalies.

There is, however, one statistic, Δ (Edwards, 1960b), which can be used to differentiate between a correlation and a varying probability in very large sets of data. Δ is defined as the sum of the terms of a distribution for a given family size that contain an even number of girls, less the sum of terms containing an odd number. In the simple binomial distribution this is clearly

$$p^N - Np^{N-1}q + \binom{N}{2}p^{N-2}q^2 - \binom{N}{3}p^{N-3}q^3 + \cdots \pm q^N = (p - q)^N$$
$$= (-1)^N G_N(-1),$$

where $p = 1 - q$ is the probability of a male birth, N is the number in the family, and $G_N(s)$ is the probability generating function of the distribution. For more complex distributions it is easy to see that the expression $(-1)^N G_N(-1)$ satisfies the original definition.

For the simple, Lexian, and Poisson binomial distributions it transpires that Δ is negligible so long as p is near one-half and variations in the probability are small. But in the correlated, or Markov, binomial distribution, as described by Edwards (1960a), Δ_N is zero when N is odd, and $r^{N/2}$ when N is even, where r is the correlation coefficient between successive births, provided that p is near one-half.

There are thus three possible types of data presented solely in the form of binomial distributions for each family size: that which fits some sort of binomial distribution, and in which Δ does not differ significantly from zero; that which fits a modified binomial but in which Δ does differ significantly from zero in some family sizes; and that which does not fit a modified binomial satisfactorily.

B. Early Investigations

Before describing more modern analyses of data in terms of these three types, it is of interest to record some earlier work. At some stage in the eighteenth century, or earlier, it must have been realized that the determination of sex was a chance phenomenon capable of being treated in terms of probabilities, and Laplace (1820) and Poisson (1830) certainly appreciated this when discussing whether the sex ratios in several parts of France were noticeably different. An important contribution was made by Newcomb (1904), who set out "to discover a criterion by which we may distinguish between inequalities in the division of a family between the two sexes which are simply the result of chance and those which are the result of a unisexual tendency on the part of the parents," and was thus the first person to appreciate the exact nature of the problem. However, he found no evidence for a "unisexual tendency."

Monographs on the sex ratio were written in Germany (Düsing, 1884), Italy (Gini, 1908), and France (Worms, 1912). Gini's book, which will be referred to below, contains many references to earlier sets of data, but it is not intended to review any of these works systematically here.

C. Data Which Conform to the Simple Binomial Distribution

In spite of the labor that has gone into the collection and analysis of data of the first type mentioned above there is little point in considering them further. Such data have been described by Parkes (1924a), Rife and Snyder (1937), Myers (1949), and Taylor (1954), and their simple binomial form is almost certainly due to the small size of the samples.

D. Geissler's Data

1. Early Analyses

The only set of data large enough to be of the second type is Geissler's (1889). This remarkable collection of data, comprising some four million births, has dominated the field of the human sex ratio for nearly three-quarters of a century. It has been analyzed many times, but still remains enigmatic. Geissler himself had all the sequences of the sexes available as well as the final distributions. He fitted the simple binomial distributions to the data, and, in the absence of tests of significance, thought they fitted well. But when he investigated the sequences themselves he found a slight negative correlation between the sexes of successive children in a family, but a positive correlation when all the previous children were of the same sex.

Subsequent workers have confined their attention to the distributions themselves, and Gini (1908) was the first to note that the observed distributions had consistently greater variances than those expected on the simple binomial hypothesis. He ascribed these to the variability of the probability between families. He also noted the curious anomalies that have plagued analyses of the data ever since: in families with an even number of children the statistic Δ_N is frequently significantly different from zero, being positive when N is a multiple of four, but otherwise negative. He proposed his "inversion" theory to explain these, in which the population consists of two parts, one having a preponderance of males in the early births of a family, changing to a preponderance of females in the later births, and the other having the same change in reverse. Apart from the inherent unlikeliness of such a dichotomy, this theory must be rejected because it gives rise to two superimposed Poisson binomial distributions in which, as has been mentioned above, the statistic Δ is negligible. In 1911 Gini published a paper in which he proposed the use of the β-binomial distribution, which will be further mentioned below, but he did not fit it to the data.

Not knowing of Gini's work on Geissler's data, Waaler (1928) tried to explain the anomalous values by supposing that the distribution of p between families was irregular, but since this improbable theory requires a different distribution of p for each family size, it must also be rejected.

Harris and Gunstad (1930a,b) analyzed the data at some length, but did not notice the anomalies, and concluded nothing new.

In successive editions of *Statistical Methods for Research Workers* Fisher (1925–1958) uses Geissler's data for families of eight in an example of fitting the binomial distribution, but does not attempt an explanation of the anomalies, stating that their unexplained presence "detracts from the value of the data." Stern (1949) investigated the data in his "Principles of Human Genetics," and inclined to the view that the probability is not continuously variable between families, but that only a small proportion of families has abnormal probabilities. He revised this view, in the light of recent research, for the second edition of his book (Stern, 1960).[2]

Slater (1944) believed that he had observed a similar excess of families with even numbers of both boys and girls in his own data, and he supposed that both men and women have a bias toward the production of one sex or the other. Thus three values of the probability would be expected: a high one when both parents were biased toward the production of males; a value of one-half when there was one parent of each kind; and a low value when both parents were biased toward the production of females. With this theory a trimodally inclined distribution would result, and it is unsatisfactory for the same reason as Waaler's theory.

But in fact Slater's statistical methods are erroneous: in considering the distributions he decides that "in order to give the infrequent larger families their proper weight, it seems right to make the calculation in terms of individuals rather than in terms of families," and he proceeds to multiply each distribution by the relevant number of children in a family before conducting the χ^2 analysis. This multiplication invalidates the test of significance: one must not multiply a body of data by a factor in order to get more data! Slater commits a further error in his Table 3, where he sums five χ^2's that are not independent. The true values of the significance levels are well below those that he quotes, and there is no need to regard his data as anomalous. The variances of the distributions do, however, exceed their simple binomial expectations.

In 1950 Lancaster wrote a paper expressly for the purpose of discrediting Geissler's data. He concluded that the anomalies are due to the unreliability of German mothers in declaring the sexes of their children, but a much more specific theory is required if the regular deviations from expec-

[2] In his review of this edition Penrose (1961) wrote "it might have been hoped that Geissler's worthless data would have been omitted." We prefer the more balanced comments of Fisher (quoted on this page) and Gini (quoted on p. 251).

tation are really to be ascribed to irregularities in the collection of the data. More detailed criticisms of Lancaster's paper have been made by Gini (1951) and Edwards (1958). The former, who is widely experienced in the analysis of data on the human sex ratio, comments:

> If Lancaster had known the literature on the subject better, he would have found that the supposed inconsistencies were discussed in it and explained. This is particularly the case for the significantly low probability of a male birth in families which had but one female with an excess of males.
>
> To sum up, it is difficult to see any serious reason for doubting the reliability of Geissler's data, all the more so as the general accuracy of German statistics is well recognized. In any case, to throw doubt on the accuracy of figures which you cannot explain is a very risky proceeding, only admissible in the case of scholars who have studied the subject thoroughly.

Although the explanations of the inconsistencies to which Gini refers are unsatisfactory, I endorse the rest of this comment. Gini's paper is a review of work on human sex ratio data, especially on Geissler's data, and contains references to many of the older investigations. With respect to Geissler's data it is largely a restatement of his earlier theories, which have been referred to, but in other respects it is a useful summary.

Robertson (1951) concluded that the probability is very significantly variable between families but, quoting Lancaster, he dismissed the data as being unreliable. His hypothesis is that the probability is normally distributed between families, in spite of the fact that the normal distribution does not confine p to the limits 0 to 1. The resulting mathematics is extremely cumbersome, and a number of approximations are necessary in order to arrive at any results.

2. *Edwards' Analysis*

An alternative distribution for p was suggested by Edwards (1958). If p is assumed to be a β-variate, the resulting modified binomial distribution and the analysis of heterogeneity are relatively simple. At the time when I used this distribution I was under the impression that it had first been given by Skellam (1948), and hence I referred to it as "Skellam's distribution." But in fact Gini (1911) assumed that p was a β-variate, and the Appendix to my paper adds nothing to Gini's treatment. I therefore propose that this modified binomial, or β-binomial, be referred to as "Gini's distribution."

In this most recent analysis of Geissler's data the following conclusions were drawn:

(1) Provided there is no correlation between the sexes of successive children in a family, and neglecting the anomalous deviations referred to above, the variance of p between families is of the order of 0.0025.

(2) There is no evidence that parents only capable of producing unisexual families exist.

(3) It may be possible to explain the anomalous deviations on the hypothesis of a negative correlation.

Further research into the effect of a negative correlation showed that it could indeed explain the anomalies qualitatively, leaving no inexplicable deviations from expectation in the data (Edwards, 1960b). But the magnitude of the correlation coefficient necessary was much too large for the theory to be acceptable (values of the order of -0.3 are required, and these are quite unsubstantiated by other data, as will be shown below). Yet it is striking that such a simple hypothesis can explain the complex variations of the statistic Δ, and it may well contain a germ of truth. The estimated values of the correlation coefficients increase with family size, as might be expected since larger families have, on the average, smaller birth intervals, and therefore the influence of the sex of a child on the sex of the succeeding one may be expected to be stronger. The biological mechanism of such a correlation is discussed below.

3. *Conclusion*

Thus after many years of analysis Geissler's data remain enigmatic. It is not possible to draw any conclusions of biological importance from them, and this situation must remain until the anomalies of the distributions can be satisfactorily explained.

E. THE DATA OF OUNSTED AND HEWITT, WEBB AND STEWART

The third type of data, that which cannot be explained on any binomial hypothesis, is perhaps the most interesting. Ounsted (1953) found that there was an excess of all-male families in his material, which was collected through propositi with convulsive disorders. Hewitt *et al.* (1955) found a similar excess in their sample of families, which were originally collected for an investigation into perinatal deaths. In the latter case the excess is heterogeneous in that it only occurs in half the sample. It is tempting to dismiss these findings as being due, in some way, to the methods of ascertainment, but there is no evidence for this, and it may well be that there are a few families with abnormal numbers of males in the English Midlands: Ounsted's families come from Oxford and the anomalous ones of Hewitt and associates' data from Northamptonshire. Both sets of data are small, and the excess of all-male families is not reflected in any other well-known data.

F. DATA IN WHICH THE ORDER OF BIRTH IS RECORDED

1. *Introduction*

Much more use may be made of data in which the actual sequence of the sexes in each family is recorded. Their analysis need not be ambiguous as is the case with data recorded solely in the form of binomial distributions. Three sets of such data have recently been analyzed: 14,230 French families of five or more children (Turpin and Schützenberger, 1948; Schützenberger, 1949); 5,477 Swedish families (Edwards and Fraccaro, 1958, 1960), and 60,334 Finnish families of two, three, and four children (Renkonen, 1956; Edwards, 1961a).

2. The Swedish Data

The conclusion drawn from the Swedish data is easy to record, for there is no evidence for any departure from a chance determination of sex. Tests were made for a variation in sex ratio with parity; for an excess in the variances of the distributions over the simple binomial expectations; for a correlation between consecutive births in a family, births separated by one other birth, and births separated by two other births; and for the influence of the sexes of previous births on whether or not procreation was continued. None of these tests gave significant results. The estimated value of the correlation coefficient between the sexes of successive births is 0.0044 ± 0.0074.

3. The French Data

On the other hand, Schützenberger's data give the following estimates of the correlation coefficients between successive births, births separated by one other birth, and births separated by two other births (Turpin and Schützenberger, 1948):

$$r_0 = 0.029 \pm 0.0035$$
$$r_1 = 0.007 \pm 0.0039$$
$$r_2 = 0.004 \pm 0.0043$$

Edwards (1959) has criticized Schützenberger's analysis, mainly because of its frequent numerical inaccuracies, but the above estimates seem to be valid. Further, Schützenberger finds that a correlation coefficient of 0.029 between the sexes of successive children adequately explains the variances of his distributions. Since r_1 does not differ significantly from r_0^2, nor r_2 from r_0^3 (see Edwards, 1959, 1960b), and the variances of the distributions are adequately explained by the correlation alone, these data are entirely compatible with the theory that sex is determined by chance except in so far as the sex of a child is correlated with that of its predecessor. Schützenberger's value of the correlation coefficient is clearly significantly different from the value found in the Swedish data, so that the two sets of data must be judged incompatible.

Schützenberger attempted to explain the correlation which he observed by postulating the existence of "périodes gynophiles et androphiles" in the life of the parents, during which the probability of bearing a boy is $p - r/2$ and $p + r/2$, respectively. But (as Edwards, 1959, pointed out), he failed to notice that such a variation in the probability would give rise to Poisson's generalized binomial distribution, the variance of which is less than the comparable simple binomial variance by $Nr^2/4$ in this case, and, further, that the simulated correlation would in fact be negative. This hypothesis is therefore untenable.

It is simpler to suppose that after a male birth the chance of survival of the next embryo is greater if the embryo is male than if it is female, and vice versa after a female birth. It is reasonable that the sex of a child should make some difference to the constitution of the mother, which in turn could influence the viability of the next embryo. This theory leads to two expectations: first, that the interval between births of like sex should be shorter than that between births of unlike sex, because the latter births would have a higher probability of being separated by an abortion; and, second, that the majority of abortions, miscarriages, and stillbirths should be of opposite sex to the preceding live birth. The first expectation is upheld by some observations of Turpin and Schützenberger (1952a,b), who found it in the case when one of the pair of births was a twin birth, while the second has never been investigated.

4. *The Finnish Data*

Renkonen's very extensive Finnish data confirm Schützenberger's findings. The original analysis was short and superficial, but Edwards (1961a) has subjected the data to an extensive analysis based on the technique of factorial experimentation.[3] In retrospect, it is clear that this method is much superior to previous methods: it is simple to use and understand, and presents an integrated account of the data such has not previously been possible.

In the case of Renkonen's data the resulting values of χ^2 are so striking that it is worthwhile repeating them. Table 1 gives the values summed over the three family sizes and divided according to the interactions being tested. It will be seen that the only significant entry concerns the association between the sexes of consecutive births, and that this is very significant. That the remaining values of χ^2 are so low is remarkable, considering the size of the data.

It may be concluded that the only departure from a chance distribution of the sexes is due to the sex of one child influencing that of the next. The mean value of the correlation coefficient is 0.0261 ± 0.0032, which does not differ significantly from Schützenberger's value. In the figures for the correlation coefficients for the various pairs of births, given by Edwards, two trends are evident: the association is weaker in later pairs

[3] See Edwards (1962) for a correction to this analysis.

of births in a family, and is stronger in the larger family sizes. The latter trend might have been anticipated, for the reason given when discussing Geissler's data, but a similar hypothesis for the association being

TABLE 1
Renkonen's Data: Factorial Analysis of χ^2

	χ^2	Degrees of Freedom
Main effects		
(Testing deviations of sex ratio from one-half)	11.3489	9
Consecutive two-factor interactions		
(Testing associations between consecutive births)	83.6294***	6
Non-consecutive two-factor interactions		
(Testing associations between births separated by at least one other birth)	7.2245	4
Three- and four-factor interactions	9.2142	6
	111.4170***	25

*** Standard notation to indicate significance at the 0.1% level.

weaker in later pairs of births in a family would require the interval between births to grow longer as the size of the family increased.

G. The Effects of Birth Control

It may be wondered whether the various authors whose work has been reviewed have taken into account the effects of birth control. In their analyses of Geissler's data both Gini and Edwards considered it, but its effects cannot be expected to be straightforward. Goodman (1961) has recently written a paper on the subject, but as there is a danger of his results being misinterpreted, it is as well to examine the problem anew.

In the first place, it is clear that, provided the sex of a birth is independently determined, no degree of birth control can change the over-all population sex ratio. Only if it is supposed that the probability of a birth being male varies between families can birth control influence the over-all sex ratio, as Goodman has shown. However, the present problem is somewhat different from that which Goodman considered, for in it the population is broken down into distributions for each family size. In this case, even if the births are independent, birth control can affect the data: for example, if couples with just one boy never had another child, but couples with one girl invariably did, the sex ratio in complete families of one child would be one, because families with two or more children would not be included in the distribution. There would, however, be no comparable disturbance in the distribution for the first child in families of one or more children. It is for this reason that Edwards and Fraccaro analyzed the distributions of the first N children in families of N or more children from the Swedish data, but this procedure has the disadvantage that succeeding distributions are then not independent, so that it is inappropriate if a complete factorial analysis is made, such as Edwards' analysis of the Finnish data.

Whether the frequency with which procreation is continued is dependent on the sexes of the children born previously can be tested, for a given family size, by setting the distribution for complete families of size N against that for the first N children in families of more than N in a contingency table. On the Swedish data this test provided no evidence for such an effect.

If there were a positive correlation between the sexes of successive children in a family, such as has been found in the French and Finnish data, it might just be possible to construct a complex theoretical model in which birth control leads to the observed results. But such a hypothesis would hardly be credible without additional evidence, especially since, in complete families of two children, a negative correlation would be anticipated due to those families with two children of like sex being more frequently enlarged than those with two children of opposite sexes, as Thomas (1951) has shown occurs. Indeed, this situation was found to hold more generally in Geissler's data by Gini (1951).

H. Miscellaneous Investigations

It is a platitude to say that familial data extending over several generations are required in order to demonstrate the heritability of the human sex ratio, and it is perhaps unfortunate that recently such data have not received the attention accorded to the types of data reviewed above. This relative disinterest has been largely due to the fact that investigators have been interested in their own data per se, and not in any biological conclusions that may be drawn from them. My own entry into this field was precipitated by a purely statistical interest in Geissler's data, and the biological relevance of the analysis only appeared at a later stage.

Two well-known family trees with a marked excess of one sex must, however, be mentioned. Lienhart and Vermelin (1946) described one which contained seventy-two female births and no male, and put forward several alternative hypotheses of how this could happen, and Harris (1946) described a family tree with a marked preponderance of males, which he ascertained through a member with microspermia. The males throughout the line are relatively infertile, and the only adult female has many male characteristics, and is infertile.

Slater's (1944) data have already been mentioned. As well as investigating the siblings of his propositi Slater also recorded the sexes of their more distant relatives, and showed that there was an association between the sex ratios in different branches of each family. However, even such a direct investigation does not show that the sex ratio contains a heritable component, for the association might be due to a similarity of environment between different branches of the same family, if differing environments gave rise to differing sex ratios by influencing, for example, the mortality rates *in utero*.

One isolated contribution to the present problem was made by Hemphill *et al.* (1958), who studied a small population of male homosexuals and found that, of the 24 who were married, 15 had children, there being 21 boys and 22 girls in all. It might have been supposed that men of abnormal sexual development of this nature would have an unusually high probability of begetting girls, but there is no evidence for this.

Racial variability of the sex ratio is a subject which attracted many of the earlier writers, and there is no doubt as to the reality of the differences [for example, the Korean sex ratio is well known to be much higher than European ones, Kang (1959) quoting a value of 0.570]. But whether these differences are genetic is a subject for enquiry: it would be interesting to follow up the marriages of Koreans with other nationalities.

I. Conclusion

Of the major analyses that have been briefly reviewed here, only those of data in which the order of birth has been recorded can have much influence on a general hypothesis, for it is clear that sex-ratio data available solely in the form of binomial distributions are of little use, their analysis being difficult and their interpretation uncertain.

The French and Finnish data are compatible, and provide strong evidence that there is a positive association between the sexes of successive children in a family. Of the French data Edwards (1959) wrote "that this hypothesis [of a correlation] should not be accepted without confirmation on independent data." In view of the Finnish data, this reservation may now be withdrawn. In the Swedish data, however, the determination of sex appears to be entirely fortuitous, and in none of these samples is there any evidence for the variability of the probability of a birth being male between families, and thus there is no indication that the sex ratio might be heritable.

Indeed, apart from the excess variances in Geissler's anomalous data, the only clues to the heritability of the sex ratio are the differences between races, Slater's familial data, and the two family trees referred to above. The latter may, however, be due to the action of rare major genes or inherited chromosome abnormalities, and it cannot be inferred that the sex ratio in the majority of families is genetically controlled, as presumably such control would be polygenic.

VI. Evidence on the Variability and Heritability of the Sex Ratio in Other Species

Knowledge of the heritability of the human sex ratio is necessarily limited by the restricted types of investigations that can be carried out on human populations. But it is reasonable to suppose that the genetic situation with regard to the sex ratio in man is similar in some respects to the situations in other species, and in particular in other mammals. Inquiries into the variability and heritability of the sex ratio in other species may therefore throw some light on the problem in man, and the findings of some such inquiries are reviewed in this section.

Some genetic control has been found in species of *Drosophila* (see, for example, Morgan et al., 1925; Gershenson, 1928; Sturtevant and Dobzhansky, 1936; Novitski, 1947; Wallace, 1948; Gowen and Fung, 1957; Malogolowkin, 1958), although Falconer (1954) failed to change the sex ratio by direct selection. Most of the cases described involve the production of unisexual, or nearly unisexual, progenies. The continued existence of genetic mechanisms causing such progenies in natural populations is difficult to explain, and the problem is taken up in the next section.

King (1918) succeeded in selecting for high and low sex ratios in the Norwegian rat, demonstrating by crossing that the doe was the important parent, but Falconer (1954) failed to do so in the house mouse. Weir (1953; Weir and Clark, 1955) selected for high and low blood pH in the mouse and found that the sex ratio responded simultaneously. In a further experiment (Weir and Wolfe, 1959; Wolfe, 1961) the sex ratio responded in the reciprocal manner in the first three generations of selection, so the results are not clear-cut. In this connection McWhirter (1956, 1960) has suggested that blood pH and sex ratio may normally be associated in mammals.

Weir (1955, 1958, 1960; Weir et al., 1958) followed up his findings with experiments that showed that the male parent "controls" the sex ratio and that the mortality rates *in utero* are not significantly sex-differential. If these results are true they indicate that it is the primary sex ratio that has been modified. On the other hand, Howard et al. (1955) showed that the sex ratio in the mouse is to a certain extent genetically controlled, but by differential mortality, since the influential genotype seems to be that of the zygote itself. They found no heterogeneity between different litters of the same mating type, parity, and size.

The problem of whether mortality *in utero* is sex-differential is a difficult one, as has been observed in the case of man. MacDowell and Lord (1925, 1926) agree with Weir *et al.* (1958) that in mice it is not, but Lindahl and Sundell (1958) found, by nuclear sexing, a high sex ratio before uterine implantation in the golden hamster, which implies differential mortality since the secondary sex ratio is about one-half.

Parkes investigated the sex ratios in various mammals in the 1920's and summarized his, and other workers', findings in a review paper "The Mammalian Sex-Ratio" (1926a). This comprehensive review is very valuable, but so much work on sex ratios has been published since its time that a modern equivalent is needed. Data from different species are scattered throughout biological literature, and the present review, with its accent on the human sex ratio, is no place for their evaluation.

But enough has been said for it to be apparent that there is evidence for genetic control of the sex ratio in some cases, with the result that the following work on the influence of natural selection on the sex ratio, which is the main thesis of this review, is not without relevance to practical situations. The existence of genetic control of the sex ratio in some cases does not, of course, imply its existence in man, although it shows that there is no biological reason to doubt that genetic control in man occurs.

VII. Natural Selection and the Sex Ratio

A. Introduction

The remarkable thing about the sex ratio, regarded as a genetic segregation subject to natural selection, is its resilience: however many individuals of one sex are killed in one generation, provided enough remain for the continued existence of the species, the sex ratio is back to normal in the next generation. This is due, of course, to the fact that the sex chromosomes are not fully subject to random assortment, as are the autosomes. Realization of this seems to have inhibited mathematical geneticists from considering the sex ratio as a segregation subject to evolutionary forces.

In 1871 Darwin considered whether the sex ratio was subject to natural selection. After surveying the available statistics and investigating the selective effects of infanticide at some length he concluded by saying that as far as he could see there was no selective advantage attached to a particular sex ratio, and that "I now see that the whole problem is so intricate that it is safer to leave its solution to the future."

B. Fisher's Theory

There the matter lay until it was taken up by Fisher (1930), who outlined the solution to the problem. His approach depends on essentially economic arguments which take into account the parental expenditure involved in the rearing of offspring to maturity. This expenditure will be a function of the time and energy which the parents are induced to spend in rearing their offspring, and must therefore depend on the mortality during the period of parental expenditure. Here is a shortened account of Fisher's argument:

Let us consider the reproductive value, or the relative genetic contribution to future generations, of some offspring at the moment when parental expenditure on them has just ceased. It is clear that the total reproductive value of the males in a generation of offspring is equal to the total value of the females, because each sex must supply half the ancestry of future generations. From this it follows that the sex ratio will so adjust itself under the influence of natural selection that the total parental expenditure incurred by children of each sex shall be equal; for if this were not so and the total expenditure incurred in producing males, for instance, were less than that incurred in producing females, then parents genetically inclined to producing males in excess would, for the same expenditure, produce a greater amount of reproductive value, with the result that in future generations more males would be reared. Selection would thus raise the sex ratio until the expenditure upon males became equal to that upon females.

In 1953 Shaw and Mohler tackled the problem, but they consider that "Fisher's treatment is phrased in non-genetical terms and does not lend itself to further development." They therefore "ignore instances involving parental care" and, working in terms of the reproductive value alone, come to the anticipated conclusion that the equilibrium sex ratio at conception should be one-half. Since the production of offspring by a sexual organism always involves expenditure by the parents, and since Fisher has shown that this expenditure is relevant, their treatment is necessarily incomplete.

Shaw (1958) followed this up with some generation-by-generation calculations of gene frequencies in models involving sex-ratio loci. But he only considered genes causing unisexual progenies, and cases where there are no sex chromosomes, so that his work is irrelevant to the human problem. Of Fisher's "principle" he says it "merits special attention since it may hold for man. Nevertheless, it cannot be considered here."

C. The Analytic Extension of Fisher's Theory

Bodmer and Edwards (1960) have attempted to put Fisher's theory on an analytic basis, and have investigated the population behavior of the sex ratio. They concluded that the important sex ratio is that at the end of the period of parental expenditure (the tertiary sex ratio), and that this has an equilibrium value of $1-h$, where the expected expenditure on a male child that outlives the period of parental expenditure to that on a female is as h to $1-h$. They showed that when the mean sex ratio of the population is at the equilibrium value all sex ratios are equally advantageous, but when it is not at equilibrium those sex ratios tending to return it to equilibrium are at a selective advantage over others.

In human populations, where h is presumably little different from one-half, the equilibrium state is reached when the tertiary sex ratio is about one-half. Writers have often commented upon the fact that the human sex ratio at the reproductive age is about a half (although higher before that age and lower after it), and Crew (1937) supposed that it was advantageous to have equal numbers of the sexes available for mating. Since the ages of reproduction and parental independence cannot differ much in man, this numerical equality of the sexes is predicted by Fisher's theory.

Bodmer and Edwards (1960) go on to show that the genetic variance in sex ratio in a population will, on certain simplifying assumptions, decrease slowly compared with the rate of approach of the population to equilibrium, and that this rate will be directly proportional to the genetic variance. If V is the genetic variance, assumed constant, then the deviation of the sex ratio of a population from its equilibrium value near one-half will decrease by a factor $e = 2.718$ in $1/(4V)$ generations. Even if the genetic variance of a human population is as large as 0.0025 this is 100 generations, or about 2000 years. However, since the variance is certainly much less, the time must be proportionately longer; thus changes in the human sex ratio due to natural selection are too slow to be detected over the period for which data are available, even if the equilibrium values have been changing, although they may not be slow compared with other evolutionary processes. Kolman (1960) has also expounded Fisher's theory mathematically, though in less detail than Bodmer and Edwards.

D. Kalmus and Smith's Theory

Kalmus and Smith (1960) have recently put forward the view that the most important reason for the selective advantage of a sex ratio of one-half over other ratios is that it maximizes the chance of encounter between the sexes, as Crew (1937) suggested. Further, they state that, in small populations, it will be advantageous if the "effective size" of the population, in terms of Sewall Wright's parameter, is as large as possible so that the genetic variance of the population shall be as large as possible, and, for a given population size, the effective size is at a maximum when the sex ratio is one-half.

Of Fisher's theory, Kalmus and Smith remark: "In the case of species which rear their young, a further reason for expecting equality of sex

ratio at the time of sexual maturity [sic] has been given by Fisher . . . ."
These views have been criticized by Edwards (1960c) on the following grounds:

Consider a population with a sex ratio of one-third, whose Fisherian equilibrium sex ratio is one-half. On Fisher's theory that population, even if isolated, will evolve toward a sex ratio of one-half. Kalmus and Smith, on the other hand, suggest that it will be replaced by a population with a sex ratio nearer to one-half because the latter population will be more prolific on account of the greater chance of its members of opposite sex meeting, or because, in a changing environment, it has greater genetic flexibility. Their suggestions contain no mechanism for intrapopulation selection. While it must be admitted that, in the absence of any other reason for a sex ratio of one-half, populations with great numerical inequality of the sexes would be at a disadvantage, it is clear that such inequalities will not arise owing to Fisherian selection operating within populations. Indeed, selection of the kind that Kalmus and Smith envisage could only be important in small populations struggling to produce enough offspring for survival, in which case one might have supposed that it would have been advantageous to have an excess of females. In this connection it is interesting to note that Pitt-Rivers (1929) claimed to have shown that decreasing human populations tend to have an excess of males, but increasing ones an excess of females, at the reproductive age.

Kalmus and Smith's theory is also difficult to reconcile with the situation in very polygamous species in which a single male, after defeating his competitors, fertilizes a great number of females. It might have been supposed that such species would contain a marked excess of females, but this has not been observed. To state that Fisher's theory is relevant "in the case of species which rear their young" is to misrepresent it: the theory is relevant whenever reproduction demands expenditure by the parents, that is, in all sexually reproducing species.

Thus it seems that the proposed theory, in spite of its formal correctness, is irrelevant to a consideration of the sex ratio in natural populations. The magnitude of variation in the sex ratio between populations that Fisherian selection allows must mean that variation in the selective advantage between populations due to their sex ratios is infinitesimal compared with variation due to other factors.

E. Further Examination of the Concept of Parental Expenditure

In view of these recent contributions to the theory of natural selection and the sex ratio, it is now possible to appreciate the part played by parental expenditure.

Shaw and Mohler's model assumed no parental expenditure and a heritable primary sex ratio, which they demonstrated has an equilibrium value of one-half. Further, they showed that sex-differential mortality does not affect this conclusion. It is easy to see that this must be so, for, as Fisher remarked, the equilibrium sex ratio "will not be influenced by differential mortality during a self-supporting period," the reason being that such mortality does not influence the reproductive value per unit parental expenditure.

Shaw and Mohler's work, and Shaw's following it, is irrelevant to the human problem simply because it takes no account of parental expenditure, with the result that it can only concern the primary sex ratio, and it is not this which is most likely to be genetically variable. The variables most probably involved are the early mortality rates, which, since they are very large, offer much scope for selection. In point of fact Fisher's theory does not specify how the equilibrium tertiary sex ratio is attained, and thus does admit of a heritable primary ratio.

The concept of parental expenditure involves that of replacing an offspring that has died early, and it may be argued that, if this concept is necessary in a particular case, it should be relevant to other situations. To take an example, which happens to involve the sex ratio, Owen (1953) has published a model for selection between two alleles at a locus where the genotypic viabilities are sex-differential. The primary sex ratio is assumed to be a constant one-half, but the adult sex ratio is, in general, not one-half at equilibrium. Introducing the concepts of parental expenditure and replacement, Fisher's theory applies, and it is clear that the equilibrium tertiary sex ratio must be one-half. Since the introduction of replacement can make such a difference, it is important to consider its relevance in any natural situation, whether involving the sex ratio or not.

This has been done by Edwards (1961b) in his consideration of the population genetics of a sex-ratio condition in *Drosophila*. This condition, which is sex-linked, and causes males to produce daughters almost exclusively, has been described by Morgan *et al.* (1925), Gershenson (1928), and Sturtevant and Dobzhansky (1936). Edwards considers that the nature of the organism makes it reasonable to omit parental expenditure from the formulation of the problem, and he goes on to review previous attempts at constructing a model for this condition (Bennett, 1958; Barker, 1958; Shaw, 1959), and to construct his own, which fits the experimental findings of Wallace (1948) satisfactorily.

It is known that the unusual segregation ratio obtained in the above case is due to irregularities in spermatogenesis (Sturtevant and Dobzhansky, 1936), and an analogous case of a disturbed autosomal ratio has been described by Dunn and his collaborators (see Dunn, 1953, 1956; Lewontin and Dunn, 1960). The alleles concerned are those at the T locus of the mouse, and their existence in natural populations presents a similar problem to the case of "sex-ratio"; it may transpire that its solution requires the application of concepts of parental expenditure and replacement.

Sandler and Novitski (1957) suggested that "where such a force, potentially capable of altering gene frequencies, is a consequence of the mechanics of the meiotic divisions . . . the name *meiotic drive* be applied." In another connection Novitski and Sandler (1957) commented on the fact that all the products of spermatogenesis may not be functional.

F. Artificial Selection for the Sex Ratio

The analysis of this chapter is, of course, but an approximation to reality. In particular, just as the experimentalist trying to select for the sex ratio finds that the genetic variance, on which he is trying to select, is hidden in the far larger binomial variance, so does natural selection: the next step in the analysis must be to take into account the finiteness of family sizes.

The lack of success in sex-ratio selection experiments (one suspects that several which have been undertaken have not been reported) may be due not only to the smallness of the genetic variance compared with the binomial variance, but also to the fact that throughout such an experiment Fisherian selection will be trying to return the sex ratio to its original value. Successful selection may, in any case, be due to the collection of sex-linked lethal factors, and King's (1918) experiment needs examining afresh with this in mind.

It should be possible to mimic the effect of natural selection on the sex ratio by artificially changing the differential infant mortality. Suppose that, in a random-mating mammalian population at equilibrium for the sex ratio, half the males born were removed from the population at birth. Then in order to return to the Fisherian equilibrium condition in which the total parental expenditure on males equals that on females, more males must be born, and the secondary sex ratio raised. Unfortunately such an experiment is only possible in a species in which the sex of the young can be determined before they become independent of their parents, as in mammalian species. With faster-breeding species such as those of the genus *Drosophila* the experiment is impracticable, and in view of the smallness of the genetic variance in sex ratio, the use of a mammal such as the mouse would lead to a very long experiment. A further difficulty is that selection might be for a combined change in the primary sex ratio and the differential infant mortality, which together may have been stabilized by natural selection.

Another limitation of the above analysis is that it makes no allowance for sex-linked loci affecting the sex ratio. If, for example, the only "sex-ratio" loci were on the X-chromosome a ratio of two-thirds would be expected. The fact that the tertiary sex ratio is about one-half in man may reflect the limited importance of sex-linked loci in determining the sex ratio.

VIII. Conclusion

Whereas theoretical considerations of the action of natural selection on the sex ratio have advanced rapidly in the last few years, little progress has been made in establishing the heritability of the sex ratio, in man at least, on which these considerations rely.

New data of the distribution of the sexes in families have been collected, and new techniques of analysis devised, but no unequivocal demonstration of the variability of the probability of a birth being male between families has followed. It transpires that there is, however, good evidence for a correlation between the sexes of successive children in some samples.

Other methods of searching for heritability and variability in human populations have proved hardly more successful, and it must be concluded that, if genetic variability exists, it is of a very low order of magnitude. Among species other than man there is stronger evidence for the heritability of the sex ratio, although in most cases the point is by no means settled.

On the theoretical side, simple genetic arguments show that the primary sex ratio may be expected to have an equilibrium value of one-half, should it be variable. In order to explain the stabilization of the subsequent mortality rates, and thus the achievement of a sex ratio of one-half at later stages, it is necessary to postulate the replacement of children who die before they have become independent of their parents. This more complex theory has been developed in general terms to show that the tertiary sex ratio is of selective importance, and that the genetic variance in this sex ratio may be expected to remain constant while the sex ratio of the population approaches its equilibrium value at a known rate.

Future theoretical developments may be expected to include models for specific loci involving sex-differential viabilities and the concept of replacement, or "compensation."

References

Allen, G. (ed.) 1960. Symposium on cytology and cell culture genetics of man. *Am. J. Human Genet.* **12**, 95–138.

Aristotle (ca. 350 B.C.) On the Genesis of Animals, Δ. (In Greek.)

Barker, J. S. F. 1958. Simulation of genetic systems by automatic digital computors. IV. Selection between alleles at a sex-linked locus. *Australian J. Biol. Sci.* **11**, 613–625.

Barr, M. L., and Bertram, E. G. 1949. A morphological distinction between neurones of the male and female, and the behaviour of the nucleolar satellite during accelerated nucleoprotein synthesis. *Nature* **163**, 676–677.

Bateson, W. 1913. "Mendel's Principles of Heredity." Cambridge Univ. Press, London and New York.

Bennett, J. H. 1958. The existence and stability of selectively balanced polymorphism at a sex-linked locus. *Australian J. Biol. Sci.* **11**, 598–602.

Bodmer, W. F., and Edwards, A. W. F. 1960. Natural selection and the sex ratio. *Ann. Human Genet.* **24**, 239–244.

Bridges, C. B. 1916. Non-disjunction as proof of the chromosome theory of heredity. *Genetics* **1**, 1–52, 107–163.

Bridges, C. B. 1925. Sex in relation to chromosomes and genes. *Am. Naturalist* **59**, 127–137.

Bunak, V. V. 1934. Maternal Influence on the Human Still-Birth Sex Ratio. (In Russian.) *Proc. Maxim-Gorky Med. Biol. Research Inst.* **3**, 195–212.

Colombo, B. 1957. On the sex ratio in man. *Cold Spring Harbor Symposia Quant. Biol.* **22**, 193–202.

Crew, F. A. E. 1937. The sex ratio. *Am. Naturalist* **71**, 529–559.

Crew, F. A. E. 1954. "Sex-Determination," 3rd ed. Methuen, London.

Dahlberg, G. 1951. The primary sex ratio and its ratio at birth. *Acta genet.* **2**, 245–251.

Darlington, C. D. 1958. "Evolution of Genetic Systems," 2nd ed. Oliver & Boyd, Edinburgh.

Darwin, C. 1871. "The Descent of Man, and Selection in Relation to Sex." John Murray, London.

Davidson, W. M. 1960. Sex determination: diagnostic methods. *Brit. Med. J.* **II**, 1901–1906.

Davidson, W. M., and Smith, D. R. 1954. A morphological sex difference in the polymorphonuclear leucocytes. *Brit. Med. J.* **II**, 6–7.

Doncaster, L. 1914. "The Determination of Sex." Cambridge Univ. Press, London and New York.

Doncaster, L. 1921. "Heredity in the Light of Recent Research," 3rd ed. Cambridge Univ. Press, London and New York.

Düsing, C. 1884. "Die Regulierung des Geschlechtsverhältnisses." Fischer, Jena.

Dunn, L. C. 1953. Variations in the segregation ratio as causes of variations in gene frequency. *Acta genet.* **4**, 139–151.

Dunn, L. C. 1956. Analysis of a complex gene in the house mouse. *Cold Spring Harbor Symposia Quant. Biol.* **21**, 187–195.

Edwards, A. W. F. 1958. An analysis of Geissler's data on the human sex ratio. *Ann. Human Genet.* **23**, 6–15.

Edwards, A. W. F. 1959. Some comments on Schützenberger's analysis of data on the human sex ratio. *Ann. Human Genet.* **23**, 233–238.

Edwards, A. W. F. 1960a. The meaning of binomial distribution. *Nature* **186**, 1074.

Edwards, A. W. F. 1960b. The human sex ratio. Ph.D. dissertation. Cambridge University Library, Cambridge.

Edwards, A. W. F. 1960c. Natural selection and the sex ratio. *Nature* **188**, 960–961.

Edwards, A. W. F. 1961a. A factorial analysis of sex ratio data. *Ann. Human Genet.* **25**, 117–121.

Edwards, A. W. F. 1961b. The population genetics of "sex-ratio" in *Drosophila pseudoobscura*. *Heredity* **16**, 291–304.
Edwards, A. W. F. 1962. A factorial analysis of sex-ratio data: a correction. *Ann. Human Genet.*, **25**, 343–346.
Edwards, A. W. F., and Fraccaro, M. 1958. The sex distribution in the offspring of 5477 Swedish Ministers of Religion, 1585–1920. *Hereditas* **44**, 447–450.
Edwards, A. W. F., and Fraccaro, M. 1960. Distribution and sequences of sexes in a selected sample of Swedish families. *Ann. Human Genet.* **24**, 245–252.
Falconer, D. S. 1954. Selection for sex ratio in mice and *Drosophila*. *Am. Naturalist* **88**, 385–397.
Fisher, R. A. 1925–1958. "Statistical Methods for Research Workers." Oliver & Boyd, Edinburgh.
Fisher, R. A. 1930. "The Genetical Theory of Natural Selection," 1st ed. Oxford Univ. Press, London and New York.
Fisher, R. A. 1958. The nature of probability. *Centen. Rev. Arts and Sci.* **2**, 261–274.
Ford, C. E. 1961. Die Zytogenese der Intersexualität des Menschen. *In* "Die Intersexualität" (C. Overzier, ed.), pp. 90–121. Thieme, Stuttgart.
Geissler, A. 1889. Beiträge zur Frage des Geschlechtsverhältnisses der Geborenen. *Z. K. Sächs. Stat. Bureaus* **35**, 1–24.
Gershenson, S. 1928. A new sex-ratio abnormality in *Drosophila obscura*. *Genetics* **13**, 488–507.
Gini, C. 1908. "Il Sesso dal Punto di Vista Statistico." Sandron, Milano-Palermo-Napoli.
Gini, C. 1911. Considerazioni sulle probabilita a posteriori e applicazioni al rapporto dei sessi nelle nascite umane. *Studi Economico-Giuridici* **3**, 5–41.
Gini, C. 1951. Combinations and sequences of sexes in human families and mammal litters. *Acta genet.* **2**, 220–244.
Goldschmidt, R. 1934. Lymantria. *Bibliographia genet.* **11**, 1–186.
Goodman, L. A. 1961. Some possible effects of birth control on the human sex ratio. *Ann. Human Genet.* **25**, 75–81.
Gordon, M. J. 1957. Control of sex ratio in rabbits by electrophoresis of spermatozoa. *Proc. Natl. Acad. Sci. U.S.* **43**, 913–918.
Gowen, J. W., and Fung, S. C. 1957. Determination of sex through genes in a major sex locus in *Drosophila melanogaster*. *Heredity* **11**, 397–402.
Harris, H. 1946. Microspermia in an individual from a family of unusually high sex ratio and low fertility. *Ann. Eugen. (London)* **13**, 156–159.
Harris, J. A., and Gunstad, B. 1930a. The correlation between the sex of human siblings. *Genetics* **15**, 445–461.
Harris, J. A., and Gunstad, B. 1930b. The problem of the relationship between the number and the sex of human offspring. *Am. Naturalist* **64**, 495–508.
Hart, D., and Moody, J. D. 1949. Sex ratio: experimental studies demonstrating controlled variations. Preliminary report. *Ann. Surg.* **129**, 550–571.
Hemphill, R. E., Leitch, A., and Stuart, J. R. 1958. A factual study of male homosexuals. *Brit. Med. J.* **I**, 1317–1323.
Hewitt, D., Webb, J. W., and Stewart, A. M. 1955. A note on the occurrence of single-sex sibships. *Ann. Human Genet.* **20**, 155–158.
Hippocrates (ca. 420 B.C.). On Diathesis, I, XXVII. (In Greek.)
Howard, A., McLaren, A., Michie, D., and Sander, G. 1955. Genetic and environmental influences on the secondary sex-ratio in mice. *J. Genet.* **53**, 200–214.
Kalmus, H., and Smith, C. A. B. 1960. Evolutionary origin of sexual differentiation and the sex ratio. *Nature* **186**, 1004–1006.

Kang, Y. S. 1959. A study of Korean population genetics: Vital statistics of different occupational groups in the Korean population. *Proc. Intern. Congr. Genet.* 10th *Congr. Montreal* 1958 **2**, 141.

King, H. D. 1918. The effect of inbreeding, with selection, on the sex ratio of the albino rat. *J. Exptl. Zool.* **27**, 1–35.

Kolman, W. A. 1960. The mechanism of natural selection for the sex ratio. *Am. Naturalist* **94**, 373–377.

Lancaster, H. O. 1950. The sex ratio in sibships, with special reference to Geissler's data. *Ann. Eugen. (London)* **15**, 153–158.

Laplace, Marquis de, 1820. "Théorie Analytique des Probabilités," 3rd ed. Introduction: Essai philosophique sur les probabilités.

Lewontin, R. C., and Dunn, L. C. 1960. The evolutionary dynamics of a polymorphism in the house mouse. *Genetics* **45**, 705–722.

Lienhart, R., and Vermelin, H. 1946. Observation d'une famille humaine à descendance exclusivement féminine. Essai d'interprétation de ce phénomène. *Compt. rend. soc. biol.* **140**, 537–540.

Lindahl, P. E., and Sundell, G. 1958. Sex ratio in the golden hamster before uterine implantation. *Nature* **182**, 1392.

McClung, C. E. 1901. Notes on the accessory chromosome. *Anat. Anz.* **20**, 220–226.

McClung, C. E. 1902. The accessory chromosome—sex determinant? *Biol. Bull.* **3**, 43–84.

MacDowell, E. C., and Lord, E. M. 1925. Data on the primary sex ratio in the mouse. *Anat. Record* **31**, 143–148.

MacDowell, E. C., and Lord, E. M. 1926. The relative viability of male and female mouse embryos. *Am. J. Anat.* **37**, 127–140.

McKeown, T., and Lowe, C. R. 1951a. The sex ratio of stillbirths related to cause and duration of gestation. An investigation of 7,066 stillbirths. *Human Biol.* **23**, 41–60.

McWhirter, K. G. 1956. Control of sex ratio in mammals. *Nature* **178**, 870–871.

McWhirter, K. G. 1960. The possibility of controlling sex-ratio at conception. III. Male serum pH and sex-ratio performance. *Mem. Soc. Endocrinol.* **7**, *Sex differentiation and development*, 98–99.

Malogolowkin, C. 1958. Maternally inherited "sex-ratio" conditions in *Drosophila willistoni* and *Drosophila paulistorium. Genetics* **43**, 274–286.

Mendel, G. J. 1865. Versuche über Pflanzen-Hybriden. *Verhandl. naturforsch. Ver. Brünn* **10**, 1–47.

Moore, K. L., Graham, M. A., and Barr, M. L. 1953. The detection of chromosomal sex in hermaphrodites from a skin biopsy. *Surg. Gynecol. Obstet.* **96**, 641–648.

Morgan T. H. 1914. "Heredity and Sex." Columbia Univ. Press, New York.

Morgan, T. H., Bridges, C. B., and Sturtevant, A. H. 1925. The genetics of *Drosophila*. *Bibliographia genet.* **2**, 1–262.

Myers, R. J. 1949. Same-sex families. *J. Heredity* **40**, 268–270.

Newcomb, S. 1904. A statistical enquiry into the probability of causes of the production of sex in human offspring. *Carnegie Inst. Wash. Publ. No.* **11**.

Novitski, E. 1947. Genetic analysis of an anomalous sex-ratio condition in *Drosophila affinis. Genetics* **32**, 526–534.

Novitski, E., and Sandler, I. 1957. Are all products of spermatogenesis regularly functional? *Proc. Natl. Acad. Sci. U.S.* **43**, 318–324.

Ounsted, C. 1953. The sex ratio in convulsive disorders with a note on single-sex sibships. *J. Neurol. Neurosurg. Psychiat.* **16**, 267–274.

Owen, A. R. G. 1953. A genetical system admitting of two distinct stable equilibria under natural selection. *Heredity* **7**, 97–102.
Painter, T. S. 1924. The sex chromosomes of man. *Am. Naturalist* **58**, 506–524.
Parkes. A. S. 1924a. The frequencies of sex combinations in human families. *Eugen. Rev.* **16**, 211–216.
Parkes, A. S. 1926a. The mammalian sex ratio. *Biol. Revs. Biol. Proc. Cambridge Phil. Soc.* **2**, 1–51.
Penrose, L. S., 1961. "Principles of Human Genetics" (Review). *Ann. Human Genet., Lond.* **25**, 165–166.
Pitt-Rivers, G. H. L-F. 1929. Sex ratios and marriage: their relation to population growth and decline. *Eugen. Rev.* **21**, 21–28.
Poisson, S. D. 1830. Mémoire sur la proportion des naissances des filles et des garçons. *Mém. acad. sci.* **9**, 239–308.
Renkonen, K. O. 1956. Is the sex ratio between boys and girls correlated to the sex of precedent children? *Ann. Med. Exptl. et Biol. Fenn. (Helsinki)* **34**, 447–451.
Rife, D. C., and Snyder, L. H. 1937. The distribution of sex ratios within families in an Ohio city. *Human Biol.* **9**, 99–103.
Robertson, A. 1951. The analysis of heterogeneity in the binomial distribution. *Ann. Eugen. (London)* **16**, 1–15.
Rothschild, Shettles, L. B., and Bishop, D. W. 1960. X and Y spermatozoa. *Nature* **187**, 253–256.
Sandler, L., and Novitski, E. 1957. Meiotic drive as an evolutionary force. *Am. Naturalist* **91**, 105–110.
Schützenberger, M. P. 1949. Résultats d'une enquête sur la distribution du sexe dans les familles nombreuses. *Semaine hôp.* **25**, 2579–2582.
Shaw, R. F. 1958 The theoretical genetics of the sex ratio. *Genetics* **43**, 149–163.
Shaw, R. F. 1959. Equilibrium for the sex-ratio factor in *Drosophila pseudoobscura*. *Am. Naturalist* **93**, 385–386.
Shaw, R. F. and Mohler, J. D. 1953. The selective significance of the sex ratio. *Am. Naturalist* **87**, 337–342.
Skellam, J. G. 1948. A probability distribution derived from the binomial distribution by regarding the probability of success as variable between the sets of trials. *J. Roy. Statist. Soc.* **B10**, 257–261.
Slater, E. 1944. A demographic study of a psychopathic population. *Ann. Eugen. (London)* **12**, 121–137.
Stern, C. 1949. "Principles of Human Genetics." W. H. Freeman, San Francisco, California.
Stern, C. 1960. "Principles of Human Genetics," 2nd ed. W. H. Freeman, San Francisco, California.
Stevens, N. M. 1905. Studies in spermatogenesis with especial reference to the "accessory chromosome." *Carnegie Inst. Wash. Publ. No.* **36**.
Sturtevant, A. H., and Dobzhansky, T. 1936. Geographical distribution and cytology of "sex ratio" in *Drosophila pseudoobscura* and related species. *Genetics* **21**, 473–490.
Sutton, W. S. 1902. On the morphology of the chromosome group in *Brachystola magna*. *Biol. Bull.* **3**, 24–39.
Taylor, W. 1954. A note on the sex distribution of sibs. *Brit. J. Prevent. & Social Med.* **8**, 178.
Thomas, M. H. 1951. Sex pattern and size of family. *Brit. Med. J.* **I**, 733–734.
Thomson, J. A. 1908. "Heredity." John Murray, London.

Turpin, R., and Schützenberger, M. P. 1948. Recherche statistique sur la distribution du sexe à la naissance. *Compt. rend. acad. sci.* **226**, 1845–1846.
Turpin, R., and Schützenberger, M. P. 1952a. Progenèse et gémellité. *Acta Genet. Med. et Gemellol.* **1**, 159–169.
Turpin, R., and Schützenberger, M. P. 1952b. Sexe et gémellité. *Semaine hôp.* **28**, 1844–1848.
Waaler, G. H. M. 1928. Om seksualproportionen ved födslen. *Nord. statist. Tidskr.* **7**, 65–84.
Wallace, B. 1948. Studies on "sex-ratio" in *Drosophila pseudoobscura*. 1. Selection and "sex-ratio." *Evolution* **2**, 189–217.
Weir, J. A. 1953. Association of blood pH with sex ratio in mice. *J. Heredity* **44**, 133–138.
Weir, J. A. 1955. Male influence on sex ratio of offspring in high and low blood-pH lines of mice. *J. Heredity* **46**, 277–283.
Weir, J. A. 1958. Sex ratio related to sperm source in mice. *J. Heredity* **49**, 223–227.
Weir, J. A. 1960. A sex ratio factor in the house mouse that is transmitted by the male. *Genetics* **45**, 1539–1552.
Weir, J. A., and Clark, R. D. 1955. Production of high and low blood-pH lines of mice by selection. *J. Heredity* **46**, 125–132.
Weir, J. A., and Wolfe, H. G. 1959. Modifications of the sex ratio of mice through genetic and environmental causes. *Proc. Intern. Congr. Genet. 10th Congr. Montreal 1958* **2**, 310.
Weir, J. A., Haubenstock, H., and Beck, S. L. 1958. Absence of differential mortality of sexes in mice. *J. Heredity* **49**, 217–222.
Welshons, W. J., and Russell, L. B. 1959. The Y-chromosome as the bearer of male determining factors in the mouse. *Proc. Natl. Acad. Sci. U.S.* **45**, 560–566.
Wilson, E. B. 1905. Studies on chromosomes. *J. Exptl. Zool.* **2**, 507–545.
Wilson, E. B. 1911. The sex chromosomes. *Arch. mikroskop. Anat. u.* **77**, 249–271.
Wolfe, H. G. 1961. Selection for blood-pH in the house mouse. *Genetics* **46**, 55–75.
Worms, R. 1912. "La Sexualité dans les Naissances Françaises." Giard & Brière, Paris.

Additional References

Allan, T. M. 1958. Rh blood groups and sex ratio at birth. *Brit. Med. J.* **II**, 248.
Allan, T. M. 1959. ABO blood groups and sex ratio at birth. *Brit. Med. J.* **I**, 553–554.
Allan, T. M. 1960. British stillbirths and first week deaths, 1950–57. *Brit. J. Prevent. & Social Med.* **14**, 35–38.
Bennett, J. H., and Walker, C. B. V. 1956. Fertility and blood groups of some East Anglian blood donors. *Ann. Human Genet.* **20**, 299–308.
Ciocco, A. 1938. Variation in the sex ratio at birth in the United States. *Human Biol.* **10**, 36–64.
Cohen, B. H., and Glass, B. 1956. The ABO blood groups and the sex ratio. *Human Biol.* **28**, 20–42.
Cohen, B. H., and Glass, B. 1959. Further observations on the ABO blood groups and the sex ratio. *Am. J. Human Genet.* **11**, 274–278.
Dahlberg, G. 1949. Do parents want boys or girls? *Acta genet.* **1**, 163–167.
Edwards, A. W. F. 1960d. On the size of families containing twins. *Ann. Human Genet.* **24**, 309–311.
Edwards, J. H. 1957. A critical examination of the reputed primary influence of ABO phenotype on fertility and sex ratio. *Brit. J. Prevent. & Social Med.* **11**, 79–89.

Fancher, H. L. 1956. The relationship between the occupational status of individuals and the sex ratio of their offspring. *Human Biol.* **28,** 316–322.

Fisher, R. A. 1928. Triplet children in Great Britain and Ireland. *Proc. Roy. Soc.* **B102,** 286–290.

Fisher, R. A., and Roberts, J. A. F. 1943. A sex difference in blood-group frequencies. *Nature* **151,** 640.

Heath, C. W. 1954. Physique, temperament and the sex ratio. *Human Biol.* **26,** 337–342.

Holt, H. A., Thompson, J. S., Sanger, R., and Race, R. R. 1952. Linkage relations of the blood group genes of man. *Heredity* **6,** 213–216.

Johnstone, J. M. 1954. Sex ratio and the ABO blood group system. *Brit. J. Prevent. & Soc. Med.* **8,** 124.

Kloepfer, H. W. 1946. An investigation of 171 possible linkage relationships in man. *Ann. Eugen. (London)* **13,** 35–71.

Kohn, H. I. 1960. The effect of paternal X-ray exposure on the secondary sex ratio in mice (F_1 generation). *Genetics* **45,** 771–778.

Lejeune, J., and Turpin, R. 1957. Influence de l'âge des parents sur la masculinité des naissances vivantes. *Compt. rend. acad. sci.* **244,** 1833–1835.

Lejeune, J., Turpin, R., and Rethoré, M. O. 1959. Sur les variations de la masculinité dans la descendance de parents irradiés. *Proc. Intern. Cong. Genet. 10th Congr. Montreal 1958* **2,** 163.

Lowe, C. R., and McKeown, T. 1950. The sex ratio of human births related to maternal age. *Brit. J. Social Med.* **4,** 75–85.

Lowe, C. R., and McKeown, T. 1951. A note on secular changes in the human sex ratio at birth. *Brit. J. Social Med.* **5,** 91–97.

MacMahon, B., and Pugh, T. F. 1954. Sex ratio of white births in the United States during the Second World War. *Am. J. Human Genet.* **6,** 284–292.

Martin, W. J. 1943. Sex ratio during war. *Lancet* **ii,** 807.

Mathew, N. T. 1941. The influences of seasons on human reproduction. *Sankhyā* **5,** 261–268.

Mathew, N. T. 1947. Factors influencing the relative proportion at birth of the two sexes. *Sankhyā* **8,** 277–281.

McKeown, T., and Lowe, C. R. 1951b. Sex ratio of stillbirths related to birth weight. *Brit. J. Social Med.* **5,** 229–235.

McKeown, T., and MacMahon, B. 1956. Sex differences in length of gestation in mammals. *J. Endocrinol.* **13,** 309–318.

Neel, J. V., and Schull, W. J. 1956. "The Effect of Exposure to the Atomic Bombs on Pregnancy Termination in Hiroshima and Nagasaki." Natl. Acad. Sci., Natl. Research Comm., Washington, D.C.

Novitski, E., and Kimball, A. W. 1958. Birth order, parental ages, and sex of offspring. *Am. J. Human Genet.* **10,** 268–275.

Novitski, E., and Sandler, L. 1956. The relationship between parental age, birth order and the secondary sex ratio in humans. *Ann. Human Genet.* **21,** 123–131.

Parkes, A. S. 1924b. Studies on the sex-ratio and related phenomena. 2. The influence of the age of the mother on the sex-ratio in man. *J. Genet.* **14,** 39–46.

Parkes, A. S. 1926b. The physiological factors governing the proportion of sexes in man. *Eugen. Rev.* **17,** 275–293.

Record, R. G. 1952. Relative frequencies and sex distribution of human multiple births. *Brit. J. Social Med.* **6,** 192–196.

Russell, W. T. 1936. Statistical study of the sex ratio at birth. *J. Hyg.* **36,** 381–401.

Sanghvi, L. D. 1951. ABO blood groups and sex ratio at birth in man. *Nature* **168**, 1077.
Schull, W. J. 1958. Empirical risks in consanguineous marriages: sex ratio, malformation, and viability. *Am. J. Human Genet.* **10**, 294–349.
Shield, J. W., Kirk, R. L., and Jacobowicz, R. 1958. The ABO blood groups and masculinity of offspring at birth. *Am. J. Human Genet.* **10**, 154–163.
Synder, R. G. 1961. The sex ratio of offspring of pilots of high performance military aircraft. *Human Biol.* **33**, 1–10.
Szilard, L. 1960. Dependence of the sex ratio at birth on the age of the father. *Nature* **186**, 649.
Turpin, R., Lejeune, J., and Rethoré, M. O. 1958. Les effets génétiques chez l'homme des radiations ionisantes. *Bull. schweiz. Akad. med. Wiss.* **14**, 571–579.
Turpin, R., and Schützenberger, M. P. 1950. Sur la masculinité à la naissance dans les grossesses multiples. *Compt. rend. acad. sci.* **231**, 1098–1099.

THE GENETICS OF *Streptomyces coelicolor*

D. A. Hopwood[*] and G. Sermonti

Botany School, University of Cambridge, Cambridge, England
and Istituto Superiore di Sanità, Roma, Italy

	Page
I. Introduction	273
II. Strains and Methods of Culture	276
A. Strains	276
B. Culture Media	277
C. Incubation and Maintenance of Stock Cultures	278
D. Preparation of Spore Suspensions and Plating	279
E. Isolation and Characterization of Mutants	279
III. Morphology of the Organism	285
A. Structure and Development of the Colony	285
B. Nuclear Cytology	287
IV. Studies of Recombination by Selective Analysis	290
A. Discovery of Gene Recombination	290
B. Crossing Techniques	291
C. Analysis of Recombination	293
D. Simplified Selective Analysis	299
E. Crosses Giving Unequal Frequencies of Complementary Genotypes	301
F. Considerations on the Fertility of *Streptomyces coelicolor*	303
V. Heteroclones	304
A. Discovery of Heterogeneous Clones (Heteroclones)	304
B. Principles for the Selection of Heteroclones	305
C. Techniques Used in the Selection and Analysis of Heteroclones	308
D. Estimation of Linkage by the Analysis of Heteroclones	310
E. The Nature of the Heteroclones	321
VI. Analysis of Short Chromosome Regions	332
VII. Speculations on Nuclear Behavior during the Life Cycle	335
VIII. Summary and Conclusions	336
Acknowledgments	339
References	339

I. Introduction

The last fifteen years have seen the development of bacterial genetics as a flourishing field of study, which has contributed much to the science of genetics as a whole, as well as revealing processes apparently charac-

[*] Present address: Department of Genetics, University of Glasgow, Glasgow, Scotland.

teristically bacterial. Gene recombination is now known to occur in several of the bacteria, a group of organisms at one time thought to reproduce entirely asexually: *Diplococcus pneumoniae* (Hotchkiss, 1951); *Escherichia coli* (Lederberg and Tatum, 1946); *Salmonella typhimurium* (Zinder and Lederberg, 1952); *Haemophilus influenzae* (Leidy et al., 1953); *Pseudomonas aeruginosa* (Holloway, 1955); *Bacillus subtilis* (Spizizen, 1958); *Vibrio cholerae* (Bhaskharan, 1960). Various different means of transfer of genetic material from one parent cell to the other have been found in these organisms: by conjugation, by the agency of bacteriophages and episomes, and as free DNA. These processes, however, all lead to the same situation: the zygote which gives rise to recombinant progeny does not contain a complete set of genetic material from each parent, but is a "merozygote" (Wollman et al., 1956). The phenomenon of merozygosis appears to distinguish the bacteria from higher organisms with chromosomal nuclei (eucaryotes).

All the species mentioned above belong to the same small group within the bacteria, the sub-order Eubacteriineae (Breed et al., 1948). Consequently "bacterial genetics" as understood today is really the genetics of only one group within the large and heterogeneous collection of organisms which constitute the bacteria; moreover this group contains the bacteria with the simplest morphology.

The order Actinomycetales contains some of the bacteria with the most complex morphology, so complex, in fact, that certain authors have been led to consider the actinomycetes as a group intermediate between the bacteria and the fungi (Waksman, 1950), or even as fungi (e.g., Carvajal, 1947). The differentiation of the colony reaches its maximum in the genus *Streptomyces*, in which the whole colony consists of an interconnected system of hyphae, certain of which complete their development by becoming transformed into chains of aerial spores. Notwithstanding a superficial resemblance to fungi, there is no doubt that the streptomycetes are bacteria (see, for example, Glauert and Hopwood, 1960). The comparatively complex morphology of the streptomycetes, which suggests that they are not very closely related to the simple rod-shaped or spherical Eubacteriineae, made them a group worthy of genetic study. Such a study would sample more fully the range of different genetic systems in the bacteria as a whole, and help to discover the extent to which the features which seem to distinguish the genetic systems of bacteria from those of eucaryotes are common to all bacteria.

The streptomycetes have several characteristics which make them suitable material for a genetic investigation. A very important feature is the presence of abundant spores, easily collected and separated from one another, and highly uniform. They contain a single nuclear body, which usually seems to carry a single set of genetic information (see later). For this reason, members of the genus *Streptomyces* have been used for many radiogenetic studies, for example: Kelner (1948), Newcombe (1953), Wainwright and Nevill (1955), Saito and Ikeda (1959).

Most streptomycetes grow on extremely simple defined media, a feature which allows nutritional mutants to be isolated readily. Unfortunately, they have not so far furnished any reliable morphological markers, but it is possible that such markers may be found in the future. Finally, *Streptomyces* is worthy of consideration for the fact that all widely used antibiotics, except the penicillins, are produced by members of this genus. Genetic recombination may be useful, on the one hand, for the production, by cross-breeding, of improved antibiotic-producing strains, and, on the other, for the study of the biosynthesis of these singular chemical compounds.

Studies of genetic recombination in *Streptomyces* were begun independently at about the same time in different laboratories. It is a curious coincidence that the first two positive results reported (Sermonti and Spada-Sermonti, 1955; Hopwood, 1957) were obtained with different isolates of *Streptomyces* classified as *S. coelicolor*. Results obtained with other species soon followed: *S. fradiae* (Braendle and Szybalski, 1957); *S. rimosus* (Alikhanian and Mindlin, 1957); *S. griseoflavus* (Saito, 1958); *S. griseus* (Braendle and Szybalski, 1959); *S. scabies* [Gregory, discussion to Braendle and Szybalski (1959)]; *S. aureofaciens* (Alikhanian and Borisova, 1961). All these species showed recombination, and as far as the phenomenon was explored, the situation appeared to be similar to that in *S. coelicolor*. Only a few negative results have been reported: in *S. griseus* and *S. cyaneus* (Bradley and Lederberg, 1956), in *S. griseus*, *S. albus*, and *S. venezuelae* (Braendle and Szybalski, 1957), and in *S. coelicolor* (Bradley, 1957, 1958), where only heterokaryosis was observed. Later, Bradley (1959, 1960) suggested the possible occurrence simultaneously of various processes of genetic interaction in *S. coelicolor*, with the constant exception of a normal recombination.

Interest in the genetics of *Streptomyces* seems to have reached its maximum at a conference on The Genetics of *Streptomyces* and Other Antibiotic-Producing Microorganisms, held in January 1959 at the New York Academy of Sciences. After this, most of the authors turned to the study of other topics, or to the practical application of recombination in *Streptomyces*. Analysis of the processes of recombination has been almost confined to *S. coelicolor*. This review aims to describe the present state of knowledge of the genetic system of *S. coelicolor*, with the object of showing the relevance of the genetics of *Streptomyces* to problems of fundamental genetic interest.

II. Strains and Methods of Culture

A. STRAINS

1. *Note on Nomenclature*

Kutzner and Waksman (1959) have shown that the various strains described in the literature as *Streptomyces coelicolor* do not all have the same combination of characteristics, and suggest that two distinct "species" are involved: *Streptomyces coelicolor* Müller, and *Streptomyces violaceoruber* Waksman and Curtis. It appears that all the strains referred to as *S. coelicolor* that have been used in genetic investigations in fact have the characteristics of *S. violaceoruber* as given by Kutzner and Waksman (1959). In this review, however, we propose to retain the specific epithet *coelicolor*, since this was the most legitimate when our studies of recombination were begun. In "Bergey's Manual" (Breed et al., 1948), the name *Streptomyces coelicolor* was used, in bold type, for a group of doubtfully homogeneous isolates, and *S. violaceoruber* was given as a synonym for, or as a sub-group of, *S. coelicolor*. Thus without wishing to enter into questions of classification, the use of the name *S. coelicolor* seemed desirable. Moreover, the possession of different morphological characteristics does not guarantee that two isolates belong to two different "species" in the genetic sense, and genetic studies may lead in the future to a further reclassification of the group. The retention of the name *S. coelicolor* seems desirable in order that a particular strain used in a series of studies should continue to be called by the same name, although some exceptions to this rule have already been made (Glauert and Hopwood, 1961).

2. *Strains of Streptomyces coelicolor Used in Genetic Studies*

At least five wild type strains of *S. coelicolor* have been used, by various workers, as starting strains in studies of genetic recombination:

I.S.S.—This strain was obtained from A. Tonolo, Istituto Superiore di Sanità, Roma, Italy (Tonolo et al., 1954). Gene recombination was found between marked derivatives of this strain by Sermonti and Spada-Sermonti (1955, 1956).

A3(2)—This is a derivative of Waksman's strain 3443 (Stanier, 1942). Erikson (1955) obtained a culture from Stanier, and isolated a single spore with a micromanipulator; this gave rise to the starting culture A3(2) from which all the stocks used in genetic work have been derived. Gene recombination between marked derivatives of A3(2) was found by Hopwood (1957, 1959). Derivatives of A3(2) gave no recombination with derivatives of I.S.S. (Braendle and Szybalski, 1959; Sermonti and Spada-Sermonti, unpublished).

NI9021—This strain was obtained by Braendle and Szybalski (1957) from K. Tsubaki, Nagao Institute, Tokyo, Japan. They selected prototrophs from mixed cultures of auxotrophic derivatives of this strain, and from mixed cultures of such derivatives with derivatives of I.S.S. Braendle and Szybalski (1959) obtained prototrophs from mixed cultures of one out of three combinations of derivatives of NI9021 and A3(2).

S16—This is strain NRRL-B-1257 of the United States Department of Agriculture, Peoria, Illinois. Bradley (1957) obtained prototrophs from mixed cultures of auxotrophic derivatives of this strain. Prototrophs were obtained from mixed cultures of derivatives of S16 and I.S.S. (Bradley, personal communication). Genetic recombination between derivatives of S16 and A3(2) or of S16 and NI9021 has not been sought.

S199—This strain, studied by Bradley (1960), was obtained from a Wisconsin soil sample. Bradley obtained prototrophs from mixed cultures of auxotrophic derivatives of this strain and from mixed cultures of derivatives of S199 and I.S.S. Bradley and Anderson (1958) obtained prototrophs from mixed cultures of derivatives of S199 and S16. Derivatives of S199 gave no recombination with derivatives of A3(2) (Hopwood, unpublished). Genetic recombination between derivatives of S199 and NI9021 has not been sought.

It appears that the strain I.S.S. of Sermonti and Spada-Sermonti is closely related to Bradley's S16 and S199, since derivatives of the three strains interact to give prototrophs, while A3(2) is less closely related to these three. NI9021 is possibly intermediate between A3(2) and the other three strains.

Nearly all the results described in this paper have been obtained by us with marked derivatives of strain A3(2), since a satisfactory series of genetic markers and a workable basis for genetic analysis were obtained relatively early in this strain by Hopwood (1957, 1959). Some of the results obtained with marked derivatives of strain I.S.S. in the early work of Sermonti and Spada-Sermonti (1955, 1956, 1959a) are also referred to.

B. CULTURE MEDIA

1. *Minimal Medium (MM)*

Various minimal media have been used at different stages of our work. The following glucose-asparagine medium (modified from Waksman, 1950) has been found to be the simplest and most satisfactory (%, w/v): glucose 1.0; asparagine, 0.05; K_2HPO_4, 0.05; $FeSO_4.7H_2O$, 0.001; KOH, 0.03; Difco Bacto-agar, 1.5–2.0. After autoclaving pH = 7.0.

Czapek-Dox agar has also been used.

2. Complete Media

Sporulation medium —(%, w/v): sucrose, 3.0; corn steep-liquor (ca. 50% solid), 1.0; yeast extract, 0.3; methionine, 0.005; KH_2PO_4, 0.1; $MgSO_4.7H_2O$, 0.05; KCl, 0.05; $FeSO_4.7H_2O$, 0.001; agar, 2.0; (%, v/v): nucleic acid hydrolyzate (from yeast), 0.3.

Reproductive medium 1—(Sermonti and Spada-Sermonti, 1959b) (%, w/v): glucose, 1.0; bacto-peptone, 0.2; yeast extract, 0.2; hydrolyzed casein (Difco), 0.2; $MgSO_4.7H_2O$, 0.05; agar, 2.0.

Reproductive medium 2—(Hopwood, 1960) (%, w/v): glucose, 2.5; bacto-peptone, 0.2; yeast extract, 0.1; K_2HPO_4, 0.5; NaCl, 0.05; $MgSO_4.7H_2O$, 0.05; agar, 1.5; (%, v/v): nucleic acid hydrolyzate (from yeast), 0.3; nucleic acid hydrolyzate (from thymus), 0.2; casein hydrolyzate, 0.5; vitamin solution, 0.1. Adjusted to pH 7.2 with N HCl, before autoclaving.

The component solutions, *casein hydrolyzate, nucleic acid hydrolyzate,* and *vitamin solution* are prepared as described by Pontecorvo (1953).

The sugars are autoclaved separately as 50% solutions and added to the media after autoclaving.

Although these three complete media have been found to be satisfactory, they may not be the best; improvements could almost certainly be made by an exhaustive study of the effects of the various components. The sporulation medium is normally used for the maintenance of stock cultures and for the plating of spore suspensions in the recovery of mutants and segregants. The fertility of crosses is much lower on the sporulation medium than on the reproductive media, sometimes by a factor of 1000; consequently one or other of the reproductive media is used for making mixed cultures.

3. Growth Factor and Antibiotic Supplements to Culture Media

Amino-acids are added to minimal medium at a final concentration of 1.0 mM; adenine and uracil at 0.1 mM; vitamins at 0.01 mM; streptomycin at 50 µg/ml.

Sometimes, particularly when cultures of *Neurospora* are being grown in the same laboratory, *actidione* is added to culture media at the rate of 10 µg/ml to prevent contamination.

C. INCUBATION AND MAINTENANCE OF STOCK CULTURES

All cultures are incubated at 28–30°C.

Stock cultures are allowed to grow for a week on slopes of complete medium in order to produce vigorously sporulating cultures, and are then stored in a refrigerator at ca. 4°C. They are subcultured every 6 to 12 months. Sometimes it is necessary to subculture poorly sporulating cultures every 3 months.

D. Preparation of Spore Suspensions and Plating

The spores of *S. coelicolor* may be suspended in distilled water; suspension is facilitated if a very dilute detergent solution is used (e.g., lauryl sulfonate 1:10,000). Water or dilute detergent is added to a well-sporulating culture and the spores are suspended by gently scraping the surface of the culture with a loop. The suspension is shaken vigorously to break up chains of spores, and is then filtered through cotton wool to remove large fragments of mycelium and aggregations of spores. Alternatively, suspensions may be filtered through Whatman No. 2 filter paper.

Rough estimates of the number of plating units per unit volume of spore suspension may be made by counting in a hemocytometer. Alternatively, the number of viable units in suspensions containing 10^6 or more spores per ml may be estimated by measuring their turbidity in a nephelometer.

Spore suspensions are plated on the surface of agar medium by spreading 0.1 ml of the suspension over the surface of a Petri dish with a glass rod. Up to 300 colonies can readily be counted on and isolated from such plates. When more accurate counts of the number of plating units per unit volume are required, the suspension is plated by placing 0.5 to 1.0 ml in a Petri dish and adding the molten agar medium, held at 45°C. 1000 or more colonies may be counted accurately on such plates.

E. Isolation and Characterization of Mutants

The only mutations so far employed as genetic markers in our work are those resulting in auxotrophy and in resistance to streptomycin. Mutations conferring resistance to other growth-inhibitory agents have been used by Braendle and Szybalski (1959) in *S. fradiae* and *S. griseus*. Production or nonproduction of the characteristic pigment of *S. coelicolor* has been unsatisfactory as a marker, owing to the extreme variability in manifestation of pigment production. Bradley (1958, 1960) has described the isolation of mutants of strains S16 and S199 resistant to particular actinophages.

1. *Irradiation*

Ultra-violet irradiation has been used routinely for the induction of biochemical mutations. A spore suspension is irradiated in a quartz test tube with a low-pressure mercury lamp. Details of the apparatus were described by Catcheside (1954). The period of irradiation has varied from 1 to 2½ minutes, giving less than 0.1% survival. The viable count of the irradiated suspension is determined by serial dilution and plating on complete medium. The bulk of the suspension, meanwhile stored at 4°C for 2 days, is then diluted to contain approximately 3000 viable units per

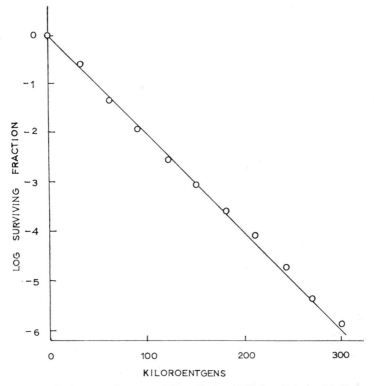

FIG. 1. Survival curve of spores of *S. coelicolor* A3(2) irradiated with X-rays.

ml, and plated on the surface of complete medium. Alternatively, a spore suspension may be plated on the surface of complete medium, and the plates irradiated, with the lids removed.

X-irradiation was used for the induction of some of the first biochemical mutations. Spore suspensions were irradiated in a glass bottle with X-rays (210 Kv, 10 mA). For the determination of a survival curve, a total dose of 300×10^3 r (60 minutes irradiation) was given, samples being removed after every 30×10^3 r and plated in glucose-asparagine-peptone medium (Newcombe, 1953) at suitable dilutions for the determination of viable counts. The survival curve of strain A3(2) (Fig. 1) was found to be exponential; a similar result was obtained by Saito and Ikeda (1959) with *S. coelicolor* strain I.S.S.

2. *Isolation of Auxotrophs*

Well-sporulating colonies derived from the spores surviving irradiation are obtained 4–5 days after plating. Replica plates are prepared on minimal medium, using a velvet pad (Lederberg and Lederberg, 1952); after 2 days incubation of the replicate plates, the original plates are scored and colonies which failed to replicate (Fig. 2) are isolated and characterized for their growth requirements (Pontecorvo, 1953).

In a series of experiments, 30,220 survivors of ultraviolet irradiation of the wild type A3(2) yielded 136 (0.45%) partially or fully characterized auxotrophic mutants. The numbers of the different phenotypes represented were (requirement of): arginine, 31; arginine, citrulline, or ornithine, 2; proline, 1; glutamic acid, 4; glutamic or aspartic acids, 2; cysteine (para-thiotrophs), 16; methionine, 7; methionine plus threonine, (homoserine), 1; threonine, 1; histidine, 13; phenylalanine, 1; tyrosine, 1; leucine, 2; valine, 2; isoleucine plus valine, 1; serine, 1; glycine, 1; serine or glycine, 1; ammonium (para-azotrophs), 15; purines, 13; uracil, 6; nicotinamide, 4; thiamine, 6; pyridoxine, 3; *p*-aminobenzoic acid, 1.

3. *Streptomycin-Resistant Mutants*

Resistant mutants are isolated by plating irradiated or unirradiated spore suspensions on plates of complete medium containing 50 µg/ml streptomycin. The gradient plate technique (Szybalski, 1952) may also be used. Streptomycin-resistant mutants are of more than one kind, as in other bacteria (Demerec, 1948). In addition to mutants with only slightly more resistance than the wild type, two kinds of high-step mutants have occurred in strain A3(2). Some, such as *str-1*, grow on 800 µg/ml streptomycin in minimal medium; others, such as *str-3*, grow only on a somewhat lower concentration of streptomycin. Mutants of the *str-3* type will utilize nitrate as sole source of nitrogen only in the absence of streptomycin; in its presence, growth occurs at a normal rate only in the presence of an ammonium salt or one of a number of amino acids as nitrogen source.

4. *List of Mutations*

Table 1 gives the symbols, characteristics, and origins of the mutant alleles obtained in strain A3(2) which have been used in recombination studies. Remarks on linkage relations and allelism, deduced from the analyses described in later sections, are also included.

Abbreviated symbols have been assigned to the eight most commonly used markers for ease of reference to them in the tables. These abbrevi-

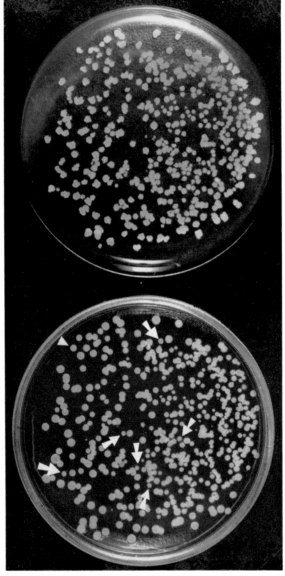

Fig. 2. Replica plating in the isolation of auxotrophic mutants. *Left*: original plate of complete medium. *Right*: replica plate on minimal medium. Arrows on the original plate indicate some of the colonies which have failed to be replicated.

THE GENETICS OF *Streptomyces coelicolor*

TABLE 1

Mutations in *Streptomyces coelicolor* Strain A3(2) Used in Crosses Up to July 1961

Symbols*		Characteristics	Mutagenic agent	Type of genetic analysis†	Remarks
				Linkage and allelism‡	

Linkage Group I

	Symbols*	Characteristics	Mutagenic agent	Type of genetic analysis†	Remarks
	amm-1	Require ammonium,	UV	(2)	4 sites very close together, left of *met-2*
	amm-2	or glutamic acid,			
	amm-3	or aspartic acid			
	amm-4				
a	arg-1	Requires arginine, or citrulline, or ornithine	UV	(1),(3)	See Tables 4 and 11
h	his-1	Requires histidine; accumulates imidazole propanediol (?)	X-rays	(1),(3)	See Tables 4, 11, and 18; "cluster" covering less than 2 units; the 3 mutations complement one another
	his-9	Requires histidine	UV	(2)	
	his-12	Requires histidine; accumulates imidazoleglycerol phosphate	UV	(2)	
m	met-2	Requires methionine	UV	(1),(3)	See Tables 4 and 11
	pdx-2	Requires pyridoxine	UV	(3)	Less than 10 units from *his-1*
	tyr-1	Requires tyrosine	UV	(3)	Less than 5 units from *his-1*

Linkage Group II

	Symbols	Characteristics	Mutagenic agent	Type of genetic analysis	Remarks
ad	ade-3	Requires adenine, or guanine, or hypoxanthine	UV	(3)	See Tables 11 and 13
	ade-8	Require adenine plus thiamine	UV	(2)	3 sites very close together, near *ade-3*
	ade-10		UV	(2)	
	ade-11		UV	(2)	

* The abbreviated symbols (indicated in this table in boldface italic type) will be used in the tables (lightface italic in the other tables) to indicate the commonest markers.

† Key to numbers in parentheses: (1) = complete selective analysis; (2) = simplified selective analysis; (3) = analysis of heteroclones.

‡ See also the linkage map in Fig. 10.

TABLE 1 (*Continued*)

Symbols*		Characteristics	Mutagenic agent	Type of genetic analysis†	Remarks
	cys-6	Requires cysteine, or thiosulfate (at 100 μg/ml)	UV	(2),(3)	Between *str-1* and *ade-3*
h3	*his-3* *his-4* *his-15*	Require histidine; accumulate histidinol phosphate	UV UV UV	(1),(3) (2) (2)	See Tables 11 and 13; 3 sites very close together
	hom-1	Requires methionine plus threonine (homoserine)	UV	(2)	Less than 10 units from *his-3*, to left
	met-3 *met-4* *met-5*	Require methionine, or homocysteine	UV UV UV	(2) (1) (2)	3 sites very close together; between *str-1* and *ade-3*
p	*phe-1*	Requires phenylalanine	X-rays	(1),(3)	See Tables 11 and 13
	thr-1	Requires threonine	UV	(2)	Less than 10 units from *his-3*, to right
s	*str-1*	Resistant to streptomycin (at ca. 800 μg/ml)	Spontaneous	(1),(3)	See Tables 11 and 13
	str-3	Resistant to streptomycin (at ca. 200 μg/m); requires ammonium in presence of streptomycin	Spontaneous	(1)	Close to *str-1*
u	*ura-1*	Requires uracil	UV	(3)	See Tables 11 and 13

* The abbreviated symbols (indicated in this table in boldface italic type) will be used in the tables (lightface italic in the other tables) to indicate the commonest markers.

† Key to numbers in parentheses: (1) = complete selective analysis; (2) = simplified selective analysis; (3) = analysis of heteroclones.

‡ See also the linkage map in Fig. 10.

ated symbols, indicated in bold-face type in Table 1, are as follows:
a = *arg-1*; ***ad*** = *ade-3*; ***h*** = *his-1*; ***h3*** = *his-3*; ***m*** = *met-2*; ***p*** = *phe-1*;
s = *str-1*; ***u*** = *ura-1*.

III. Morphology of the Organism

The streptomycetes have been the subject of morphological study for several decades, and certain general features of the organization of their colonies have been known for some time (Waksman, 1950). However, many details of their life cycle and cytology have remained obscure. The

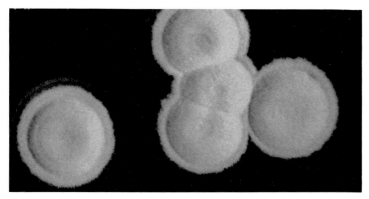

Fig. 3. Colonies of *S. coelicolor* growing on complete medium (× 3).

following description of *Streptomyces coelicolor* is based on a series of recent observations of the living organism with the phase-contrast microscope (Hopwood, 1960), of preparations stained with nuclear stains examined in the light microscope (Hopwood and Glauert, 1960a), and of thin sections examined in the electron microscope (Glauert and Hopwood, 1959, 1960, 1961; Hopwood and Glauert, 1960b).

A. Structure and Development of the Colony

A colony of *Streptomyces* (Fig. 3), like that of a mold, consists of a continuous system of interconnected hyphae. Within the colony the mycelium is differentiated into two rather distinct layers, the substrate mycelium and the aerial mycelium (Fig. 4a). The substrate mycelium develops by the elongation and branching of the one or more germ-tubes produced by a spore when it germinates. On an agar medium, the hyphae of the substrate mycelium grow over the surface of the agar and penetrate into it; they are much-branched (Fig. 4b) and vary in diameter from 0.3 to 1.0 μ. The colony increases in circumference by the elongation of a number of major hyphae in a radial direction, the spaces between these hyphae

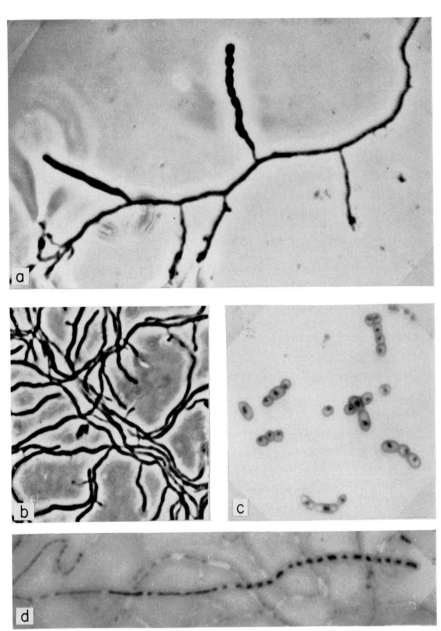

Fig. 4. (a) Microcolony of *S. coelicolor* growing on a "starvation" medium, observed with phase-contrast; a spore (*bottom left*) has produced a main substrate hypha bearing two aerial hyphae, the right hand one forming spores (× 2000). (b) Substrate hyphae at the edge of a colony on complete medium; phase-contrast (× 2000). (c) Spores stained with Azure A-SO$_2$ (DeLamater, 1951) showing chromatinic bodies (× 2500). (d) Aerial hypha stained with Thionin-SO$_2$ (DeLamater, 1951); basal segments (*left*) contain rod-shaped chromatinic bodies; rounded bodies are seen within developing spores (*right*) (× 2500).

being colonized by successively smaller side-branches. Evidence for anastomosis between the hyphae is difficult to obtain, owing to the small size of the hyphae (Hopwood, 1960). Septa divide the protoplasm of the hyphae into compartments; the septa, unlike those of many fungi, do not have a central perforation, so that the migration of large structures from compartment to compartment is not possible. When the colony has reached a certain size, specialized side-branches of the substrate hyphae grow upwards and develop into aerial hyphae (Fig. 4a). These side-branches are recognizable as soon as they appear as small budlike outgrowths of the substrate hyphae. In *S. coelicolor* there is no evidence that the aerial mycelium originates from the substrate mycelium by a process other than simple branching, although unconfirmed claims have been made that in other streptomycetes a process of hyphal fusion occurs at this stage (see discussion in Hopwood, 1960). The time of onset of the development of the aerial hyphae is strongly influenced by the composition of the medium; precocious development of the aerial hyphae is in general favored by cultivation on "starvation" media (Hopwood, 1960) or on media with a high sugar:organic nitrogen ratio. The origin of the aerial mycelium may be detected with the naked eye, since the aerial mycelium, being hydrophobic, appears as a whitish, powdery layer on the surface of the colonies; colonies consisting entirely of substrate mycelium appear smooth and slightly shiny. Colonies showing the powdery layer are usually said to be "sporulating," although the spores may not have begun to form. The hyphae of the aerial mycelium have a greater and less variable diameter than those of the substrate mycelium, and are straighter and less highly branched. They become transformed into chains of spores through a progressive division of the protoplasm into shorter and shorter units by the development of new septa between the existing ones. In general sporulation begins at the apex of a hypha and proceeds toward the base, but the basipetal gradient is frequently irregular. The spores are surrounded by a wall that is somewhat thicker than that of the hyphae of the substrate and aerial mycelium (Glauert and Hopwood, 1961), but differs from the thick, many-layered spore wall of eubacteria (Robinow, 1953). The spores of *Streptomyces* are only slightly more resistant to heat than the mycelium.

B. Nuclear Cytology

The nuclear material of *S. coelicolor* resembles that of eubacteria in its appearance after staining by methods such as those of Feulgen, Piekarski, or DeLamater (Hopwood and Glauert, 1960a). As in eubacteria, the cytoplasm is so rich in ribonucleic acid that differential staining of the nuclear material with dyes of the Romanofsky series is possible only by the use of

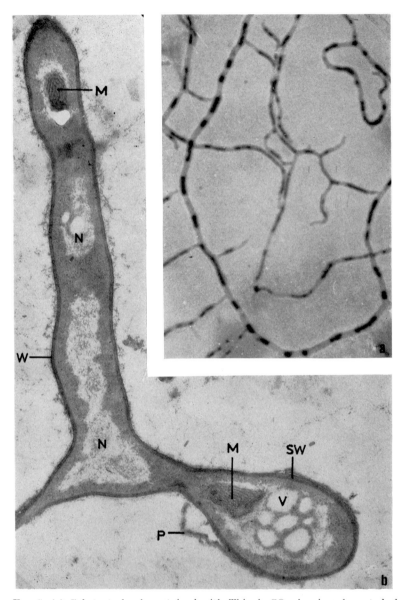

Fig. 5. (a) Substrate hyphae stained with Thionin-SO$_2$ showing elongated chromatinic bodies ($\times$ 2500). (b) Electron micrograph of a thin section of a germinated spore: N, nuclear regions; V, vacuoles in nuclear region of spore; M, membranous bodies; W, hyphal wall; SW, spore wall; P, wall of parent hypha surrounding spore.

carefully balanced mixtures of stains (Piéchaud, 1954), unless the ribonucleic acid is first removed by acid hydrolysis (as in the above-mentioned staining procedures) or by treatment with ribonuclease (Tulasne and Vendrely, 1947). The chromatinic bodies of *S. coelicolor*, like those of eubacteria, remain stainable at all stages of their development (Figs. 4 and 5), divide without the manifestation of a spindle or of normal chromosomes, and are not separated from the cytoplasm by a

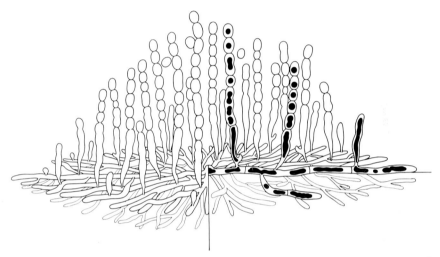

Fig. 6. Schematic drawing of a *Streptomyces* colony on agar medium. Substrate and aerial hyphae are shown extremely shortened and thickened. The right half of the colony is sectioned to show nuclear bodies (black) and septa.

nuclear membrane (Fig. 5b). However, in the substrate mycelium of *S. coelicolor*, the chromatinic bodies are more variable in shape and size than in a typical eubacterium; in addition to small rodlike or dotlike bodies, large irregularly lobed bodies are present (Fig. 5a), which in electron micrographs of thin sections (Hopwood and Glauert, 1960b) are seen to be single nuclear regions rather than aggregates of separate bodies unresolvable in the light microscope (Fig. 5b).

In the aerial mycelium a rather regular sequence of changes in the appearance of the chromatinic bodies precedes the delimitation of the spores. In the young aerial hyphae, and in the basal segments of older ones, each compartment between adjacent septa contains a single large rod of chromatinic material, which appears as a single nuclear region in electron micrographs of thin sections. Each compartment corresponds in length to up to about eight spores. During the transformation of the aerial hyphae into spores, the rodlike chromatinic bodies appear to undergo a process of gradual subdivision which produces shorter and shorter chromatinic bodies (Fig. 4d), ending with the formation of the dotlike spore nuclei (Fig. 4c). One may postulate that the chromatinic rods already contain the nuclear material corresponding to a number of spore nuclei and that the subdivision of the rods reflects the segregation of units already present in the rods.

Figure 6 is a schematic drawing of a colony of *S. coelicolor*, showing some of the main features of its organization.

IV. Studies of Recombination by Selective Analysis

A. Discovery of Gene Recombination

The demonstration of genetic recombination in an organism previously considered to reproduce only asexually requires the preparation of marked strains which differ in at least two characteristics, and the selection, from mixed cultures of the two strains, of phenotypes showing markers from both parents. In the absence of information on the ploidy of the species, the nature of the mutations, and the phases of the life cycle, it is difficult to assess the significance of the recovery of recombinant phenotypes. This information may be deduced provisionally from a consideration of related organisms with known genetic systems, and later studies may confirm or contradict the working assumptions.

In the first attempts to make "crosses" in *Streptomyces coelicolor* (Sermonti and Spada-Sermonti, 1955, 1956; Hopwood, 1957) it was assumed, by analogy with the situation in bacteria and molds, that the induced mutations used to mark the parent strains (biochemical deficiencies and resistance to streptomycin) were gene mutations, recessive to wild type alleles. The discovery of recombinant phenotypes containing presumed recessive alleles from both parents suggested that recombination between the genomes of the parent strains had given rise to new haploid genotypes containing non-homologous segments of the genomes of the two parents. The first published example was the following (Sermonti and Spada-Sermonti, 1955):

parents: *met his* + + and + + *pro glu*
recombinant: *met* + *pro* +.

Later the complementary recombinant (+ *his* + *glu*) was isolated repeatedly. Hopwood (1957) obtained a similar result with different markers in a different strain of *S. coelicolor*, and further observed that the frequencies of complementary recombinant phenotypes were usually equal. These results did not exclude the possibility that other recombinant phenotypes, in particular those showing the wild type characters, might result from the superposition of genomes (diploidy, disomy, etc.) instead of from true recombination. However, in an analysis of the linkage relations of several loci in *S. coelicolor* A3(2) (Hopwood, 1959), the internal consistencies of the data definitely excluded the occurrence of false or phenotypic recombination, at least as the major process. On the other hand, the cooperation of entire genomes in heterokaryons or heterozygous diploids to give false recombination has been repeatedly affirmed by Bradley (1957, 1958, 1960). In strain A3(2), mixed cultures of a given pair of phenotypes always gave a particular pattern of recombination, no matter whether the parent strains had been obtained by successive mutation from the wild type, or as recombinants in previous crosses (Hopwood, 1959). This indicated that the majority, at least, of the recombinants were stable haploid segregants, and not merely heterokaryons.

In this section we shall describe the methods we have used for the preparation of mixed cultures, for the selection and characterization of recombinants, and for the elaboration of the data. The rationale of the linkage tests has already been described briefly by Hopwood (1959). The data presented and elaborated in Section IV,C come from the same source.

B. Crossing Techniques

1. *Preparation of Mixed Cultures*

Mixed cultures are prepared by streaking inocula of the two parent strains in turn on the surface of a slope of complete agar. The streaks are made with a loopful of a suspension of spores or with a dry inoculum of spores and mycelium, in such a way as to give growth over the entire surface of the agar after the inocula have been thoroughly mixed together on the slope. After 3–4 days incubation, the surface of the cultures is normally seen to be sporulating. A spore suspension is prepared (see Section II,D), and is usually centrifuged and resuspended in sterile water to remove traces of growth factors dissolved from the mixed culture.

2. *Plating on Selective Media*

The markers so far employed in crosses of *S. coelicolor* are biochemical deficiencies and resistance to streptomycin. Selective media for the recovery of recombinants are supplemented in such a way that the parent strains cannot grow, owing to lack of a growth factor from the medium or to the presence of streptomycin; thus the selectable alleles are the wild type alleles of the biochemically deficient mutants and the mutant allele for streptomycin resistance. The media are chosen to allow the growth of the maximum number of recombinant classes on the minimum number of media. A four-point cross with two selectable alleles in each parent strain may be represented as $ABcd \times abCD$, where the capital letters indicate the selectable alleles. In such a cross four types of selection are possible, and so four media are employed, which select for the combinations of markers AC, AD, BC, or BD.

Supplements are added to the molten agar at the concentrations indicated in Section II,B,3. The spore suspension is plated on the surface of the selective media as described in Section II,D. The suspension is normally plated undiluted and at tenfold and hundredfold dilutions in order to obtain a suitable density of recombinant colonies. When more accurate counts of the numbers of recombinants per unit volume of spore suspension are required, the suspension is plated by mixing it with the molten agar medium.

3. Isolation of Recombinants

Forty-eight or 72 hours after plating the spore suspension on selective media, the recombinant colonies are counted. At 72 hours they are normally sporulating vigorously and are readily isolated.

In a four-point cross, on each selective medium, the alleles at two loci are unselected and therefore four genotypes are recoverable (see Section IV,C,1). Although the colonies may be classified by replica plating from the original selective media to two diagnostic media, this is inadvisable since scoring the replica plates is generally laborious. Random samples are therefore isolated to master plates containing the same medium as that on which they were selected, in order to prevent the growth of parental spores picked up from the surface of the selective plates. The colonies are isolated with a small loop or a needle, depending on their density on the selective plates, and streaked in regions of the master plate corresponding to the fifty squares of a grid over which the plate is placed during the operation. Two possible procedures may be adopted in isolating recombinants: (1) to isolate from each selective medium a number of colonies proportional to the total count per plate on that medium; (2) to isolate an equal number of colonies from each medium. The first method is preferable for the comparison of the frequencies of analogous or complementary genotypes on various selective media, the second for the determination of the allele ratios at unselected loci.

4. Classification and Purification of Recombinants

When the cultures on the master plates are sporulating vigorously (2–3 days) the master plates are replicated to three media: two diagnostic media and a control plate of the same composition as the master plate. The recombinants are classified after a further 2 days.

The origin of certain recombinant colonies from groups of plating units or from heterogenous plating units may lead to their misclassification. This results in an apparent excess of prototrophs, since a mixed colony can be replicated on all media. Thus apparently prototrophic recombinants, especially when they are rare, should be re-tested by transferring them to minimal medium. If they grow they may be considered as true prototrophic recombinants, if not, as mixed colonies, which are excluded from the classification. This precaution was not taken in our first crosses, and misclassification may consequently explain the excess of prototrophs which disturbs certain of the data (see Table 5).

When a recombinant colony has to be "purified" in order to be used in further studies, the object of the exercise must be quite clear. When we need a clone with constant characteristics, a strain may be re-streaked and single colonies reisolated until they show themselves to be stable and homogenous. Alternatively, we may be interested in the genetic information contained in the smallest viable unit, even if this is genetically heterogenous. In this case, successive reisolation may be undesirable and lead to the loss of the information we are seeking, if the unit is highly unstable (see later: heteroclones). A more suitable procedure is a good mechanical dispersion of the original inoculum, a good and repeated filtration of the spore suspension through closely packed cotton wool, and plating at a sufficiently high dilution.

C. Analysis of Recombination

1. *Information Obtainable from the Selective Analysis*

The information obtainable from a four-point cross of the type $ABcd \times abCD$ is summarized in Table 2. Genotypes containing all four selectable markers (i.e., $ABCD$, simplified to . . . . in Table 2) are recoverable on all four media, those with three selectable alleles (e.g., a . . .) on two selective media (selecting for BC and BD), and those with two selectable alleles (e.g., a . c .) on only one medium (selecting for BD). In a cross of this kind, one recombinant . . . . is recoverable, four of the type a . . . , and four of the type a . c . , making a total of nine genotypes. The last four form two pairs of complementary genotypes (a . c . and . b . d; . b c . and a . . d). Out of a total of 16 values (4 genotypes × 4 selective media), the 15 degrees of freedom (d.f.) are distributed thus: 7 for the variance between different media, 2 for the variance between complementary genotypes, and 6 for the variance between non-complementary genotypes. This last variance may be taken as an index of the existence of linkage. There are insufficient degrees of freedom for the estimation of the effects of differential viability.

TABLE 2
Recombinant Genotypes Recoverable in a Four-Point Cross on Four Selective Media

Medium selective for:	Cross: $ABcd \times abCD$*†				Number of media on which each genotype is recoverable
	A C	B D	B C	A D	
	Recoverable genotypes (simplified symbols)				
	. . . .	. . . .	. . . .	. . . .	4
	. . . d	—	. . . d	—	2
	. b . .	—	—	. b . .	2
	—	. . c .	—	. . c .	2
	—	a . . .	a . . .	—	2
‡{	. b . d	—	—	—	1
‡{	—	a . c .	—	—	1
‡{	—	—	a . . d	—	1
‡{	—	—	—	. b c .	1
Number of genotypes per medium:	4	4	4	4	16

* Simplified symbols: . . c d $\times$ a b . .
† Capital letters (or dots in simplified symbols) indicate the selectable alleles.
‡ Complementary genotypes.

An example of the information obtained in a four-point cross involving the markers *met-2*, *his-1*, *arg-1*, and *str-1* is shown in Table 3. About a hundred recombinant colonies from each selective medium were classified. The relative frequencies of the various genotypes were obtained by multiplying the proportion of each genotype among the total recombinants classified on a particular medium by the total number of recombinant colonies growing on that medium after plating of a unit volume of spore suspension. These figures appear in the body of Table 3. To obtain the seven estimates of the frequencies of pairs of complementary genotypes (corresponding to the seven recombination patterns) the frequencies of the members of the two available pairs of complementary genotypes are averaged, and the mean frequencies of the other classes are taken as estimates of the remaining complementary pairs. This procedure is justified by the finding that the frequencies of complementary genotypes are approximately equal.

The same four markers may be used in different combinations in two further four-point crosses in which each parent contains two selectable alleles: $mh+s \times ++a+$ and $m+as \times +h++$. In each of the three

THE GENETICS OF *Strepotmyces coelicolor* 295

TABLE 3

Frequencies of Recombinant Genotypes Recovered in a Four-Point Cross on Four Selective Media*

Cross A54: $+ h\ a\ s \times m + + +$

	Supplements in the selective medium				Average fre- quencies per unit volume	Recombination between the loci†	
	Methionine, arginine, streptomycin	Histidine	Histidine, methionine, streptomycin	Arginine		Pattern	Relative frequency
Selected alleles:	$h^+\ s$	$m^+\ a^+$	$a^+\ s$	$m^+\ h^+$			
Recombinant genotypes							
$m + + s$	61	—	71	—	66	mha/s	51
$+ + + s$	1	4	7	3	4	ms/ha	3
$+ + + +$	—	7	—	3	5	m/has	4
$+ h + s$	—	44	18	—	31	mhs/a	24
$m + a\ s$	18	—	—	—	18 ⎫ 22	mh/as	17
$+ h + +$	—	26	—	—	26 ⎭		
$+ + a\ s$	1	—	—	1	1	h/mas	1
$m\ h + s$	—	—	0	—	0 ⎫ 0	ma/hs	0
$+ + a +$	—	—	—	0.1	0.1 ⎭		
Colonies per unit volume:	81	81	96	7			100

	Recombination between pairs of loci†					
	m/h	m/a	h/a	m/s	h/s	a/s
	3	24	24	51	51	51
	4	17	17	17	17	24
	1	3	1	4	3	3
	0	4	0	0	1	0
Tentative estimates of recombination percentages‡:	8	48	42	72	72	78
	5	32	28	*50*		

* About 100 colonies from each medium tested.

† Loci are indicated by the symbols of the mutant alleles.

‡ Assuming no linkage between s and other loci (recombination % = 50).

crosses, of course, a different genotype or pair of genotypes corresponds to a particular pattern of recombination. In order to compare the frequencies of the corresponding genotypes in the three crosses, the relative frequencies of the seven pairs of complementary genotypes in each cross have to be expressed as percentages of the total recombinant progeny. A comparison of the apparent frequencies of recombination in the same region in the different crosses gives us a considerable amount of information on the effect of differential viability of the various genotypes on the estimation of recombination. In Table 4 the variance in the rows reflects variation due to differences in the viability of different genotypes, while variance in the columns reflects variation due to the different frequencies of the various patterns of recombination. It is clear that the latter is very much larger than the former.

A more direct analysis of the variability due to linkage and that due to differential viability (and other factors formally indistinguishable from it) of the different genotypes based on the actual numbers of recombinants classified may be obtained from the data elaborated in Table 5. Here, the numbers of recombinants selected on the same medium in the two crosses in which the markers are in different coupling arrangements are compared. The numbers of recombinants corresponding to the same pattern of recombination may be compared ignoring the genotype (χ^2B), or else the numbers of recombinants with the same genotype may be compared, ignoring the pattern of recombination (χ^2A). If viability effects are absent, the relative frequencies of recombination patterns should be the same in the two crosses, irrespective of the genotypes; if, instead, there is no linkage effect the relative frequencies of different genotypes should be the same in the two crosses. It is quite clear from a comparison of chi-squares A and B (Table 5) that there is linkage, but that viability effects are also sometimes present. It is noteworthy that the biases leading to significant chi-squares B are usually due to an excess of prototrophs; the excess prototrophs may have resulted from misclassification of mixed colonies (see Section IV,B,4).

2. *Estimation of Linkage*

Any analysis of linkage carried out on a single selective medium must be based on the multiplication of values which include those corresponding to multiple crossover classes. The latter are inevitably small, and, further may be strongly influenced by interference (Bailey, 1951). When however, all the markers are employed as nonselected markers on various media as in the type of analysis outlined above, it is possible to estimate the relative frequencies of recombination between pairs of markers by summing the observed values, and thus avoid the placing of undue weight

TABLE 4
Analysis of Linkage in Three Four-point Crosses: Relative Frequencies Of Recombination*

Pattern of recombination between the loci†	Crosses			Means	Crossover in regions (see map below)
	A54 $+ h\ a\ s$ × $m + + +$	A42 $m\ h + s$ × $+ + a +$	A55 $m + a\ s$ × $+ h + +$		
mhas/ mha/s	— 51	— 78	— 55	— 61	—
ms/ha m/has	3 4	2 0.4	3 3	2.5 2.4	1
mhs/a mh/as	24 17	16 4	9 27	16 16	2
h/mas ma/hs	1 0	0 0	3 0.1	1.3 0.1	1,2
	100	100	100	100	

	Recombination between pairs of loci†					
	m/h	m/a	h/a	m/s	h/s	a/s
Cross A54	8	48	42	72	72	78
Cross A42	2	22	20	83	84	95
Cross A55	9	42	40	84	88	67
Means:	6.5	38	34	80	81	80
Estimates of recombination percentages‡:	4	24	21		50	

Map:

```
        |--------24--------|
        m  h               a              s
        |  |               |              |
           4      21
```

Regions: 1   2

* Data from Hopwood (1959).
†,‡ See notes to Table 3.

TABLE 5

Tests for the Presence of Linkage in Three Four-Point Crosses*

Crosses:	A55 m + a s × + h + +		A42 m h + s × + + a +		A54 + h a s × m + + +	
Pattern of recombination between loci	Genotype	No.	Genotype	No.	Genotype	No.
	Medium: MM + histidine, arginine and streptomycin		Medium: MM + methionine arginine and streptomycin		Medium: MM + methionine, histidine, and streptomycin	
mha/s	$+ h + s$	80	$+ + a\ s$	90	$m + + s$	73
m/has	$+ + a\ s$	1	$m + + s$	0	$+ + + a\ s$	1
ma/hs	$+ + + s$	0	$m + a\ s$	2	$m + a\ s$	1
mh/as	$+ h\ a\ s$	13	$+ + + s$	3	$m + a\ s$	21
	χ^2A = 143.04; $P \ll 0.01$		χ^2A = 176.74; $P \ll 0.01$		χ^2A = 91.14; $P \ll 0.01$	
	χ^2B = 3.86; P: 0.3–0.2		χ^2B = 16.62; $P < 0.01$		χ^2B = 0.92; P: 0.9–0.8	
	Medium: MM + methionine		Medium: MM + histidine		Medium: MM + arginine	
mhs/a	$m + + +$	18	$+ + + +$	66	$+ + + +$	44
h/mas	$+ + + +$	3	$+ h + s$	3	$+ + a\ s$	9
ma/hs	$+ + + s$	0	$+ + h +$	19	$+ + a +$	1
mh/as	$m + + +$	72	$+ h + +$	8	$+ + + s$	38
	χ^2A = 177.44; $P \ll 0.01$		χ^2A = 109.56; $P \ll 0.01$		χ^2A = 62.67; $P \ll 0.01$	
	χ^2B = 68.12; $P \ll 0.01$		χ^2B = 9.72; P: 0.02–0.01		χ^2B = 18.05; $P < 0.01$	

Pattern of recombination between loci						
	mha/s		mhs/a		has/m	
	h/mas		m/has		h/mas	
	ma/hs		mh/as		ma/hs	
	mh/as		ms/ha		ms/ha	

* χ^2A (3 d.f.): hypothesis of no linkage—same genotype, same frequency in the two crosses.
χ^2B (3 d.f.): hypothesis of no viability effect—same pattern of recombination, same frequency in the two crosses.

on the smallest values. An estimate of linkage is obtained by treating the data of a four-point cross in the way shown in the lower half of Table 3. As stated above, it is possible to estimate the frequencies of all the pairs of recombinant genotypes, but not of the parental genotypes. Thus the recombination frequencies are only relative (Hopwood, 1959). The relative frequency of recombination between two loci is obtained by summing the relative frequencies of the pairs of recombinant genotypes (four out of seven) which have the alleles at these loci in nonparental combinations. Thus the relative recombination frequency between *met-2* and *arg-1* (m/a) is equal to: $24 + 17 + 3 + 4 = 48$. The loci *met-2* and *his-1* are the most closely linked; *arg-1* is at a distance from them of about five times their distance apart, and nearer *his-1* than *met-2*. *Str-1* is far from the other three loci, and equally distant from all three, suggesting that it is unlinked with them. If we tentatively assign a value of 50% to the recombination between *str-1* and the other markers, we can convert the relative frequencies of recombination between the other loci into absolute values (bottom of Table 3).

More reliable estimates of linkage relations are clearly obtainable by considering the combined data of the three crosses involving the same markers. In Table 4, the relative frequencies of the genotypes corresponding to the same pattern of recombination in the three crosses are compared, and mean values are calculated. The conclusions drawn from the analysis of the single cross in Table 3 are confirmed. The absence of linkage between *str-1* and the other markers is indicated by the mean recombination frequencies between pairs of loci (lower half of Table 4) calculated from the data of the three crosses, and is confirmed by the equal frequencies of recombination patterns differing only with respect to the *str-1* locus. The order *met-2–his-1–arg-1*, deducible from the recombination percentages, is confirmed by the observation that the classes produced by double crossover are the least frequent.

The discovery of a marker unlinked to the others in the early stages of the genetic analysis of *S. coelicolor* A3(2) was particularly fortunate since it allowed the relative frequencies of recombination between pairs of loci to be converted immediately into absolute terms. Proceeding from the location of these four markers in two linkage groups it has been possible to map other genes with respect to the original loci by means of the analysis of four-point crosses on four selective media (Hopwood, 1959).

D. SIMPLIFIED SELECTIVE ANALYSIS

A simplified method for determining the relative frequencies of recombination between pairs of loci in four-point crosses was described by

Hopwood (1959). The spores from the mixed culture are plated on four selective media as in the complete analysis. Each medium selects the recombinants between two loci, and so the number of colonies that grow on each medium is proportional to the frequency of recombination between a certain pair of loci. The analysis requires only the counting of colonies without the isolation and classification of recombinants. However, only four of the six relative recombination frequencies in a four-point cross can be estimated, the other two being the distances between non-selectable alleles in coupling. When the absolute frequency of recombination between one pair of loci is already known, the others can be determined by comparison with it. This type of analysis theoretically gives results equivalent to those of the complete analysis when the frequencies of complementary genotypes are equal. This condition, which as we shall see is not always true, cannot however be verified in the simplified analysis.

Hopwood (1959) used this method of analysis for the tentative location of several loci. The internal consistency of the data is verified approximately when two of the four deducible recombination frequencies are already known and their relative magnitudes are in agreement with expectation. Table 6 shows a simplified analysis of the linkage relations of various markers (x) with markers situated in the two linkage groups ($arg-1$, and $his-3$); the recombination frequencies of these loci with a third

TABLE 6
Simplified Selective Analysis in the Detection of Linkages Involving New Markers (x); Counts per Plate

		Crosses: his-3 arg-1 str-1 × x						
		Medium selective for				Reference linkage $h3$-s	New linkage indicated (% recombination*)	
Cross	New marker (x)	$a^+ s$	$h3^+ s$	$a^+ x^+$	$h3^+ x^+$	(% recombination*)	with a	with $h3$
A 133	hom-1	114	26	120	11	11.4	—	ca. 5
A 139	met-3	84	45	125	74	26.8	—	—
A 140	met-4	1233	417	2800	1800	16.9	—	—
A 141	met-5	198	78	216	85	19.7	—	ca. 20
A 138	cys-6	376	152	578	325	20.2	—	—
A 282	phe-1	1183	480	1693	1194	20.3	—	—

* Recombination between a and s made equal to 50%.

(*str-1*) are already known, and can be compared in the various crosses in order to test the internal consistency of the data. Two of the six markers appear to be linked to *his-3*: *hom-1* at about five units, and *met-5* less closely. However, the internal consistency of the data is poor, the chi-square for homogeneity of the ratio of counts on the media which reflect the two previously known recombinations in the six crosses being 33.1, with 5 degrees of freedom. Thus this method of analysis can be considered reliable only when it reveals a close linkage. Table 7 shows an example of a preliminary investigation of the linkage between three methionine markers. In each cross only a single recombination percentage is known, so that there is no test of the internal consistency of the data. Nevertheless it is clear that two of the methionine loci (*met-3* and *met-4*) are very closely linked, if not allelic, and distant from a third (*met-2*). Fuller analysis has confirmed that *met-3* and *met-4* are in linkage group II, *met-2* being in linkage group I. A more detailed analysis of the linkage relations of some histidine loci which form a cluster is described in Section VI.

TABLE 7

Simplified Selective Analysis of Recombination between Three Methionine Loci (*met-2*, *met-3*, *met-4*)

	Crosses					
	A303		A304		A318	
	met-2 arg-1 str-1 × *met-3*		*met-2 arg-1 str-1* × *met-4*		*met-4 his-3 str-1* × *met-3*	
Medium selective for:	a^+ s	$m2^+$ $m3^+$	a^+ s	$m2^+$ $m4^+$	$h3^+$ s	$m3^+$ $m4^+$
Counts per plate:	131	133	235	244	68	0
Linkage between selected markers*:	*absent*	loose or absent	*absent*	loose or absent	present	close

* Known reference linkages in italics.

E. Crosses Giving Unequal Frequencies of Complementary Genotypes

In some crosses the selective analysis described in the previous sections becomes unreliable because the basic requirement on which it depends is not fulfilled; complementary recombinant genotypes have unequal frequencies. The differences in frequency were small in the first crosses of strains of *S. coelicolor* A3(2), some of which were reported by Hopwood

(1959), so that they did not present a serious obstacle to the genetic analysis. Later, such differences began to appear rather frequently, and in certain crosses were very large, occasionally one of the two genotypes being up to a hundred times more frequent than the complementary type. We do not know whether the absence of asymmetries in the earlier crosses was a fortunate coincidence, or whether the phenomenon of asymmetry is related in some way to the successive re-isolation of strains or to the introduction of new markers into the stocks.

The phenomenon of asymmetry may provide a clue to the understanding of the genetic system of *S. coelicolor*. The equality or inequality of the frequencies of the complementary classes depends on the strains employed in the crosses, and is reproducible to a large extent in experiments carried out at different times and by different workers. Table 8 summarizes the results of two crosses each carried out three times. The same markers are employed in the two crosses, and one of the parent strains (HG1-7 *his-1*) is the same; the strains *met-2 phe-1 str-1* with which it is crossed differ in

TABLE 8
Strain-Dependence of Segregation Patterns in Crosses with the Same Combinations of Markers: Relative Frequencies (Per Cent) of Different Genotypes among Total Recoverable Recombinants

Genotypes	Cross:	HG4-1 $m + s\ p$ × HG1-7 $+ h + +$			Cross:*	HG22-16 $m + s\ p$ × HG1-7 $+ h + +$			Cross-over in region
	A	B	C	Mean	D	E	F	Mean	
$+ h\ s\ p$	42	39	24	35	6	3	6	5	—
$m + + +$	20	13	22	19	59	40	61	53	
$+ h\ s +$	11	16	14	14	5	9	6	7	2
$m + s +$	14	14	29	19	19	41	23	28	
$+ + + +$	4	5	3	4	2	2	2	2	1
$+ + s\ p$	3	7	1	3	6	2	1	3	
$m\ h\ s +$	0	0	0	0	0	1	0	0	
$+ + + p$	5	3	4	4	2	0	0	1	1,2
$+ + s +$	1	3	3	2	1	2	1	1	

Map: m h s p

Regions: **1**   **2**

*In crosses HG22-16 × HG1-7 (D, E, and F), there is a large excess of recombinants carrying $m +$ over those carrying the complementary combination $+ h$.

origin, having been isolated as recombinants from different crosses. The most frequent classes of complementary recombinants have the genotypes $+ h s p$ and $m + + +$, and arise by exchange between unlinked markers. Their frequencies are similar in the cross HG1-4 × HG1-7 although $+ h s p$ is always rather more frequent; in the cross HG22-16 × HG1-7, on the other hand, $m + + +$ is about ten times more frequent than $+ h s p$. An excess of the $m +$ over the $+ h$ genotype is also shown in the frequencies of recombinants produced by crossing-over in region **2** in the HG22-16 × HG1-7 crosses, but not in the HG4-1 × HG1-7 crosses. Clearly the simplified form of selective analysis (Section IV,D) is totally inapplicable when complementary genotypes have very different frequencies: in fact the two distances *his-1–phe-1* and *met-2–str-1* determined from the counts on the two selective media, which are nearly equal in the more symmetric crosses (A, B, and C), differ by a factor of ten in the others (D, E, and F). This obviously prevents the use of the simplified selective analysis for the location of new markers; when the locations are known, however, this analysis provides a rapid test for the identification of parent strains which give rise to asymmetrical progenies. Using this test, recombinants have been identified which derived from the same cross, had the same phenotype, but behaved differently in crosses with a reference strain, some producing symmetrical and others asymmetrical progenies. Although it is not yet possible to describe this behavior quantitatively, it thus appears that the capacity to give rise to symmetrical or asymmetrical patterns of recombination segregates in a cross. Since strains that produce symmetrical or asymmetrical progenies are indistinguishable in their growth and sporulation, it seems that the difference between them is due to some inheritable factor that influences the crossing behavior of a strain rather than affecting its viability.

F. Considerations on the Fertility of *Streptomyces coelicolor*

Bradley and Anderson (1958) claimed that certain combinations of auxotrophic strains of *S. coelicolor* did not give rise to prototrophic colonies. In particular, they stated that a strain S207 formed prototrophs with two other strains (S125, S206) but not with a further group of strains (S202, S203, etc.). Even though this incompatibility was not attributable to allelism between the markers of the strains which formed "sterile" combinations, it could, however, have been attributed to interactions between different markers. Unfortunately the methods used by Bradley and Anderson did not allow the study of true genetic recombination. Before building a hypothesis of a compatibility system it would have been desirable at least to obtain two strains bearing the same markers but showing different types of fertility.

A fertility system might be manifested quantitatively rather than qualitatively. Unfortunately it has been impossible to devise a good parameter for measuring fertility in *Streptomyces*, in contrast to the situation in *Escherichia coli*, where the number of recombinants can be expressed in relation to the input of cells of the minority parent. For practical reasons we usually express the number of recombinants as a fraction of the spores in the suspension harvested from the mixed culture, although such estimates vary widely even in different crosses of the same strains. In a cross in which the inocula were adjusted in order to secure an equal output of spores of the two parents, the number of recombinants of a type which required only recombination between linkage groups was nearly equal to the number of spores of each parental type. In the same cross, when one parental type was ten times more frequent than the other among the harvested spores, recombinants represented only 1% of the total spores. When no attempt is made to balance the inoculum of the two parents so as to achieve an equal output, the fraction of spores which give rise to recombinant colonies on a medium which selects for recombination between unlinked markers usually varies between 10^{-2} and 10^{-5}.

Certain rare crosses have turned out to be regularly sterile, whatever the ratio of spores of the two parental genotypes in the mixed culture. Thus a small group of strains has been identified which give no recombination among themselves, but are fertile when crossed with strains not belonging to this group. All the other strains studied have been fertile in all combinations. A certain number of strains which carry the same markers differ only in their fertility in crosses, and so fertility must be controlled by a factor independent of the markers; nothing is yet known of the mechanism of control. The situation shows some similarities with the fertility system of *Escherichia coli* K12 (see Hayes, 1960), the group of self-sterile strains perhaps being comparable with F^- strains of *E. coli* and the fertile strains with F^+ or Hfr strains.

V. Heteroclones

A. Discovery of Heterogeneous Clones (Heteroclones)

Sermonti and Spada-Sermonti (1959a, b), working with strain I.S.S. of *Streptomyces coelicolor*, found that certain of the colonies produced by plating spore suspensions derived from mixed cultures on selective media were heterogeneous: two or more different recombinant phenotypes were recovered from a single colony. Sermonti *et al.* (1960) investigated this phenomenon further in crosses of strains of *S. coelicolor* A3(2) containing the two closely linked markers *met-2* and *his-1* in repulsion. A high proportion of the colonies arising on minimal medium were found to be heterogeneous. A spore suspension prepared from any such colony and plated on complete medium gave rise to segregants with a variety of phenotypes, up to twelve of the sixteen possible combinations of four markers being recovered from a single colony. The colonies were named *heteroclones* (heterogeneous clones) and were postulated to arise from plating units containing a highly unstable heterozygous nucleus which underwent segregation to produce stable haploid segregants with parental and recombinant genotypes.

An interesting feature of the heteroclones studied by Sermonti and associates (1960) was the imbalance of some allele pairs. Among the segregants from heteroclones derived from four-point crosses containing the markers *met-2*, *his-1*, *arg-1*, and *str-1*, the two alleles at each of the loci *met-2*, *his-1*, and *arg-1* (in linkage group I) were present in approximately equal numbers. The allele *str-1* (in linkage group II), however, was present with a very low frequency in most of the heteroclones, the great majority of the segregants being sensitive to streptomycin. It was later found (Hopwood et al., 1961) that some alleles were apparently missing altogether from many of the heteroclones, and sometimes this incompleteness of the heteroclones affected groups of linked loci simultaneously.

The heteroclones promised to be a powerful tool in the genetic analysis of *S. coelicolor*, since they offered the hope of carrying out genetic analysis without recourse to selective methods, with their inevitable limitations. Since the heteroclones derived from heterozygous units, and their growth on selective media required the cooperation of two genomes, they also offered a *cis/trans* test of allelism. Moreover, the study of the nature of the heteroclones promised to contribute to the understanding of the mechanism of recombination.

B. Principles for the Selection of Heteroclones

For the selection of heteroclones, three conditions must be satisfied which are also applicable to the recovery of mixed clones deriving from unstable heterozygous units (heterogenotes) in other microorganisms. (1) The selection must be carried out on a medium which allows the growth of the minimum number of the haploid recombinant cells with which the heterozygotes are mixed, without prejudicing the growth of the heterozygotes. (2) The medium must not allow the preferential growth of some haploid segregants from the heterozygotes, which could emerge as sectors, altering the composition of the mixed clones. (3) A method must be devised to detect the heterogeneous clones among the colonies present on the selective medium.

The first condition is realized by using, for the synthesis of the heterozygotes, strains carrying two closely linked nutritional markers in repulsion. The wild alleles of these markers will be present in every heterozygous nucleus (unless the heterozygote is incomplete for that region), but one or other will be lacking from the great majority of the haploid recombinants, that is, from those in which recombination has not occurred in the short region between the two loci. Thus, on a medium which selects for the two wild alleles, only a small fraction of the haploid recombinants can grow, while all, or at least a large proportion of the heterozygotes can develop into colonies. Such systems of selection have been used for the recovery of persistent heterogenotes in *Escherichia coli* (Lederberg, 1949) and of heterozygous diploid ascospores in *Aspergillus nidulans* (Pritchard, 1954; Arditti and Strigini, 1961). In *Neurospora* selection for recombination between closely linked nutritional markers has led to the discovery of the *pseudowilds*—disomics, heterozygous for markers on a single chromosome (Pittenger, 1954). In the recovery of the heterozygotes it may be important to allow for their possible incompleteness, which may be due to partial homozygosity (Arditti and Strigini, 1961) or partial hemizygosity (Lederberg, 1949). If alleles determining the need for a growth factor are present in the homozygous or hemizygous region, the growth factor must be present if the partial heterozygotes are to be recovered.

The second requirement is achieved by the same strict selective conditions that reduce the number of haploid recombinants recoverable from the original mixed culture. By the same mechanism the probability is reduced that a segregant capable of growth on the selective medium will arise within and emerge from the mixed clone.

The third requirement is easily satisfied when the heterozygote is so unstable that it dissociates almost completely into haploid units (spores) during the life span of the colony. Since in the haploid spores there is no longer the possibility of cooperation between the different genomes present in the original heterozygote, the heterogeneous colony can be recognized by its inability to be perpetuated by means of haploid spores. Thus the conidia of a heterokaryon of *Aspergillus* containing complementary auxotrophic nuclei cannot perpetuate the colony on unsupplemented medium; the same is true of the heteroclones of *S. coelicolor* (Sermonti et al., 1960). In *Streptomyces* this can be determined by replica plating of the colonies onto a medium of the same composition as that on which they arise; the stable recombinant clones are transferred, while the heteroclones are not.

In *S. coelicolor* heteroclones can be selected satisfactorily in crosses containing a pair of nutritional markers in repulsion at about 5 units apart. Two such pairs of markers have been used extensively: *met-2* and *his-1* in linkage group I and *ade-3* and *phe-1* in linkage group II. Using

THE GENETICS OF Strepotmyces coelicolor 307

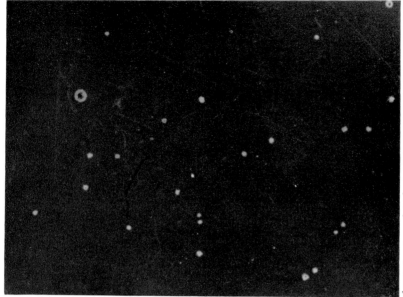
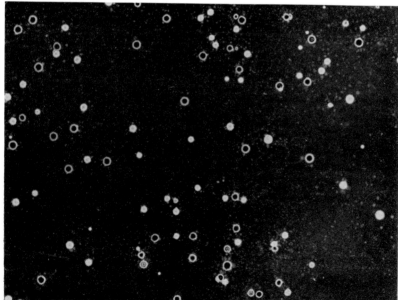

FIG. 7. Cross: *his-9 met-2 str-1* × *his-1*. The same number of spores were plated on each of two selective media. *Left*: MM + histidine + streptomycin (selection for recombination between the unlinked loci *met-2* and *str-1*); most of the colonies are normal sized stable recombinants. *Right*: MM + methionine (selection for recombination between the two closely linked *his* loci); all but one of the colonies are small heteroclones.

one such pair, heteroclones represent about 10% of the colonies arising on selective media. This percentage rises to nearly 100 in crosses involving very closely linked markers in repulsion, as, for example, pairs of histidine markers in the *his-1* region (Fig. 7).

C. Techniques Used in the Selection and Analysis of Heteroclones

The methods of preparing and plating spore suspensions from mixed cultures are the same as those used in selective analysis (see Section IV,B). The plating media lack the growth factors corresponding to the pair of markers which serve to select the heteroclones; the growth factors corresponding to some or all of the remaining nutritional markers are added to selective media when heteroclones containing only the mutant allele at one or more loci are to be recovered. When the colonies are sporulating vigorously, and before a possible background growth of the parental spores appears on the selective medium (3–4 days), replica plates are prepared on a medium of the same composition. The original plates are kept in the refrigerator during the 2 days required for the colonies on the replica plates to grow, in order to avoid the development of a background growth which would contaminate the spore suspensions prepared from the heteroclones. The heteroclones are identified by their failure to give rise to growth on the replica plate (Fig. 8). Frequently there is a clear-cut size difference between the heteroclones and the stable recombinant colonies, as in Fig. 8. However, replica plating should always be carried out in order to recognize those heteroclones in which stable segregant sectors able to grow and replicate on the selective medium have arisen.

No means of altering the proportion of plating units giving rise to heteroclones has been found. Centrifugation, heat inactivation, and repeated filtration of spore suspensions from mixed cultures do not significantly alter the ratio between heteroclones and haploid recombinants. This incidentally suggests that the heteroclones arise from units resembling normal spores.

The heteroclones are designated with a symbol of the type C28/2U, where C28 is the code number of the cross, and 2U indicates heteroclone number 2 isolated on a medium containing uracil. The supplements to selective media are indicated by the following letters: A = arginine; Ad = adenine; H = histidine; M = methionine; P = phenylalanine; S = streptomycin; U = uracil; the sign − indicates unsupplemented minimal medium.

Spores are harvested from a heteroclone by rubbing the surface of the colony with a small loop carrying a drop of water, and are suspended in 0.5 ml of water in a test tube. The spore suspension is plated on com-

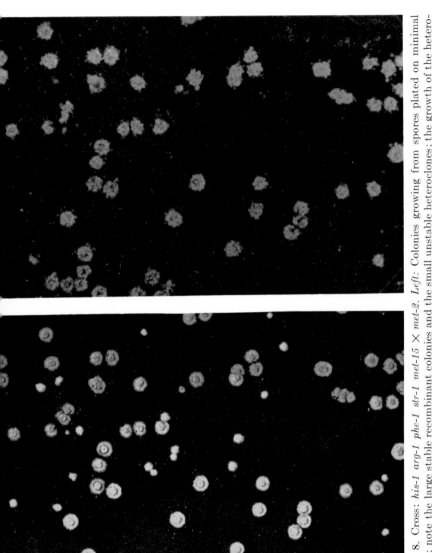

FIG. 8. Cross: *his-1 arg-1 phe-1 str-1* × *met-15 met-2*. *Left*: Colonies growing from spores plated on minimal medium; note the large stable recombinant colonies and the small unstable heteroclones; the growth of the heteroclones is made possible by the complementation of the parental genomes in heterozygous nuclei. *Right*: replica plate on minimal medium; the recombinant colonies have been replicated; practically all the heteroclones have failed to be replicated because of segregation of the heterozygous nuclei before spore formation.

plete medium in an approximate tenfold dilution series in order to obtain plates from which single segregant colonies are readily isolated. The following rapid method may be used: 0.1 ml of the spore suspension is placed on a plate of complete medium and 0.1 ml of water on two further plates; the spore suspension is spread on the first plate with a glass rod and then, without being sterilized, the rod is used to spread the drop of water on the second and third plates in turn. After incubation of the plates for 3–4 days, segregant colonies are transferred to master plates of complete medium, 50 per plate. Their genotypes are determined by replica plating from the master plate to a series of diagnostic media, capable of characterizing all the markers (Fig. 9). Care must be taken to re-test rare phenotypes apparently showing few or no growth requirements, in order to avoid misclassification. A mixture of different auxotrophic clones may be misclassified by replication on a series of media each lacking one growth factor, and appear as a phenotype with only the requirements common to the components of the mixture. Re-testing is done on a medium containing only those supplements apparently required.

A time-table of the stages in the recovery and analysis of the heteroclones is shown in the tabulation.

Stage	Time (in days)
Growth of the mixed culture	4
Development of heteroclones on selective media	4
Recognition of heteroclones by replica plating	2
Growth of segregants from heteroclones	3
Growth of isolated segregants on master plates	3
Characterization of segregants on diagnostic media	2
Re-testing of doubtful segregants	2
Complete analysis	20

D. ESTIMATION OF LINKAGE BY THE ANALYSIS OF HETEROCLONES

1. *Disturbed Allele Ratios in Heteroclones*

The segregants from the heteroclones allow the genetic analysis of *S. coelicolor* without recourse to selective methods; compared with the recombinants isolated on selective media they have the great advantage that no genotypes are excluded. The analysis of heteroclones is complicated, however, by the fact that some allele ratios among the segregants may be disturbed. Hopwood *et al.* (1961) put forward the hypothesis that the disturbances observed in certain heteroclones might be due to the lack from the heterozygous units of a segment of chromosome, starting from a breakage point, which we may call zero (0). The frequency of the alleles coupled with the 0 point would be reduced in proportion to their proximity to that point. This hypothesis, which we shall regard for the purposes of this section simply as a formal model, leads to certain conclusions that can be treated quantitatively.

2. *Estimation of Linkage between Pairs of Loci Showing Single Disturbances*

If we consider two pairs of alleles, we have four classes of segregants, two parental (AB and ab) and two recombinant (Ab and aB). The two

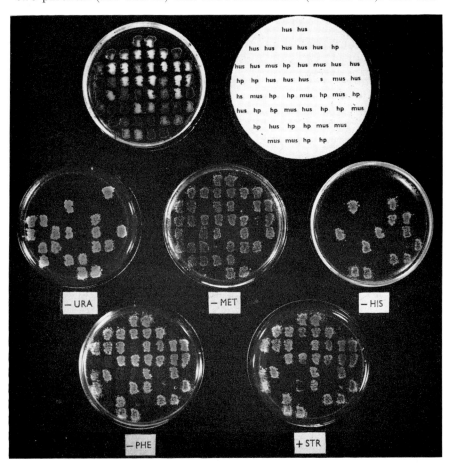

Fig. 9. Replica plating in the characterization of segregants from heteroclones. *Top left:* the master plate. *Top right:* phenotypes of the segregants as deduced from the replica plates below; u, m, h, and p stand for inability to grow on media lacking uracil (−URA), methionine (−MET), histidine (−HIS), and phenylalanine (−PHE), respectively; s stands for ability to grow on media containing streptomycin (+STR).

alleles at each locus may be present with equal frequency, in which case the complementary segregant classes must be equal in frequency; thus the estimation of linkage presents no problems. This situation, however, occurs rarely in heteroclones; usually the ratio of at least one of the pairs of alleles deviates significantly from 1:1.

Let us postulate a 0 point linked to one or both of the two loci in question. As a corollary of the theory involving a single breakage point, the 0 point cannot lie between two linked loci. There are now eight possible types of segregant instead of four, in the following complementary pairs:

$$
\begin{array}{cccc}
A\,B & a\,b & a\,B & A\,b \\
a\,b\,0 & A\,B\,0 & A\,b\,0 & a\,B\,0
\end{array}
$$

All the segregants containing 0 are supposed to be inviable, and therefore absent, but the frequencies of the other four classes still allow the estimation of the frequencies of all the segregants, assuming the equality of complementary products. The 0 point is nearer to that allele having the lowest frequency (say b). The four recoverable classes therefore have the following significance:

AB parental
ab recombination between b and 0
aB recombination between a and b
Ab recombination between a and b and between b and 0.

The sum of the percentages of the two classes produced by recombination between a and b is an estimate of the distance between the two loci, while the sum of the percentages of the two classes produced by recombination between b and 0 (i.e., the percentage frequency of the allele b) is an estimate of the distance between b and the 0 point. In the absence of interference, the following relation should be true:

$$\text{no. }(AB):\text{no. }(ab) = \text{no. }(aB):\text{no. }(Ab)$$

The significance of any deviation from this relation may be assessed by means of a 2×2 independence test with one degree of freedom. If there is no significant deviation, the estimate of linkage between a and b may be considered reliable. A large deviation would require the assumption of other factors disturbing the analysis; a second breakage point could be postulated linked to the less asymmetric locus. Under these circumstances the straightforward estimation of linkage is not possible (see Section V,D,3).

The segregants obtained from two heteroclones (C28/2U and C28/1−) derived from the same cross are listed in Table 9. The two heteroclones

TABLE 9
Segregants Recovered in Two Heteroclones from the Same Cross: C28/2U and C28/1−

$$\text{Cross (C28):} \quad \frac{+\ h\ +\ p\ +}{m\ +\ s\ +\ u}$$

	No recombination between h-s			Recombination between h-s		
	Segregants			Segregants		
		Number in heteroclone			Number in heteroclone	
Crossover in region	Genotype	C28/2U	C28/1−	Genotype	C28/2U	C28/1−
—	$+\ h\ +\ p\ +$	37	79	$m\ +\ +\ p\ +$	44	21
—	$m\ +\ s\ +\ u$	10	6	$+\ h\ s\ +\ u$	9	17
1	$m\ h\ +\ p\ +$	0	0	$+\ +\ +\ p\ +$	5	1
1	$+\ +\ s\ +\ u$	0	0	$m\ h\ s\ +\ u$	0	0
2	$m\ +\ s\ p\ +$	12	9	$+\ h\ s\ p\ +$	5	26
2	$+\ h\ +\ +\ u$	0	4	$m\ +\ +\ +\ u$	1	1
3	$m\ +\ s\ +\ +$	13	5	$+\ h\ s\ +\ +$	13	44
3	$+\ h\ +\ p\ u$	0	0	$m\ +\ +\ p\ u$	0	0
1, 2	$+\ +\ s\ p\ +$	1	1	$m\ h\ s\ p\ +$	1	0
1, 2	$m\ h\ +\ +\ u$	0	0	$+\ +\ +\ +\ u$	0	0
2, 3	$+\ h\ +\ +\ +$	5	11	$m\ +\ +\ +\ +$	1	3
2, 3	$m\ +\ s\ p\ u$	0	0	$+\ h\ s\ p\ u$	0	1
	Total	78	115		79	114

Frequencies of alleles (per cent)

	$\dfrac{m}{+}$	$\dfrac{+}{h}$	$\dfrac{s}{+}$	$\dfrac{+}{p}$	$\dfrac{u}{+}$
Heteroclone C28/2U	$\dfrac{52}{48}$	$\dfrac{55}{45}$	$\dfrac{41}{59}$	$\dfrac{33}{67}$	$\dfrac{13}{87}$
Heteroclone C28/1−	$\dfrac{20}{80}$	$\dfrac{21}{79}$	$\dfrac{47}{53}$	$\dfrac{40}{60}$	$\dfrac{13}{87}$

Map: m h s p u

Regions: 1   2   3

show different types of disturbances, although both contain all the markers present in the parents; the differences are reflected in significantly different ratios between the frequencies of pairs of alleles in linkage group I (lower part of Table 9).

In Table 10 the data for certain pairs of loci in the heteroclone C28/2U are elaborated. For each pair of loci, the frequencies of the four classes of segregants are used to estimate the percentage of recombination between the loci. The 2×2 chi-square indicates the reliability of the linkage estimates. It is significant at the 5% level of probability only for two pairs of loci; in these cases it is not significant at the 2% level. Thus all the recombination percentages may be considered as tolerable estimates.

This analysis can give no direct information on differential viability of the various genotypes, since a difference in the frequencies of the two alleles at a locus due to differential viability could not be distinguished from one due to linkage of the locus with a 0 point. On the hypothesis that the more disturbed allele ratio is due to differential viability, a nonsignificiant chi-square would indicate that the alleles at the other locus did not differ significantly in viability. Since all the markers so far studied have appeared in at least one heteroclone as the more balanced locus in linkage tests giving a nonsignificant 2×2 chi-square, viability effects cannot be serious. Moreover the ratio of the frequencies of the two alleles at a given locus varies widely from heteroclone to heteroclone even in the same cross, a situation not easily reconcilable with differential viability. In any case, a viability effect at one locus would not vitiate linkage estimates, while bias due to viability effects at both loci would be detected by the 2×2 chi-square test.

Linkage between several pairs of loci has been estimated by the analysis of the segregants from many heteroclones. Table 11 lists a series of reliable estimates. The estimates have been considered reliable when the 2×2 chi-squares give probabilities greater than 0.05. Estimates of recombination percentages between loci in different linkage groups have not been included in the list; they have always been near 50% (see Table 10), or else unreliable (see Table 12). Different estimates of the recombination frequency between the same pair of loci are homogeneous (Table 11). The distances show satisfactory additivity (see map in Table 11) and are in fair agreement with those deduced from selective analysis.

As an illustration of linkage estimates considered as unreliable on the basis of the 2×2 test, the data for some pairs of loci in the heteroclone C28/1− are elaborated in Table 12. For certain pairs of loci, the chi-square is highly significant; these are the pairs for which the linkage estimates are inconsistent with the conclusions derived from other data. In particular, the apparent linkage between *met-2* and *ura-1* is in contra-

TABLE 10

Frequencies of Combinations of Alleles at Pairs of Loci in Heteroclone C28/2U*

Pairs of loci†

	m, h	m, s	m, p	m, u	s, p	p, u	s, u
Parental allele combinations	+ h 69 m + 81	+ + 47 m s 36	+ p 48 m + 25	+ + 66 m u 11	+ p 86 s + 45	p + 105 + u 20	+ + 92 s u 19
Recombinant allele combinations	m h 1 + + 6	m + 46 + s 28	m p 57 + + 27	m + 71 + u 9	s p 19 + + 7	+ + 32 p u 0	s + 45 + u 1
Per cent recombination	4.5	47	53	51	17	20	29
χ^2‡: Probability:	2.72 0.05–0.1	0.50 0.3–0.5	0.08 0.7–0.8	0.33 0.5–0.7	0.54 0.3–0.5	4.35 0.02–0.05	5.25 0.02–0.05

* From data in Table 9.
† Loci are indicated by the symbols of the mutant alleles.
‡ Test of independence (1 d.f.): equal ratios within parental and recombinant allele combinations.

TABLE 11
Recombination between Pairs of Loci in Heteroclones

Pair of loci	Hetero- clones	Combinations of alleles								Per cent recom- bination	Hetero- geneity χ^2
		Parental				Recombinant					
		Geno- type	No.	Geno- type	No.	Geno- type	No.	Geno- type	No.		

Linkage group I

Pair of loci	Hetero- clones	Geno- type		No.	Geno- type		No.	Geno- type		No.	Geno- type		No.	Per cent recom- bination	Hetero- geneity χ^2
met-2 his-1	C28/1−	+	h	182	m	+	45	+	+	2	m	h	0	0.87	
	C28/1P			171			35			11			0	5.07	
	C28/2P			234			33			2			0	0.74	9.89
	C28/2U			69			81			6			1	4.46	5 d.f.
	C36/1−			260			100			16			2	4.76	
	C36/2−			129			60			2			4	3.08	
	pooled													3.18	
his-1 arg-1	C36/2−	h	+	121	+	a	55	+	+	7	h	a	12	9.74	
	C36/3−			71			59			11			5	11.0	
	C47/3−			87			38			4			14	12.6	3.87
	C47/2M			76			53			9			2	7.86	5 d.f.
	C47/3H			91			28			2			10	9.16	
	R14/3−	h	a	49	+	+	32	h	+	9	+	a	5	14.7	
	pooled													10.6	
met-2 arg-1	C36/2−	+	+	119	m	a	55	m	+	9	+	a	12	10.8	
	C36/3−			72			56			10			8	12.3	
	C47/4−			62			57			18			7	17.4	3.28
	R14/3−	+	a	49	m	+	29	+	+	12	m	a	5	17.9	5 d.f.
	R27/1−			141			82			17			16	12.9	
	R27/5−			78			64			18			7	13.9	
	pooled													13.9	

Linkage group II

Pair of loci	Hetero- clones	Geno- type		No.	Geno- type		No.	Geno- type		No.	Geno- type		No.	Per cent recom- bination	Hetero- geneity χ^2
his-3 str-1	R27/4−	+	+	56	h3	s	11	+	s	6	h3	+	2	10.7	0.49
	R27/5−			123			20			19			4	13.9	1 d.f.
	pooled													12.9	
str-1 ade-3	C36/1−	+	+	247	s	ad	84	s	+	41	+	ad	6	12.4	
	C36/2−			134			29			27			5	16.4	2.39
	C47/3H			82			27			16			6	16.8	2 d.f.
	pooled													14.4	

THE GENETICS OF *Streptomyces coelicolor*

TABLE 11 (*Continued*)

Pair of loci	Hetero- clones	Combinations of alleles								Per cent recom- bination	Hetero- geneity χ^2
		Parental				Recombinant					
		Geno- type	No.	Geno- type	No.	Geno- type	No.	Geno- type	No.		
ade-3 phe-1	C36/1−	+ p	273	ad +	87	+ +	15	ad p	3	4.76	
	C36/2−		155		34		6		0	3.08	
	C36/3−		104		33		6		3	6.16	2.33
	C47/4−		122		17		4		1	3.47	5 d.f.
	C47/2M		92		47		2		4	3.87	
	C47/3H		90		31		8		2	7.63	
	pooled									**4.74**	
phe-1 ura-1	C47/4−	p u	122	+ +	10	p +	11	+ u	1	8.33	0.08
	C47/3H		91		27		1		12	9.92	1 d.f.
	pooled									**9.13**	
str-1 phe-1	C28/1−	+ p	101	s +	72	s p	37	+ +	19	24.5	
	C28/2U		86		45		19		7	16.6	
	C36/1−		232		81		44		21	17.2	8.32
	C36/2−		129		30		26		10	18.5	4 d.f.
	C47/1H	+ +	77	p s	47	+ s	10	p +	10	13.9	
	pooled									**18.4**	
ade-3 ura-1	C47/4−	+ u	124	ad +	9	ad u	9	+ +	2	7.64	1.16
	C47/3H		93		23		10		5	11.5	1 d.f.
	pooled									**9.45**	
str-1 ura-1	C28/2P	+ +	164	s u	11	s +	87	+ u	7	34.9	3.87
	C47/3H	+ u	79	s +	19	s u	24	+ +	9	25.2	1 d.f.
	pooled									**31.8**	

Summary of linkages

```
          group I                              group II
met-2   his-1   arg-1           his-3   str-1   ade-3   phe-1   ura-1
  |       |       |               |       |       |       |       |
      3      11                       13      14      5       9
  |---------------|               |---------------|
         14                              18
                                         |---------------|
                                                 9
                                         |-----------------------|
                                                    32
```

TABLE 12

Frequencies of Combinations of Alleles at Pairs of Loci in Heteroclone C28/1—*

	Pairs of loci†						
	m, h	m, s	m, p	m, u	s, p	p, u	s, u
Parental allele combinations	$+\ h$ 182 $m\ +$ 45	$+\ +$ 95 $m\ s$ 20	$+\ p$ 108 $m\ +$ 15	$+\ +$ 162 $m\ u$ 7	$+\ p$ 101 $s\ +$ 72	$p\ +$ 137 $+\ u$ 28	$+\ +$ 115 $s\ u$ 24
Recombinant allele combinations	$+\ +$ 2 $m\ h$ 0	$m\ +$ 25 $+\ s$ 89	$m\ p$ 30 $+\ +$ 76	$m\ +$ 38 $+\ u$ 22	$s\ p$ 37 $+\ +$ 19	$+\ +$ 63 $p\ u$ 1	$s\ +$ 85 $+\ u$ 5
Per cent recombination	0.9	50	>46§	>26§	25	<28‖	<39‖
x^2‡: Probability:	0.49 0.3–0.5	0.75 0.3–0.5	10.18 ≪0.01	42.35 ≪0.01	1.05 0.3–0.5	8.55 <0.01	6.77 <0.01

* From data in Table 9.
† ‡ See notes to Table 10.
§ No evidence of linkage—parental allele combinations overestimated (more asymmetric—see page 319).
‖ Evidence of linkage—recombinant allele combinations overestimated (more asymmetric—see page 319).

diction to the conclusion from other heteroclones that they belong to linkage groups I and II, respectively (see heteroclone C28/2U, Table 10). The data can be interpreted by postulating two 0 points, one to the left of *met-2* and one to the right of *ura-1*.

3. Bias in Estimates of Linkage Produced by Double Disturbances

When two zero points linked to the alleles under examination have to be postulated, we can sometimes obtain some qualitative information on linkage, even though we cannot estimate a recombination percentage. If the two 0 points are in coupling: $AB/0ab0$, two extra crossovers are required to release one of the parental combinations of markers (ab) from coupling with the zero points. The two recombinant classes (Ab and aB) each require either one or the other extra crossover. The asymmetry in the frequencies of the complementary parental genotypes is therefore greater than that in the frequencies of the recombinant genotypes. Further, the sum of the frequencies of the recombinant genotypes gives an underestimate of the frequency of crossing-over between the pair of loci, for the following reason. The extra crossovers produce reductions in the frequencies of each of the recombinant genotypes (if, for example, the frequency of each is reduced by a factor of 10, the total frequency of the two becomes $0.1 + 0.1 = 10\%$ of the unreduced value), and two combined reductions in the frequency of only one of the parental genotypes ($0.1 \times 0.1 + 1 = 50.5\%$ of the unreduced value); thus the frequencies of the recombinants are more reduced than are those of the parental types.

The situation is reversed in the case of two zero points in repulsion: $AB0/0ab$. Here, the frequencies of the parental types, which are the less asymmetric, are the more reduced, and the distance between the loci is overestimated.

When the combined percentage frequency of the recombinant types, even though overestimated, turns out to be significantly less than 50, there is an indication of linkage (see Table 12). If the combined percentage frequency of the recombinant types, even though underestimated, turned out to be significantly greater than 50, there would be an indication of a change in the coupling arrangement of the alleles in question.

4. Determination of the Sequence of Loci

The analysis of the heteroclones reveals the order of triplets of linked loci, when the frequencies of the eight possible combinations of markers are examined. The observed complementary classes have never so far shown equal frequencies, a fact attributable to the presence of a 0 point outside one of the end markers. In the same way as in the analysis of pairs of loci, a single breakage point should not disturb the estimates of linkage between the markers of the triplet, and a comparable reliability test may be employed to detect disturbances. In the absence of significant disturbances, the members of the complementary segregant classes may be summed to give four values, one for the parental classes, two for the classes produced by crossover between one or the other pair of adjacent loci, and one for the double crossover classes; the latter will be the smallest. The expected value for the double crossover classes can be determined from the recombination fractions between the two pairs of adjacent loci, and a coincidence value estimated. In the four examples presented in Table 13, there is no significant interference between the two intervals defined by the triplet. The following orders of loci may be deduced: *his-3* - *str-1* - *ade-3*; *str-1* - *ade-3* - *phe-1*; *str-1* - *phe-1* - *ura-1*, which combine to give the following order of the five loci: *his-3* - *str-1* - *ade-3* - *phe-1* - *ura-1*.

TABLE 13
Determination of the Orders of Trios of Loci by Identification of the Double Crossover Class*

	Trios of loci					
	his-3 str-1 ade-3		*str-1 ade-3 phe-1*		*str-1 phe-1 ura-1*	
Heteroclone:	R27/1 −		C36/1 −		C28/2U	
Crossovers (c.o.)	Genotypes	Total number	Genotypes	Total number	Genotypes	Total number
No c.o.	+ + *ad* 150 *h3 s* + 40	190	+ + *p* 232 *s ad* + 81	313	+ *p* + 86 *s* + *u* 19	105
Single c.o.	+ *s* + 15 *h3* + *ad* 22	37	*s* + *p* 41 + *ad* + 6	47	*s* + + 26 + *p u* 0	26
Single c.o.	+ + + 14 *h3 s ad* 9	23	*s ad p* 3 + + + 15	18	*s p* + 19 + + *u* 1	20
Double c.o.	+ *s ad* 6 *h3* + + 0	6	+ *ad p* 0 *s* + + 0	0	+ + + 6 *s p u* 0	6
Expected double (no interference)	4.9		2.2		5.3	
Order of loci:	*his-3* *str-1* *ade-3* *phe-1* *ura-1*					

* Data from heteroclones.

E. The Nature of the Heteroclones

1. *Kinds of Genomic Deficiencies Shown by Heteroclones*

For the purpose of linkage studies, we have so far considered only data relative to pairs or trios of loci within different heteroclones. In this sec-

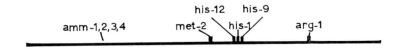
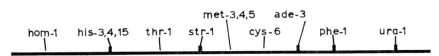

Fig. 10. Linkage map of *Streptomyces coelicolor* A3(2).

tion we shall examine the allele ratios in the heteroclones as a whole in order to obtain information about their constitution.

The most unbalanced allele ratios are always found at loci nearest to the ends of the linkage groups. This situation may be interpreted by postulating that the zero point is a point of breakage of the chromosome, with the resulting deletion of a terminal segment. By the term deletion, however, is not necessarily meant the separation and loss of a segment, but more generally its inability to contribute genetic material to the progeny. The deletion hypothesis, rather than a hypothesis of reduced viability of an allele at the most unbalanced locus, is supported by the extreme variability in the imbalances from heteroclone to heteroclone.

The real absence of certain alleles (often leading to the loss of alleles coupled to them) is shown by the effect of the composition of the selective medium on the number of heteroclones recovered (Table 14). The larger numbers of heteroclones on media with more growth factors indicate the occurrence of heteroclones in which the wild allele corresponding to a nutritional deficiency is absent from the plating unit. The occurrence, with a reduced frequency, of heteroclones on media containing streptomycin indicates the recessiveness of the mutation for resistance and the absence of the allele *str-s* from certain heteroclones.

TABLE 14
Effect of Composition of the Selective Medium on the Numbers of
Heteroclones Recovered* †

$$\text{Cross C1:} \frac{+\ h\ +}{m\ +\ a} \quad \frac{s\ p}{+\ +}$$

Constitution of chromosome II in heteroclones

Selective media (MM plus)	$\dfrac{s\ \ p}{+\ +}$	$\dfrac{s\ \ p}{}$	$\dfrac{}{+\ +}$	Number of heteroclones recovered
—	+	−	+	172
phenylalanine	+	+	+	230
streptomycin	−	−	−	6
streptomycin and phenylalanine	−	+	−	85

$$\text{Cross C21:} \frac{+\ h\ +}{m\ +\ a} \quad \frac{s\ +}{+\ p}$$

Constitution of chromosome II in heteroclones

Selective media (MM plus)	$\dfrac{s\ +}{+\ p}$	$\dfrac{s\ +}{}$	$\dfrac{}{+\ p}$	Number of heteroclones recovered
—	+	+	−	96
phenylalanine	+	+	+	172
streptomycin	−	+	−	54
streptomycin and phenylalanine	−	+	−	57

* + = growth; − = no growth.

† Increase of heteroclone number on addition of phenylalanine in both crosses indicates the occurrence of heteroclones lacking the allele p^+. Heteroclones recovered in the presence of streptomycin lack the allele s^+. When s and p are in coupling (cross C1), streptomycin practically eliminates heteroclones, if phenylalanine is absent.

THE GENETICS OF *Streptomyces coelicolor*   323

The heteroclones considered in this and the following sections have been obtained in various experiments, designed for various objects, and therefore do not represent by any means a random sample of all possible types of heteroclone. Of more than a hundred heteroclones examined, not one was undisturbed; that is, none showed a segregation in equal numbers of all the alleles introduced into the cross. The disturbances varied from the imbalance of allele frequencies at a single locus (reflected in a lesser imbalance of the alleles at linked loci), to the absence of all the alleles derived from one parent, with the exception of that used for the selection of the heteroclone.

The presence of two linkage groups in *S. coelicolor* provides us with a test that enables us to detect any gross disturbances in the segregation frequencies within a heteroclone. The test is a chi-square of independence which tests the hypothesis that the various combinations of the markers in one linkage group have the same relative frequencies in association with each combination of markers in the other linkage group (see Table 15). Nonsignificant chi-squares given by this test indicate the absence not only of linkage between markers in different linkage groups, but also of preferential combinations of unlinked markers, such as might result from a significant viability interaction between markers in different linkage groups, from sectoring within the heteroclone, or from contamination with extraneous clones.

The independence of the two linkage groups is clear in nearly all the heteroclones, including those in which several 0 points must be postulated. In fact, most of the rare cases in which the chi-square of independence is significant can readily be interpreted on the basis of contamination with an extraneous clone, usually of a parental phenotype.

Zero points have been found at one or both ends of one or both linkage groups, and no strict correlation has been detected between the various deletions. This observation, which suggests that there are four independent points in which breakage may occur, supports the idea that the two linkage groups represent two structural entities, that is, two distinct chromosomes.

Two zero points in the same pair of chromosomes are indicated by the analysis of many heteroclones. Usually they are in repulsion; that is, allele ratios at the two terminal loci deviate from 1:1 in opposite directions. Occasionally the deviations are in the same direction, possibly suggesting two zero points in coupling, but change in the coupling arrangement of the terminal markers themselves is a possible explanation.

In those heteroclones selected on a medium containing supplements which make one chromosome dispensable, it is very common for all the markers of one linkage group from one parent to be absent. This situation

TABLE 15
Tests of Independence between Linkage Groups in Heteroclones; Frequencies of Segregant Genotypes*

Heteroclone† C36/1 −: $\dfrac{0\ m\ +\ a}{+\ h\ +}$ $\dfrac{s\ ad\ +\ 0}{+\ +\ p}$

Markers of group II:	+ + p	s ad +	s + p	+ + +	+ ad +	Total
Markers of group I						
+ h +	146	53	22	7	4	232
m + a	39	16	10	6	2	73
m + +	16	3	5	2	0	26
+ h a	19	6	1	0	0	26
+ + a	7	3	3	0	0	13
Total	227	81	41	15	6	370

χ^2(16 d.f.) = 14.86; P = 0.7–0.5

Heteroclone† R27/5 −: $\dfrac{m\ +}{+\ a}$ $\dfrac{+\ +\ ad\ +}{0\ h3\ s\ +\ p}$

Markers of group II:	+ + ad +	+ + + p	h3 s + p	+ s + p	Total
Markers of group I					
+ a	39	21	6	7	73
m +	32	13	10	7	62
+ +	6	4	2	3	15
m a	2	2	2	1	7
Total	79	40	20	18	157

χ^2(9 d.f.) = 4.44; P = 0.9–0.8

Heteroclone† R14/6 −: $\dfrac{0\ m\ +\ +}{+\ h\ a}$ $\dfrac{0\ s\ +\ p\ +\ 0}{+\ ad\ +\ u}$

Markers of group II:	+ ad + u	+ + + u	s ad + u	+ + p u	+ + p +	Total
Markers of group I						
+ h a	29	10	2	1	3	45
+ h +	16	7	3	2	0	28
m + +	12	2	3	2	1	20
Total	57	19	8	5	4	93

χ^2(8 d.f.) = 7.16; P = 0.7–0.5

* Rare genotypes omitted (less than 5% of total).
† In heteroclone formulas, alleles in defect are indicated by a 0 distal to them.

has been interpreted as the lack of one chromosome (hemizygosity), but it could also be due to homozygosity of the markers in one linkage group. In certain heteroclones, a terminal marker is completely absent, and hemizygosity of the homologous allele is the simplest explanation. However, rarely, the segregation at linked loci is not disturbed in the way that it would be by a deletion including a terminal locus, and homozygosity of this locus is therefore indicated. In almost all the heteroclones so far examined, including those of higher order (see Section V,E,3), the markers have been found in the parental coupling arrangements; very rarely changes in coupling have been detected.

2. *Schematic Representation of Heteroclones*

The nature of the heterozygous unit from which a heteroclone develops is deduced from the composition of the population of segregants present in the heteroclone. This deduction assumes that processes of segregation within the heteroclone are extremely numerous, so that the segregants represent, statistically, the genetic potentialities of the heteroclone, without disturbances due to clonal effects. On this assumption a heteroclone may be represented diagrammatically by means of a kind of linkage map containing the segregating markers and one or more zero points (Fig. 11). The distance from a zero point to the least frequent allele corresponds to the percentage frequency of this allele, that is, the frequency of recombination in the region which frees the allele from coupling with the zero point. Such distances show good additivity with the intragenic distances. Thus the heteroclones are represented as two pairs of parallel lines, corresponding to the two linkage groups in the diploid condition. One or more of the four lines may end, before the segment parallel with it, in a zero point, indicated by 0; this represents a breakage point or an entity having the same effect. Chromosomes whose markers do not appear among the segregants are omitted altogether.

This purely formal representation, which serves as a model to summarize the data, does not imply that all the segregating nuclei within the heteroclone have the same composition, but rather represents the average result of the events which the heterozygous units undergo during the life span of the colony. The distance between a zero point and each locus thus gives the probability of each allele in coupling with the zero point being transmitted to the segregant progeny. The observed reduction of the frequency of such an allele would represent an average result of terminal deletions distal to the allele, and possibly also including it, which would tend to decrease the frequency of transmission of the allele, together, perhaps, with less extensive or less frequent deletions in the homologous chromosome, which would tend to increase its transmission. If, however, a

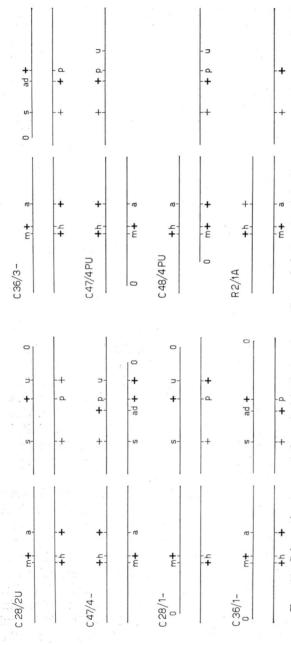

FIG. 11. Schematic representation of eight heteroclones. In each heteroclone the lines represent the two pairs of chromosomes (in the last three heteroclones one member of the second pair is missing). The distances between pairs of loci are proportional to the average recombination percentages (Table 11). The ending of some chromosome segments at a zero point (0) reflects the formal incompleteness of the zygote, postulated to explain the reduction in frequency of the alleles on the interrupted segment. The distance between a zero point and the allele nearest to it is proportional to the percentage frequency of that allele. The total lengths of the uninterrupted chromosomes are arbitrary. For detailed discussion see text. The boldface type + indicates those wild type alleles selected for by the plating medium, all the other alleles being unselected.

deletion were already established at the end of a chromosome in the plating unit from which the heteroclone derived, the homologous region in the other chromosome could clearly not be deleted in viable products.

3. *Higher-Order Heteroclones*

The data so far described have been obtained from the analysis of single whole heteroclones, the segregants representing a random sample of the spores of a whole colony. The results have been regarded as indicating the genetic potentialities of the heterozygous nucleus which gives rise to the colony, and the diagrammatic representation of the heteroclones is based on this assumption. However, this does not mean that the same potentialities are possessed by all the heterozygous units within the heteroclone up to the moment of its segregation.

Up to a million spores are obtainable from a single heteroclone. If a sample of these is plated on a medium capable of selecting heteroclones, a small proportion of new ones appears, usually representing less than 1% of the total plating units; these second-order heteroclones presumably arise from units whose nuclei have failed to undergo segregation. If the submerged mycelium of a heteroclone is broken into small pieces and plated, the proportion of plating units giving rise to heteroclones is very much higher, but some segregants are still found. The general pattern of segregation is similar to that obtained by plating the spores of the same colony, except for a reduced internal consistency of the data, as judged by the test of independence of the two linkage groups, probably due to selective effects; in fact genotypes capable of growth on the selective medium are present with abnormally high frequencies. These observations indicate that segregation of the heterozygous units is not limited to the formation of the spores, but occurs also in the vegetative mycelium, although with a much lower frequency. The formation of the spores leads to a virtually complete segregation of the heterozygous nuclei; only rarely do nuclei pass the barrier of sporulation, and it is still unknown whether they are included in normal spores. Several thousand second-order heteroclones may be obtained from a single original heteroclone; thus the heterozygous condition is conserved through at least a dozen nuclear generations.

The analysis of second-order heteroclones gives a picture of the variability between the heterozygous nuclei within the parent heteroclone. Many of the second-order heteroclones maintain the same genetic potentialities as the parent colony, but some show a very different pattern of segregation. Some show deficiencies not present in the parent, or an accentuation of its imbalances, but others show an inversion of them; an allele present with a low frequency in the parent heteroclone may be predominant in some of those of second order, or even a chromosome absent from the small sample of segregants analyzed from the parent heteroclone may appear in some of those of second order. Table 16 shows the allele ratios in a parent heteroclone and in four heteroclones derived from it; two of these show very different ratios from those of the parent. However, the allele str-s, absent from the parent, did not reappear in any of the four daughter heteroclones nor in any of eight other daughter heteroclones examined.

TABLE 16
Allele Ratios in Three Second-Order Heteroclones Compared with Those in the Parent First-Order Heteroclone

	Heteroclones			
Alleles	R/14/3− 1st order	R14/3−/ 1MHAU 2nd order	R14/3−/ 2MHAU 2nd order	R14/3−/ 4AUAdP 2nd order
$m/+$	34/61	49/0	19/29	4/45
$+/h$	37/58	49/0	20/28	5/44
$+/a$	41/54	49/0	11/37	9/40
$s/+$	0/95	0/49	0/48	0/49
$+/ad$	63/32	31/18	27/21	0/49
$p/+$	61/34	32/17	26/22	0/49
$+/u$	69/26	39/10	39/9	0/49

4. *Evolution of the Heterozygous Nucleus in Heteroclones*

As noted above, hardly any heteroclone, even of higher order, has shown a coupling arrangement of markers different from that of the parent strain. This excludes the widespread occurrence, during the multiplication of the heterozygous nuclei, of mitotic recombination or of segregation followed by re-fusion of haploid products, a phenomenon referred to as "recycling of meiosis" (Lederberg, 1957); the processes of reduction and recombination would seem, therefore, to occur simultaneously, as in a normal meiosis. Usually, however, such reduction events would not give rise to complementary products, owing to the presence of lethal deletions.

The evolution of the heterozygous nucleus within heteroclones may be studied in experiments on populations of heteroclones. The spores from a group of 20–30 heteroclones, derived either from the same mixed culture or as daughters of the same parent heteroclone, were collected, mixed, and sown on a selective medium for the recovery of new heteroclones.

The operation was repeated using spores from the second-order heteroclones, and again with spores from the third-order heteroclones to give two or three successive generations. In parallel, samples of the spores of the mixed populations of heteroclones were sown on complete medium for the study of the segregants. In these experiments heterozygous nuclei were forced to reproduce themselves for a greater number of generations than normally occur in a single heteroclone between the sowing of the original heterozygous unit and segregation. This experiment has some analogies with that of the artificial delay of maturation of bacteriophages (Visconti and Delbrück, 1953).

TABLE 17
Recombination between Some Pairs of Loci in Successive Generations of Heteroclones

Cross: R32 $\frac{+\ h\ +}{m+\ a}$ $\frac{+\ +\ p}{s\ ad\ +}$

Pairs of loci	Genotypes	Numbers		
		1st generation heteroclone	Pooled 2nd generation heteroclones	Pooled 3rd generation heteroclones
str-1 and ade-3	+ +	88	85	80
	s ad	35	15	76
	s +	33	83	69
	+ ad	3	5	13
	Per cent recombination:	23	47	34
ade-3 and phe-1	+ p	119	166	143
	ad +	35	17	54
	ad p	3	3	35
	+ +	2	2	6
	Per cent recombination:	3.2	2.7	17
str-1 and phe-1	+ p	86	84	79
	s +	32	13	44
	s p	36	85	101
	+ +	5	6	14
	Per cent recombination:	26	48	48

In successive generations, the relative frequencies of the various classes of segregants in the pooled population of spores change; the independence of the two linkage groups remains undisturbed. The frequency of recombination between the linked loci tends to increase, more rapidly if the original values are greater, sometimes reaching or even exceeding 50%; thus the linkage gradually becomes looser and may eventually disappear (Table 17). Within those single heteroclones examined in parallel with the pooled population that allow the evaluation of linkage between pairs of markers, the recombination values do not show an increase, nor are changes in coupling arrangement detected. However, many heteroclones show multiple deletions. The increase of recombination in the pooled population thus seems to be due to the increase of heteroclones with more than one deficiency, and may be due to forced recombination in chromosome pairs with two deficiencies in repulsion. This would be partly balanced by the effect of double deficiencies in coupling. Under these conditions, recombination would become largely independent of linkage.

5. Aberrant Types of Heteroclones; Heterokaryons

Heteroclones are considered aberrant when the test of independence of the two linkage groups is significant. An anomaly revealed by this test is not necessarily due to the nature of the heterozygous unit from which the heteroclone derives, but could be determined by selective or clonal effects during the development of the colony. In fact, in certain aberrant heteroclones, there is an excess of one or more of those genotypes of segregants which are able to grow on the selective medium. In others the aberration is caused by the presence in excess of a genotype, frequently parental, which is not able to grow on the selective medium and which probably derives from a contaminating clone.

A clearly definable type of aberrant heteroclone is that containing only the two parental genotypes, sometimes accompanied by one or more recombinant genotypes with a very low frequency. The parental genotypes evidently complement one another to give rise to a colony on a medium which does not allow the growth of either separately, but they do not interact genetically; therefore such colonies are referred to as *heterokaryons*, in spite of the fact that the presence of two different types of nucleus in the same cytoplasm has not been demonstrated. Such heterokaryons were described in *S. coelicolor* strain A3(2) by Sermonti and associates (1960), and are probably analogous with those observed by Bradley and Lederberg (1956) in *S. griseus* and *S. cyaneus*, by Braendle and Szybalski (1957) in *S. griseus*, *S. albus*, and *S. venezuelae*, and by Saito (1958) in *S. griseoflavus*, while having no affinity with the peculiar semi-stable "heterokaryons" described by Bradley (1957, 1958) in *S. coelicolor* (see below). The ratio between the two parental genotypes in the population of spores harvested from the heterokaryons of *S. coelicolor* strain A3(2) usually differs markedly from 1:1, frequently being as high as 10:1 or 20:1. The ratio tends to be constant in heterokaryons from the same cross selected on the same medium, and varies, with the composition of the medium, in favor of the selectively less handicapped type. The presence of a small percentage of recombinant genotypes in certain heterokaryons indicates a genetic interaction between the parental nuclei, even at a reduced level compared with that in the normal heteroclones, which is evidence for the presence of the two parental nuclei in the same cytoplasm, at least in some of the hyphae. The proportion of heterokaryons among the total heteroclones varies from cross to cross; in certain crosses a majority of heterokaryons is recovered, while in others heterokaryons are not found.

6. *Some Peculiar Kinds of "Heterokaryon" Described by Bradley*

Bradley (1957) isolated on minimal medium, from mixed cultures of various auxotrophic strains of *S. coelicolor*, prototrophic colonies which remained prototrophic even when subcultured by single spore isolation. However, these clones were stated to give rise regularly to dissociates of parental phenotype, although with a very low frequency (less than one among 2000 prototrophic colonies). According to Bradley, the occasional appearance of parental phenotypes indicated that the prototrophic colonies were heterokaryons, though clearly they would need to have been of a special type (perhaps a kind of permanent dikaryon), in which the two component genomes remained closely associated even during spore formation. Bradley (1959) supposed that the nucleus of *Streptomyces* "may be merely an amorphous association of several complete and partial genomes." In such intricate associations, true genetic recombination was, curiously enough, absent.

Certain of the auxotrophic derivatives from the so-called prototrophic heterokaryons showed the parental markers in new combinations. These auxotrophs were "quite stable," but occasionally produced parental dissociates, with a very low frequency; consequently they were considered by Bradley to be "anomalous heterokaryons." Bradley (1958) explained the manifestation of nutritional requirements by these anomalous heterokaryons containing the two complete complementary auxotrophic genomes of the parents by postulating a strong imbalance between the frequencies of the two parental types of nucleus (up to 1:1000), little selection among the diverse nuclear types in heterokaryons, and multinucleate spores. As Gregory (discussion to Bradley *et al.*, 1959) pointed out, this would imply up to 1000 nuclei in a single spore in these thoroughly anomalous heterokaryons. In his reply, Bradley stated that the imbalance could be produced by factors differentially affecting nuclear division. Starting from a binucleate heterokaryotic spore, a mycelium could be formed in which the frequencies of the two types of nucleus were in a ratio of 1:1000; it remained unexplained how this mycelium could regularly give rise anew to binucleate heterokaryotic spores. These suggestions moreover contradicted the previous assertion that "little selection occurs among the diverse nuclear types," which was needed to explain the relative stability of the anomalous heterokaryons.

Bradley (1960) sought to support the idea of the heterokaryotic nature of the prototrophic recombinant phenotypes with experiments which can be summarized as follows: in crosses between streptomycin-resistant and streptomycin-sensitive auxotrophs, the prototrophic recombinants were all sensitive to streptomycin, irrespective of the coupling of the markers in the parent strains. However, nearly half the so-called streptomycin-sensitive recombinants gave rise to vigorous colonies when streaked on a medium containing streptomycin. Since such colonies were less numerous (1–10%) than on control plates without streptomycin, in a test which was obviously only very approximately quantitative, the recombinants were considered as streptomycin-sensitive. On the contrary, the presence of colonies on a medium containing streptomycin showed the occurrence of resistant recombinants, even if perhaps of delayed origin, together with the sensitive recombinants.

We cannot offer any full explanation of Bradley's results; some of them may be due to imperfect purification or characterization of strains. For example, in Tables 6–8 of Bradley et al. (1959), isolates from mixed cultures appear to change their phenotype repeatedly during successive subculture. Moreover, in these tables, prototrophic isolates often undergo complete dissociation into parental phenotypes on subculture, in contrast to the situation reported by Bradley (1957), and "anomalous" streptomycin-sensitive prototrophs sometimes produce nearly 100% streptomycin-resistant colonies on subculture, in contrast to the result reported by Bradley (1960). Bradley et al. (1959) themselves pointed out the impossibility of reproducing data quantitatively, the unpredictability of character inheritance, and the consequent difficulty in developing an acceptable hypothesis to account for their results.

VI. Analysis of Short Chromosome Regions

The methods developed during the studies described in this review may make *Streptomyces coelicolor* an organism suitable for the study of short chromosome regions. Three main requirements are needed for such a study; a reliable test of allelism must be available, the genetic analysis must be capable of high resolving power (Pontecorvo, 1959), and the mutations studied must not result in poor fertility of the crosses. The heteroclones provide a test of allelism allowing the complementation of mutants with similar phenotype to be studied. No mutation has yet been found which results in sterility in crosses with strains carrying other mutations; the phenomenon of semi-sterility (see section IV,F) appears not to be associated with any particular mutant phenotypes. This situation may be contrasted with that in *Neurospora* and *Aspergillus* where the effect on fertility of certain combinations of markers severely limits the study of recombination in short regions. The resolving power of the genetic analysis requires fuller discussion.

Recombinant phenotypes rarely exceed 1% of the spores harvested from mixed cultures. 10^5–10^6 recombinants are obtainable from a single mixed culture, but these have to be selected from a population of 10^7–10^8 parental spores. In contrast, 100% of the spores harvested from a heteroclone are segregants. Up to 10^6 spores can be obtained from a single heteroclone, and the spores of 20–30 heteroclones can readily be harvested to form a population of ca. 10^7 segregants. The mutations in the short region to be studied may be used to select the heteroclones, provided the mutations are not allelic; by definition no heteroclones develop in crosses between alleles. When the mutations are allelic, the appropriate growth factor must be present in the selective medium, and another pair of closely linked markers must be used to select the heteroclones. The selection of rare recombinants from the heteroclones would require a combination of the methods for the analysis of heteroclones and those of the selective analysis; we have not yet attempted such a study. The data reported in this section have been obtained from preliminary crosses carried out primarily during the development of methods of genetic analysis in *S. coelicolor*; only the selective analysis has so far been employed in these studies.

By means of the simplified selective analysis, a group of four mutations (*amm-1-4*) which result in a requirement for ammonium or glutamic or aspartic acids have been located to the left of *met-2*; the four sites are very closely linked together. Three mutations (*met-3, met-4,* and *met-5*) which result in a requirement for methionine or homocysteine have been found to be very closely linked together and located close to *str-1*. Three histidineless mutations (*his-3, his-4,* and *his-15*), which result in the accumulation of histidinol phosphate are very closely linked together near the left end of linkage group II. Three mutations which result in a requirement for adenine plus thiamine are located in a short region close to *phe-1* in linkage group II. In each of these four cases, we are probably dealing with groups of alleles; mutations at different or identical sites in the same functional unit.

The best studied region lies to the right of *met-2* and comprises three loci concerned in the synthesis of histidine. The total length of the region has not been determined accurately, but probably does not exceed two units. The three mutations so far studied are phenotypically distinguishable: *his-9* shows no accumulation of Pauly-positive material identifiable in paper chromatograms (Ames and Mitchell, 1952), *his-12* accumulates imidazoleglycerol phosphate, and *his-1* accumulates a substance with the same R_f as imidazolepropanediol (Ames and Mitchell, 1955). Crosses in all combinations give rise to heteroclones on media lacking histidine, indicating that the three mutations occupy distinct and complementary loci.

TABLE 18

Determination of the Order of Three Closely Linked Histidine Loci (*his-12*, *his-1*, and *his-9*); Numbers of Recombinants Carrying Different Combinations of Outside Markers

Recombinant genotypes*	Crosses					
	m h12 + + / + + h1 a	m + h1 + / + h12 + a	m h12 + + / + + h9 a	m + h9 + / + h12 + a	m h1 + + / + + h9 a	m + h9 + / + h1 + a
m + + a	1	25	1	64	0	26
+ + + a	0	14	7	19	7	10
m + + +	17	0	5	29	8	10
+ + + +	39	8	13	8	29	4

Order of loci: met-2 | his-12 | his-1 | his-9 | arg-1

* Selected on minimal medium plus methionine and arginine.

To determine the order of the loci, six strains were prepared in which the three *his* mutations were combined either with *met-2* or with *arg-1*, which lie to the left and right, respectively, of the group of *his* loci. The six strains were crossed in all possible complementary combinations (Table 18). Recombinants were selected on minimal medium supplemented with methionine and arginine, but not histidine, and were classified for the outside markers. If each *his* mutation under examination was in coupling with the outside marker adjacent to it, the histidine-independent recombinants should have been mainly prototrophic; if in repulsion, they should mostly have required methionine and arginine. All the crosses gave one or the other result, reciprocal crosses giving complementary results (Table 18). Thus the order of the loci appears to be: (*met-2*) *his-12 his-1 his-9* (*arg-1*). This order, in contrast to the situation in *Salmonella* (Hartman *et al.*, 1960; Ames, *et al.*, 1960), would appear not to correspond to that of the steps in the biosynthesis if we assume that the accumulation of the compound tentatively identified as imidazopropanediol by *his-1* is due to a block in the pathway of synthesis after the production of imidazoleacetol phosphate. Nevertheless, the phenomenon of clustering of the three loci controlling steps in the same biosynthesis remains as a character that *S. coelicolor* has in common with other bacteria (Demerec and Hartman, 1959); in *S. coelicolor*, however, another *his* locus, represented by the mutations *his-3*, *his-4*, and *his-15*, is present in quite a different part of the genome, in linkage group II.

The accurate estimation of the frequency of recombination between the loci in question is made difficult by the low frequency of recombinants with respect to heteroclones among the products of a mixed culture. Recombinants deriving directly from haploid plating units or as early-formed lobes or sectors from developing heteroclones are difficult to distinguish. Some technique will have to be found to eliminate the heteroclones when studying closely linked genes which complement one another.

VII. Speculations on Nuclear Behavior during the Life Cycle

The relative frequencies of heteroclones and homogeneous recombinants are the same among the spores of a mixed culture and those of a heteroclone; moreover the pattern of recombination is fundamentally similar. We may therefore postulate that the recombinants obtained directly from a mixed culture arise from heteroclones formed during the preceding growth of the culture. Heterozygous nuclei would be formed in the mixed mycelium and undergo the same evolution as we observe in the isolated heteroclones, rarely passing the barrier of sporulation.

In isolated heteroclones, the heterozygous nuclei seem to undergo a number of equational divisions (up to about a dozen) before segregating in the aerial hyphae, the products of segregation being included in the spores. Segregation also occurs in the substrate mycelium, but with a lower frequency. During the equational divisions, the chromosomes appear to suffer successive deletions, which tend gradually to reduce the length of the duplicated regions. We have no indication that crossing-over takes place during the equational divisions; it probably occurs only at the time of reduction, as in a normal meiosis, although its result is complicated by the presence of the chromosomal deficiencies.

Loss of chromosome segments occurs after the formation of the heterozygotes (post-zygotic elimination), but we cannot say whether it may also precede their formation. On suitably supplemented selective media, a high proportion of the recovered heteroclones are very incomplete, but this does not prove the occurrence of pre-zygotic elimination of chromosome segments since the heterozygotes which we recover in first-order heteroclones have probably already undergone several nuclear divisions in the mixed culture.

The recovery of heterokaryons containing the two complete parental genomes proves the transfer of intact genomes from one hypha to another. However, we cannot say whether heterokaryosis is a stage in the formation of heterozygous nuclei, or whether heterokaryons and heterozygotes are formed by alternative processes. Moreover, if heterokaryosis leads to heterozygosity, loss of chromosome segments might occur after intercellular transfer and before or during the formation of zygotes.

The heteroclones have been described as a special type of colony, but we have no reason to believe that the behavior of their nuclei is specifically due to heterozygosity; rather, it may be that the heterozygosity allows a general type of behavior to be revealed which cannot be detected in colonies containing only one type of genetic information. Thus the difference between heteroclones and homogeneous colonies might simply be comparable to that between a crossed and a selfed perithecium of a homothallic *Aspergillus*. The high frequency of recombination observed in well-balanced mixed cultures of *S. coelicolor* indicates that processes of nuclear fusion must be very common, and possibly occur with regularity in the hyphae.

Cytological observations cannot give us unequivocal evidence of the occurrence of processes of nuclear fusion and segregation. The large rod-shaped chromatinic bodies of the mycelium of *S. coelicolor* probably contain material corresponding to several of the small chromatinic bodies found in the spores. It is impossible to say whether they arise by fusion of separate nuclei or by a process of multiplication not followed by separation of daughter products. A regular process of segregation of sub-units from a single large chromatinic body giving rise to the uniform spore nuclei appears to be involved in sporulation. This process parallels the almost complete segregation of the heterozygous nuclei connected with sporulation in the heteroclones.

VIII. Summary and Conclusions

For convenience of reference to the text each section will be summarized separately, using a corresponding number to that employed in the text.

i. *Introduction.* All the species of bacteria previously studied genetically belong to the sub-order Eubacteriineae, which contains the bacteria with the simplest morphology. The streptomycetes are the group of bacteria with the most complex morphological differentiation; a study of their genetics is therefore of interest to extend our knowledge of bacterial genetic systems. The possession of uninucleate spores makes *Streptomyces* particularly suitable for several types of genetic study.

ii. *Strains and Methods of Culture.* Nearly all the data discussed in this review apply to *Streptomyces coelicolor* strain A3(2) [*S. violaceoruber* according to Kutzner and Waksman (1959)]. Culture methods and techniques for the production and characterization of mutants are those used routinely in microbial genetics.

iii. *Morphology of the Organism.* Colonies of *S. coelicolor* are made up of a layer of substrate mycelium, from which develop the hyphae of the aerial mycelium; these finally become transformed into chains of spores. The hyphae of the substrate mycelium contain complex nuclear bodies of variable dimensions, while the aerial hyphae contain large rodlike chromatinic bodies which, at sporulation, become subdivided, eventually producing the dotlike spore nuclei. As in other bacteria, there is no evidence of a nuclear membrane in electron micrographs of thin sections.

iv. *Studies of Recombination by Selective Analysis.* Among the spores derived from a mixed culture of strains differing in several markers are numerous recombinants having the parental markers in new combinations. In four-point crosses, nine different genotypes can be recovered on four selective media. Two pairs of these are complementary genotypes, which usually appear with similar frequencies. The others allow the estimation of the other pairs of complementary recombinant genotypes, assuming equal frequencies for the members of each pair. Data for sets of three crosses involving four markers in different coupling arrangements have been elaborated together in order to balance viability disturbances. In the absence of the parental classes, only relative frequencies of recombination between pairs of loci can be estimated. However, the early identification of a marker unlinked to the others has allowed the conversion of these estimates into absolute values. Some information on linkage, particularly between closely linked genes, can be obtained from a simplified method of analysis. A dozen markers have been located in two linkage groups. Certain crosses give unequal frequencies of complementary classes and the phenomenon appears to be genetically determined. Preliminary evidence for the existence of a group of self-sterile strains is discussed.

v. *Heteroclones.* Among the recombinant colonies that appear on selective media after the plating of the spores of a mixed culture are heterogeneous colonies that contain a large number of genotypes. These colonies (heteroclones) are presumed to derive from heterozygous units that undergo rapid segregation during their multiplication. The frequency of the heteroclones relative to haploid recombinants is high when the medium selects for two closely linked markers in repulsion. The heteroclones can be recognized by their small size and, more important, by their inability to be perpetuated by replica plating on the same medium on which they arose. All the spores in a heteroclone contain segregant nuclei from the heterozygote, and their analysis, which does not have the limitations of a selective analysis, allows the direct estimation of recombination between pairs of loci, the determination of the sequence of trios of loci, and confirms the existence of two independent linkage groups. The heteroclones suffer from various disturbances in allele ratios, extending to the complete absence of certain alleles, and often to the absence of all the alleles of a linkage group from one parent. The disturbances in allele ratios have been interpreted as being due to terminal deletions, which may take place at each end of the two linkage groups. These deletions complicate the estimation of linkage, but some simple statistical tests have been devised for the identification of unreliable estimates. The analysis of second-order heteroclones allows the study of the evolution of the heterozygous nucleus within heteroclones; it undergoes successive deletion with only very rare changes in the coupling of the markers. This makes unlikely the occurrence of mitotic crossing-over or the recycling of meiosis. Certain heteroclones segregate only the two parental genotypes, and have been interpreted as heterokaryons.

vi. *Analysis of Short Regions*. Selective analysis, and particularly the analysis of heteroclones, offers a high resolving power for the study of short regions. The heteroclones provide a reliable test of allelism. A cluster of three closely linked loci controlling histidine synthesis has been identified.

vii. *Speculations on Nuclear Behavior During the Life Cycle*. During the growth of mixed cultures, fusions between nuclei from different hyphae probably occur frequently, giving rise to heterozygous units; these multiply equationally, until they undergo almost complete segregation at spore formation. The deductions from genetic analysis find some parallels in cytological observations. It is not yet possible to determine whether the first heterozygous nuclei are complete, or whether primary "merozygosis" occurs. Frequent post-zygotic eliminations certainly occur during the multiplication of the heterozygous nuclei.

* * *

From these results we may conclude that a regular feature of the genetic system of *Streptomyces coelicolor* is the incompleteness of the heterozygous nuclei at the time of their segregation. This phenomenon is thus common to all bacteria so far investigated. In *S. coelicolor*, the incompleteness is due, at least to a considerable extent, to losses of chromosomal material after the formation of the zygote (post-zygotic elimination); little is known about the state of the zygote at the time of its formation. On the other hand, in the conjugation of *Escherichia coli* K12, the genetic system most comparable with that of *S. coelicolor*, the incompleteness of the primary zygotes (pre-zygotic elimination), due to partial transfer of the Hfr chromosome, is well documented (Wollman et al., 1956), while information on the fate of the zygotes is difficult to obtain, and consequently very limited (Lederberg, 1957; Anderson, 1958). *E. coli*, whose isolated cells are easily manipulated, is particularly adapted to the study of the phases which precede zygote formation, while *S. coelicolor*, with its specialized structures in which the products of segregation become isolated, is adapted to the study of those phases which follow zygote formation. These features, leading to different technical limitations, on the one hand make it difficult to compare the genetic systems of *E. coli* and *S. coelicolor*, but on the other promise to extend our knowledge of the genetics of the bacteria as a whole.

An important difference between the genetic systems of *S. coelicolor* and *E. coli* is the presence of two linkage groups in the former and one in the latter. This is not a trivial difference, since the process of genetic transfer and the associated fertility system in *E. coli* are closely connected with the presence of a single continuous chromosome, and are not compatible with the presence of two chromosomes. Confirmation of the existence of two chromosomes and the study of the fertility system which is on the horizon of the genetics of *S. coelicolor* promise to contribute notably to knowledge of the comparative genetics of the bacteria.

Acknowledgments

It is a great pleasure to thank Mrs. Isabella Spada-Sermonti for her invaluable cooperation in the work described in this review. The photographs in Figures 4 and 5 come from unpublished observations by D. A. H. in collaboration with Miss Audrey M. Glauert; we are grateful to her for permission to use them, and to Dr. A. Mancinelli for some of the data presented in Table 8. Our thanks are due to members of the Cambridge Part II Botany classes of 1954–1959 for isolating most of the mutations listed in Table 1. In the course of the work reported here, financial support in the form of grants for travel and research have been received from the Italian Comitato Nazionale Energia Nucleare (C.N.E.N.) and the Rockefeller Foundation. One of us (D. A. H.) wishes to thank the Director of the Istituto Superiore di Sanità, Roma, for generous hospitality, Dr. L. C. Frost for suggesting a search for gene recombination in *Streptomyces*, and Dr. H. L. K. Whitehouse for encouragement throughout the work. One of us (G. S.) wishes to express his gratitude to Prof. C. Barigozzi and Dr. G. Magni for their continuous stimulating interest.

References

Alikhanian, S. I., and Borisova, L. N. 1961. Recombination in *Actinomyces aureofaciens*. J. Gen. Microbiol. **26**, 19–28.

Alikhanian, S. I., and Mindlin, S. Z. 1957. Recombinations in *Streptomyces rimosus*. Nature **180**, 1208–1209.

Ames, B. N., and Mitchell, H. K. 1952. The paper chromatography of imidazoles. *J. Am. Chem. Soc.* **74**, 252–253.
Ames, B. N., and Mitchell, H. K. 1955. The biosynthesis of histidine. Imidazoleglycerol phosphate, imidazoleacetol phosphate and histidinol phosphate. *J. Biol. Chem.* **212**, 687–696.
Ames, B. N., Garry, B., and Herzenberg, L. A. 1960. The genetic control of the enzymes of histidine biosynthesis in *Salmonella typhimurium*. *J. Gen. Microbiol.* **22**, 369–378.
Anderson, T. F. 1958. Recombination and segregation in *Escherichia coli*. *Cold Spring Harbor Symposia Quant. Biol.* **23**, 47–58.
Arditti, R. R., and Strigini, P. 1961. Diploid ascospores in *Aspergillus nidulans*. *Sci. Repts ist. super. sanità* **1**, 1–8.
Bailey, N. T. J. 1951. The estimation of linkage in bacteria. *Heredity* **5**, 111–124.
Bhaskharan, K. 1960. Recombination of characters between mutant stocks of *Vibrio cholerae*, strain 162. *J. Gen. Microbiol.* **23**, 47–54.
Bradley, S. G. 1957. Heterokaryosis in *Streptomyces coelicolor*. *J. Bacteriol.* **73**, 581–582.
Bradley, S. G. 1958. Genetic analysis of segregants from heterokaryons of *Streptomyces coelicolor*. *J. Bacteriol.* **76**, 467–470.
Bradley, S. G., 1959. Mechanisms controlling variation in streptomycetes. *Ann. N.Y. Acad. Sci.* **81**, 899–906.
Bradley, S. G. 1960. Reciprocal crosses in *Streptomyces coelicolor*. *Genetics* **45**, 614–619.
Bradley, S. G., and Anderson, D. L. 1958. Compatibility system controlling heterokaryon formation in *Streptomyces coelicolor*. *Proc. Soc. Exptl. Biol. Med.* **99**, 476–478.
Bradley, S. G., and Lederberg, J. 1956. Heterokaryosis in *Streptomyces*. *J. Bacteriol.* **72**, 219–225.
Bradley, S. G., Anderson, D. L., and Jones, L. A. 1959. Genetic interaction within heterokaryons in streptomycetes. *Ann. N.Y. Acad. Sci.* **81**, 811–823.
Braendle, D. H., and Szybalski, W. 1957. Genetic interaction among streptomycetes: heterokaryosis and synkaryosis. *Proc. Natl. Acad. Sci. U.S.* **43**, 947–955.
Braendle, D. H., and Szybalski, W. 1959. Heterokaryotic compatibility, metabolic cooperation, and genetic recombination in *Streptomyces*. *Ann. N.Y. Acad. Sci.* **81**, 824–851.
Breed, R. S., Murray, E. G. D., and Parker Hitchens, A. (eds.) 1948. "Bergey's Manual of Determinative Bacteriology," 6th ed., pp. 935–936. Williams & Wilkins, Baltimore, Maryland.
Carvajal, F. 1947. The production of spores in submerged cultures by some *Streptomyces*. *Mycologia* **39**, 426–440.
Catcheside, D. G. 1954. Isolation of nutritional mutants of *Neurospora crassa* by filtration enrichment. *J. Gen. Microbiol.* **11**, 34–36.
DeLamater, E. D. 1951. A staining and dehydration procedure for the handling of microorganisms. *Stain Technol.* **26**, 199–204.
Demerec, M. 1948. Origin of bacterial resistance to antibiotics. *J. Bacteriol.* **56**, 63–74.
Demerec, M., and Hartman, P. E. 1959. Complex loci in microorganisms. *Ann. Rev. Microbiol.* **13**, 377–406.
Erikson, D. 1955. Loss of aerial mycelium and other changes in streptomycete development due to physical variations of cultural conditions. *J. Gen. Microbiol.* **13**, 136–148.
Glauert, A. M., and Hopwood, D. A. 1959. A membranous component of the cytoplasm in *Streptomyces coelicolor*. *J. Biophys. Biochem. Cytol.* **6**, 515–516.

Glauert, A. M., and Hopwood, D. A. 1960. The fine structure of *Streptomyces coelicolor*. I. The cytoplasmic membrane system. *J. Biochem. Biophys. Cytol.* **7**, 479–488.

Glauert, A. M., and Hopwood, D. A. 1961. The fine structure of *Streptomyces violaceoruber* (*S. coelicolor*). III. The walls of the mycelium and spores. *J. Biochem. Biophys. Cytol.* **10**, 505–516.

Hartman, P. E., Loper, J. C., and Šerman, D. 1960. Fine structure mapping by complete transduction between histidine-requiring *Salmonella* mutants. *J. Gen. Microbiol.* **22**, 323–353.

Hayes, W. 1960. The bacterial chromosome. *In* "Microbial Genetics," *Symposium Soc. Gen. Microbiol. 10th* pp. 12–38.

Holloway, B. W. 1955. Genetic recombination in *Pseudomonas aeruginosa*. *J. Gen. Microbiol.* **13**, 572–581.

Hopwood, D. A. 1957. Genetic recombination in *Streptomyces coelicolor*. *J. Gen. Microbiol.* **16**, ii–iii.

Hopwood, D. A. 1959. Linkage and the mechanism of recombination in *Streptomyces coelicolor*. *Ann. N.Y. Acad. Sci.* **81**, 887–898.

Hopwood, D. A. 1960. Phase-contrast observations on *Streptomyces coelicolor*. *J. Gen. Microbiol.* **22**, 295–302.

Hopwood, D. A., and Glauert, A. M. 1960a. Observations on the chromatinic bodies of *Streptomyces coelicolor*. *J. Biochem. Biophys. Cytol.* **8**, 257–266.

Hopwood, D. A., and Glauert, A. M. 1960b. The fine structure of *Streptomyces coelicolor*. *J. Biochem. Biophys. Cytol.* **8**, 267–278.

Hopwood, D. A., Mancinelli, A., Sermonti, G., and Spada-Sermonti, I. 1961. Eterocloni in *Streptomyces*. *Atti assoc. genet. ital.* **6**, 71–73.

Hotchkiss, R. D. 1951. Transfer of penicillin resistance in Pneumococci by the desoxyribonucleate fractions from resistant cultures. *Cold Spring Harbor Symposia Quant. Biol.* **16**, 457–461.

Kelner, A. 1948. Mutation in *Streptomyces flaveolus* induced by X-rays and ultraviolet light. *J. Bacteriol.* **56**, 457–465.

Kutzner, H. J., and Waksman, S. A. 1959. *Streptomyces coelicolor* Müller and *Streptomyces violaceoruber* Waksman and Curtis, two distinctly different organisms. *J. Bacteriol.* **78**, 528–538.

Lederberg, J. 1949. Aberrant heterozygotes in *Escherichia coli*. *Proc. Natl. Acad. Sci. U.S.* **35**, 178–184.

Lederberg, J. 1957. Sibling recombinants in zygote pedigrees of *Escherichia coli*. *Proc. Natl. Acad. Sci. U.S.* **43**, 1060–1065.

Lederberg, J., and Lederberg, E. M. 1952. Replica plating and indirect selection of bacterial mutants. *J. Bacteriol.* **63**, 399–406.

Lederberg, J., and Tatum, E. L. 1946. Novel genotypes in mixed cultures of biochemical mutants of bacteria. *Cold Spring Harbor Symposia Quant. Biol.* **11**, 113–114.

Leidy, G., Hahn, E., and Alexander, H. E. 1953. *In vitro* production of new types of *Haemophilus influenzae*. *J. Exptl. Med.* **97**, 467–482.

Newcombe, H. B. 1953. Radiation induced instabilities in *Streptomyces*. *J. Gen. Microbiol.* **9**, 30–36.

Piéchaud, M. 1954. La coloration sans hydrolyse du noyau des bactéries. *Ann. inst. Pasteur* **86**, 787–793.

Pittenger, T. 1954. The general incidence of pseudowild types in *Neurospora crassa*. *Genetics* **39**, 326–342.

Pontecorvo, G. 1953. The genetics of *Aspergillus nidulans*. *Advances in Genet.* **5,** 141–238.
Pontecorvo, G. 1959. "Trends in Genetic Analysis." Columbia Univ. Press, New York.
Pritchard, R. H. 1954. Ascospores with diploid nuclei in *Aspergillus nidulans*. *Proc. Intern. Congr. Genet. 9th Congr. Bellagio, Italy 1953* (*Caryologia Suppl.*) p. 117.
Robinow, C. F. 1953. Spore structure as revealed by thin sections. *J. Bacteriol.* **66,** 300–311.
Saito, H. 1958. Heterokaryosis and genetic recombination in *Streptomyces griseoflavus*. *Can. J. Microbiol.* **4,** 571–580.
Saito, H., and Ikeda, Y. 1959. Cytogenetic studies on *Streptomyces griseoflavus*. *Ann. N.Y. Acad. Sci.* **81,** 862–878.
Sermonti, G., and Spada-Sermonti, I. 1955. Genetic recombination in *Streptomyces*. *Nature* **176,** 121.
Sermonti, G., and Spada-Sermonti, I. 1956. Gene recombination in *Streptomyces coelicolor*. *J. Gen. Microbiol.* **15,** 609–616.
Sermonti, G., and Spada-Sermonti, I. 1959a. Genetics of *Streptomyces coeliocolor*. *Ann. N.Y. Acad. Sci.* **81,** 854–861.
Sermonti, G., and Spada-Sermonti, I. 1959b. Preliminary results of genetic analysis in *Streptomyces coelicolor*. *Selected Sci. Papers ist. super. sanità* **2,** 437–447.
Sermonti, G., Mancinelli, A., and Spada-Sermonti, I. 1960. Heterogeneous clones (heteroclones) in *Streptomyces coelicolor* A3(2). *Genetics* **45,** 669–672.
Spizizen, J. 1958. Transformation of biochemically deficient strains of *Bacillus subtilis* by deoxyribonucleate. *Proc. Natl. Acad. Sci. U.S.* **44,** 1072–1078.
Stanier, R. Y. 1942. Agar-decomposing strains of the *Actinomyces coelicolor* species group. *J. Bacteriol.* **44,** 555–570.
Szybalski, W. 1952. Gradient plate technique for study of bacterial resistance. *Science* **116,** 46–48.
Tonolo, A., Casinovi, C., and Marini-Bettolo, G. B. 1954. Sul pigmento di uno *Streptomyces* sp. *Rend. ist. super. sanità* **17,** 949.
Tulasne, R., and Vendrely, R. 1947. Demonstration of bacterial nuclei with ribonuclease. *Nat.* **160,** 225–226.
Visconti, N., and Delbrück, M. 1953. The mechanism of genetic recombination in phage. *Genetics* **38,** 5–33.
Wainwright, S. D., and Nevill, A. 1955. Some effects on post-irradiation treatment with metabolic inhibitors and nutrients upon X-irradiated spores of *Streptomyces*. *J. Bacteriol.* **70,** 547–551.
Waksman, S. A. 1950. "The Actinomycetes." Chronica Botanica, Waltham, Massachusetts.
Wollman, E.-L., Jacob, F., and Hayes, W. 1956. Conjugation and genetic recombination in *Escherichia coli* K-12. *Cold Spring Harbor Symposia Quant. Biol.* **21,** 141–162.
Zinder, N. D., and Lederberg, J. 1952. Genetic exchange in *Salmonella*. *J. Bacteriol.* **64,** 679–699.

FINE STRUCTURE OF GENES IN THE ASCOMYCETE
Ascobolus immersus

P. Lissouba, J. Mousseau, G. Rizet, and J. L. Rossignol

Laboratoire de Génétique Physiologique du Centre National de la Recherche Scientifique,
Gif-sur-Yvette (S. et O.) France

		Page
I.	Introduction	343
II.	Reasons for Choosing the Ascomycete *Ascobolus immersus*	345
III.	Occurrence, within the Same Series, of Both Tetrads with Reciprocal and Tetrads with Non-Reciprocal Recombinants: Series 75	346
IV.	Localization of Crossing-Over and Conversion in Series 19	349
	A. Possibility of Arranging the Mutants in Groups	349
	B. Analysis of 6:2 Tetrads from A × A, B × B, C × C, B × C, A × B, and A × C Crosses	352
V.	Analysis of Series 46; Evidence for a Polarized Genetic Unit: the Polaron	355
	A. The Mutants of Series 46	355
	B. The Polaron	355
	C. Confirmation of the Polarity	360
VI.	Structure of Series 19 and the Concept of the Polaron	361
VII.	Tentative Interpretation of the Properties of the Polaron	362
VIII.	Linkage Structures and Relationships between Polarons	364
	A. Results Concerning Mutant *277* of Series 46	364
	B. Results Concerning Series 19	367
	C. Interpretation of the Results	367
IX.	Occurrence of Odd Segregations	374
X.	Conclusions	375
	Acknowledgments	378
	References	379

I. Introduction

It is generally believed that DNA is the carrier of Mendelian heredity and the knowledge of its structure, stimulated by the work of Watson and Crick (1953), has made considerable progress during the past few years. On the other hand, it is known today that the gene, unit of classical genetics, is a complex element: the functional unit (cistron) and the mutational unit (muton) are units of different size (Benzer, 1957). However, it is as yet impossible to superimpose these two kinds of knowledge and to suggest a structure which the geneticists would recognize as the classical gene, and onto which the biophysicists and biochemists could fit the details of the double helices of Watson and Crick.

For the analysis of the fine structure of the gene the geneticist is limited to the study of intragenic recombinations. The latter, however, are characterized by their rareness. This fact explains why they were not discovered earlier. The frequency of these recombinational events is sometimes below 10^{-8} and there was, therefore, little chance of observing them in higher organisms where the number of individuals studied is always low. The probability of detecting this type of recombination was higher in microorganisms; here experiments are possible involving large numbers of individuals, among whose progeny it is possible to select a chosen type even if it is extremely rare.

Indeed, when crosses are made between two mutants, both of which, according to the old terminology, belong to the same series of multiple alleles, wild type recombinants are observed to occur with a small but significant frequency. As most of the mutants are biochemical mutants (auxotrophs) such recombinations produce prototrophs for which it is easy to select.

Is it possible to use the frequencies of these recombinations in order to construct gene maps in the same way as chromosome maps were made a few decades ago? If the classical gene is represented on the chromosome as a segment, that is to say, as a linear structure, the construction of such a gene map should be possible and even easy. Indeed, the linear arrangement of genes on chromosomes is deduced chiefly from the additivity of recombination frequencies when these frequencies are small. Thus a perfect additivity was expected inside the gene and it could be hoped that such investigations would provide very precise information on the structure of the chromosomes. The real situation is, however, more complex. Briefly, two kinds of difficulties are encountered.

First, very often the additivity of recombination frequencies is not confirmed and the events leading to recombinations are not independent of each other: there is interference. But here this interference is *negative*, whereas, when chromosome maps are concerned, it is *positive*. In other words, multiple events are more frequent than expected.

Second, intragenic recombinations are the result of two different mechanisms. In addition to the classical *crossing-over*, giving rise to reciprocal recombinants in the same tetrad, a second mechanism occurs which produces only a single type of recombinant (wild type or double mutant). The latter phenomenon, called *conversion*, is characterized for a couple of *sites* by a segregation ratio of 3:1 (instead of 2:2). The most generally accepted interpretation implies a *copy-choice* mechanism during chromatid replication (Lederberg, 1955; Levinthal, 1956; Case and Giles, 1958; Stadler, 1958).

In spite of the variety of materials used so far, none was entirely satisfactory for the study of these problems, since very rare recombinants must be detected and the genetic analysis of the tetrads containing these recombinants must be carried out. Even when the study of recombination frequency is done under favorable conditions, it is very difficult, and usually impossible, to analyze 10^4 or 10^5 tetrads. The investigation of recombinant tetrads can only be efficient if they can be immediately, or very easily, detected. Such a possibility is offered by the morphologically aberrant ascospore mutants of the ascomycete *Ascobolus immersus*.

II. Reasons for Choosing the Ascomycete *Ascobolus immersus*

Ascobolus immersus is a coprophilic discomycete very widely distributed and easily isolated in pure culture from wild fructifications. Various species of the same genus have already been submitted to genetic analysis. Several species, including *Ascobolus immersus*, are heterothallic (Rizet, 1939). These molds grow well, and fructify easily, at least on certain media, and their cycles are short.

Abnormal ascospores, particularly colorless ascospores (mature wild spores are deep brown), have been described in several species (Bistis, 1956). This is not surprising owing to the complexity of the chromogenesis in these organisms. *Ascobolus immersus* produces large spores (65–70 μ) which can be easily isolated but are not arranged in linear order within the ascus. They germinate well on 1.25% bacto-peptone agar containing 1.5% NaOH after overnight incubation at 40°C. The sodium hydroxide effect (Yu, 1954) and the technique have been elaborated for *A. immersus* by Mme. Makarewicz of the Warsaw Genetics Institute (personal communication). In this way the germination of the eight ascospores of an ascus is frequently obtained. Dissection of the asci is unnecessary because the mature spores are projected in clusters and the entire tetrad can be collected, for example, on a glass slide coated with gelatin or on an inverted Petri dish cover containing 3% agar. In this way it is possible to observe over 1000 tetrads collected on a single glass slide, all the tetrads remaining isolated from each other. A single culture grown on sterilized horse dung can provide on successive "screens" from 10,000 to 40,000 asci, depending on the temperature.

When the asci projected by a wild type culture are examined, groups of 8 spores exhibiting a visible segregation (for example, 4 colored and 4 white spores) are observed with a frequency of about 10^{-3}. These abnormalities are most of the time due to mutations, and when the abnormal spores are sown separately, mutant strains are obtained. These mutant strains differ from the wild strain either by the pigmentation pattern (rough or granular ascospores in which the pigment is condensed in small drops) or by their color (pink and white ascospores). Up to this time we have collected more than 2000 mutants differing from the wild strain by a single gene. In other words, crosses with the wild strain yield immediately detectable 4:4 segregations. Crosses between them result in various types of segregations, and the analysis of the latter makes it possible to define the relationship between the genes concerned (*linkage* or *independence*).

The most striking observation during the preliminary studies was the clustering of the mutants in "series" (Rizet *et al.*, 1960c). Each series comprises a certain number of mutants (from 2 to 145) which, when intercrossed, yield no, or very few, wild recombinants (brown ascospores, usually called "black" spores). Thus each series seems to represent a complex locus, the various units behaving either as *homoalleles*, or as *heteroalleles*, according to Roman's terminology (1958). When very large progenies (10^5 tetrads or more) are analyzed, true homoalleles are only very rarely detected. In other words, it is unlikely that the same mutant will be isolated more than once.

The rough spore mutants that have been isolated can be grouped into two series. This is true also for the granular spore mutants. White spore mutants, however, can be grouped into at least twenty series, four of which are under investigation (series 12, 75, 46, and 19). The two latter also contain a few pink spore mutants.

III. Occurrence, within the Same Series, of Both Tetrads with Reciprocal and Tetrads with Non-Reciprocal Recombinants: Series 75

There are more than 140 white ascospore mutants in series 75. Crosses between two such mutants yield: (1) a very large majority of asci containing 8 white spores (parental tetrads); (2) a variable, but always low or very low, proportion of asci containing 6 white and 2 black spores, which can be considered, as a first approximation, as tetratypes resulting from a crossover; and (3) extremely few, if any, asci containing 4 black and 4 white spores. These 4:4 asci must be ascribed either to a double event or to back mutation.

Series 75 is the richest in mutants and also yields the highest frequency of 6:2 asci; if all of these resulted from crossing-over, then the series would cover a chromosome segment of about 4 Morgan units. We also know that this segment is some distance from its centromere. An attempt has been made to map these sites using not only the frequency of 6:2 asci, but also some of the mutants which appear to represent deficiencies (Benzer, 1959). Although several millions of tetrads have been observed, we met many difficulties such as reversions for certain sites, homoalleles behaving very differently, non-additivity of frequencies, the production in some crosses of a non-negligible proportion of 7:1 asci, etc. In spite of its preliminary state, this study shows the existence of regions with many sites and of regions containing few sites.

The nature of the event producing 6:2 asci has been investigated particularly in the progeny of all crosses involving the five mutants *231, 322, 278, 147,* and *1987*. These mutants do not show any particular abnormality and there are good reasons to believe that they delineate almost the entire chromosome segment involved in the series. "Distances" between sites, taken from the 6:2 frequency, are given by Fig. 1.

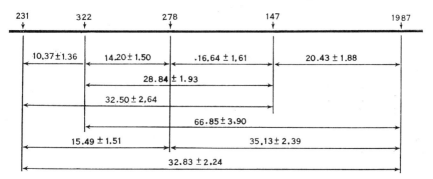

Fig. 1. Relative positions of sites *231, 322, 278, 147,* and *1987* (Series 75) determined by the frequency of 6:2 asci (average number per 1000 tetrads).

From the ten possible crosses between these five mutants 157 of the 6:2 asci have been analyzed by backcrossing each of the eight strains produced by an ascus to the two parental strains. These asci are of two kinds. All of them contain a pair of wild type recombinants, but while some of them contain, at the same time, a pair of double mutant recombinants, others contain no reciprocal recombinant. The former asci are indeed tetratypes resulting from a crossover. Out of the six white spores from the second type of asci, two pairs have the genotype of one parent (*majority parent*) and one pair has the genotype of the other parent (*minority parent*). They correspond therefore to a 2:2 segregation for the two sites of the majority parent and a 3:1 segregation for the two sites of the minority parent (conversion).

In a given cross, a conversion can take place, depending on the particular tetrad, either at the level of the mutant site located on the left, or at the level of that on the right. In the former case the majority site is the one on the right and in the latter case it is the one on the left. In Table I these two eventualities are presented separately.

In nine crosses out of ten, the 6:2 tetrads resulted either from crossing-over or from conversion. The relative frequencies of these two mechanisms are very variable; hence cross 1, where there was no crossing-over, is not necessarily an exception. On the whole, cases of conversion are much more frequent than cases of crossing-over (118 against 39). We shall return later to the fact that, clearly, conversions involve more frequently the couple of sites to the right than those to the left (84 against 34). The converse does not occur. The dissymmetry is sometimes very pronounced (crosses 3, 4, 8, and 9).

TABLE 1

Genetic Structure of 6:2 Asci from Crosses between Mutants *147*, *231*, *322*, *278*, and *1987* of Series 75

				Conversions	
	Cross	No. of asci analyzed	Crossing-over	left major	right major
1	*231* × *322*	11	0	7	4
2	*231* × *278*	14	1	9	4
3	*231* × *147*	16	4	11	1
4	*231* × *1987*	17	5	11	1
5	*322* × *278*	16	3	7	6
6	*322* × *147*	12	6	4	2
7	*322* × *1987*	17	13	3	1
8	*278* × *147*	19	3	12	4
9	*278* × *1987*	18	3	12	3
10	*147* × *1987*	17	1	8	8
	Total	157	39	84	34
				118	

This study is too preliminary to lead to precise conclusions. It does, however, allow the consideration of certain problems.

(1) In certain crosses the frequency of 6:2 asci is definitely significant. As we detect originally only the wild recombinants, it can be thought that the total frequency of recombinants is of the order of that usually found between neighboring, linked genes; it is interesting to observe in these cases a majority of conversions.

(2) If in the same cross some of the 6:2 asci result from conversion and others from crossing-over, it is clearly impossible to define a unit of "distance" between the sites involved; the distance measured is a complex value.

Since, moreover, the relative frequencies of conversion and crossing-over vary from cross to cross, following what are apparently complicated rules (compare, for example, the result of cross *322* × *1987* with that of crosses *322* × *278*, *278* × *147*, and *147* × *1987*, which involve the three fractions of the same segment), the notions of distance equality, additivity, or non-additivity are necessarily confused. The order of sites on a map built from the frequency of 6:2 asci may even appear questionable.

The segment represented in Fig. 1 is obviously very complex. This must be true also for the subsegments since most of the crosses between sites contiguous on the map yield 6:2 asci coming from either crossing-over or conversions, the latter involving either the left site or the right site. Certain relationships must, however, be emphasized. In cross 7, the 6:2 asci result almost entirely from crossing-over; in crosses 1, 2, and 10, they result almost entirely from conversions; the conversions correspond almost always to the same majority parent (crosses 3 and 4). These relationships suggested that a simpler situation will be found in the shorter series comprising fewer mutants.

IV. Localization of Crossing-Over and Conversion in Series 19

A. Possibility of Arranging the Mutants in Groups

Series 19 comprises about fifty mutants of which over thirty have been investigated in more or less detail. They are divided into three groups known as Groups A, B, and C.

Group A comprises eleven mutants (*49, 60, 65, 270, 1028, 1130, 1467, 1510, 1656, 1678*, and *2073*). Crosses between members of the group yield 6:2 asci with a frequency not exceeding 1/4000, and, in most cases, much lower or nil. To at least four of them (*49, 60, 1028, 1130*) a certain length must be ascribed because, when crossed with certain others which give 6:2 asci in crosses with one another (i.e., are heteroalleles), no such asci are produced. Table 2 shows the number of 8:0 and 6:2 asci obtained in crosses between four of these mutants which have never shown reversion.

In Group B, 15 mutants are known at present, viz.: *19, 72, 82, 158, 243, 484, 622, 865, 900, 1261, 1292, 1768, 1830, 1844*, and *1882*. Thirteen of these, crossed in all possible ways, yield frequencies of 6:2 asci lower than, or rarely equal to, 1/4000. At least five (*19, 72, 484, 1292*, and *1882*) overlap several others. If it is assumed that heteroalleles do not overlap, a map of this group can be drawn according to the Benzer technique (unpublished map).

Group C contains fewer mutants. Five of them have been investigated, viz.: *55, 218, 324, 1634*, and *1802*. The analysis of the crosses between them, carried out on more than 5×10^5 asci, showed the occurrence of 6:2 tetrads (about 1/50,000) only in cross *55* × *218*.

Some mutants belonging to this series have undoubtedly resulted from more profound events. In Group A, *49* and *1130* are homoallelic for all the sites of the group. Mutants *95* and *119* gave no wild recombinants with Group B and C sites. Mutant *176* is the homoallele of all the mutants of the series.

Except for these anomalies, Groups A, B, and C therefore correspond to three clusters of mutants: within each, the sites are either very close to each other or overlap.

TABLE 2

Number of 8:0 and 6:2 Asci Observed Following Crosses between Four Mutants Belonging to Group A of Series 19*

Mutant	Asci	1656	60	1678	270
1656	8:0	>72,300	44,000	53,800	13,300
	6:2	0	4	11	2
60	8:0	—	>95,500	40,850	128,000
	6:2		0	0	9
1678	8:0	—	—	>117,000	25,000
	6:2			0	1
270	8:0	—	—	—	>62,000
	6:2				0

* Actually only ¼ of the area of each Petri dish cover serving of "screen" was sampled. The number of 8:0 asci thus counted was multiplied by four and the result given to the nearest hundred; no such approximation was needed for 6:2 asci.

Numerous crosses have been carried out between mutants of different groups. The results of the first experiments, involving the mutants isolated first and carried out at 25–26°C, are shown in Table 3. Since the results of the various crosses A × B, B × C, and A × C were homogeneous, they have been grouped together.

TABLE 3

Summary of Results Obtained in Crosses A × B, B × C, and A × C of Series 19*

Cross	8:0 asci	6:2 asci	Frequencies of 6:2 asci
A × B	85,170	133	$1.6 \pm 0.2 \; 10^{-3}$
B × C	45,000	71	$1.6 \pm 0.3 \; 10^{-3}$
A × C	26,440	495	$19.2 \pm 0.9 \; 10^{-3}$

* First series of experiment.

It can be seen from Table 3 that the frequency of 6:2 asci is clearly larger in intergroup than in intragroup crosses. It is clear also that the frequency of 6:2 asci is much higher in A × C crosses than in A × B and B × C crosses; it is even larger than the sum of the A × B and B × C frequencies.

A number of crosses of this type, involving the more recently isolated mutants and carried out at 23–24°C, yielded much lower frequencies of 6:2 asci [possibly due to the temperature (Lissouba, 1961) or to the genotype]; and, above all, much more variable ones. This is especially true for A × B and A × C crosses. Some of these results are shown in Table 4.

TABLE 4
Comparison of the Frequencies of 6:2 Asci in Crosses A × B, B × C, and A × C*

A × B crosses	Frequency of 6:2 asci	B × C crosses	Frequency of 6:2 asci	A × C crosses	Frequency of 6:2 asci
60 × 243	0.71 ± 0.51	243 × 55	0.87 ± 0.42	55 × 60	12 ± 0.35
60 × 1844	1.4 ± 0.62	1844 × 55	0.27 ± 0.13	55 × 60	12 ± 0.35
270 × 243†	8.8 ± 1.41	243 × 324†	1.37 ± 0.61	270 × 324†	17.2 ± 2.4
270 × 622	3.8 ± 0.8	622 × 324	0.93 ± 0.55	270 × 324	11.29 ± 1.4
1028 × 19	0.89 ± 0.33	19 × 324	1.08 ± 0.42	1028 × 324	13.88 ± 3.2
1467 × 484	0.23 ± 0.14	484 × 324	0.68 ± 0.34	1467 × 324	6.74 ± 1.1
1510 × 243	2.48 ± 0.56	243 × 324	0.73 ± 0.21	1510 × 324	6.57 ± 1.1
1656 × 72	0.36 ± 0.12	72 × 324	0.73 ± 0.54	1656 × 324	3.5 ± 0.4
1678 × 72	0.36 ± 0.11	72 × 324	0.73 ± 0.54	1678 × 324	11.50 ± 1.6
1678 × 484	1.6 ± 0.52	484 × 324	1.4 ± 0.55	1678 × 324	11.50 ± 1.6
2073 × 72	0.24 ± 0.12	72 × 324	0.73 ± 0.54	2073 × 324	7.06 ± 1.3

* Except when indicated otherwise, the incubation temperature was 23°C.
† Growth temperature 25°C.

Even though in all these cases (which are merely examples representing a much larger body of data) the frequencies of 6:2 asci are much more variable than these shown in Table 3, the fact remains that the "distance" A → C is always larger than the sum of "distances" A → B + B → C. This confirms the existence of "groups" of sites and it appears that these groups are arranged along the chromosome segment corresponding to Series 19 in the order A → B → C.

A confirmation of this interpretation is given by the results of crosses between a double mutant A.C and a mutant B (*1678.324 × 484*), on the one hand, and between a double mutant A.B and a mutant C (*60.484 × 324*), on the other. The first of these crosses produced eight 6:2 asci out of nearly 470 × 10³ tetrads, the second seven 6:2 asci out of 7130 tetrads. The latter frequency is of the same order of magnitude as those obtained in cross *484 × 324* and in most B × C crosses; this is easily understandable if Group B is localized between Groups A and C. The former, much lower frequency, agrees with the view that the very rare wild recombinants result from a double event, and this argues in favor of the indicated order.

The grouping of the sites and the order of the groups being apparently established, it became possible to undertake the genetic analysis of 6:2 asci in the various kinds of crosses.

B. Analysis of 6:2 Tetrads from A × A, B × B, C × C, B × C, A × B, and A × C Crosses

The difficulty of this analysis is obviously a function of the frequency of 6:2 asci. It is easy for A × C crosses, more difficult for A × B or B × C crosses, and becomes very difficult for A × A, B × B, and C × C crosses. Tables 5, 6, and 7 therefore show results represented by very unequal numbers.

TABLE 5
Genetic Structure of 6:2 Asci from Crosses A × A, B × B, C × C, and B × C

No.	Cross			No. of asci analyzed	Cross-ing-over	Con-ver-sion	Majority parent	Minority parent
1	A × A	1656 ×	60	5	0	5	1656	60
2		1656 ×	270	8	0	8	1656	270
3		60 ×	270	8	0	8	60	270
4		1678 ×	270	5	0	5	1678	270
		Total		26		26		
5	B × B	865 ×	622	8	0	8	865	622
6		19 ×	1844	7	0	7	19	1844
7		19 ×	622	5	0	5	19	622
8		1844 ×	622	1	0	1	1844	622
9	C × C	218 ×	55	1	0	1	218	55
10	B × C	865 ×	55	3	0	3	865	55
11		19 ×	218	6	0	6	19	218
12		19 ×	55	18	0	18	19	55
13		1844 ×	218	4	0	4	1844	218
14		1844 ×	55	8	0	8	1844(7), 55(1)	55(7), 1844(1)
15		622 ×	55	2	0	2	622	55
		Total		63		63		

A fundamental difference appears between the results given in Table 5 and those given in Tables 6 and 7. While no crossing-over was detected in crosses A × A, B × B, C × C, and B × C (Table 5), most of the A × B and A × C crosses gave 6:2 asci resulting from either crossover or conversion, as though the segment corresponding to Series 19 consisted of two subsegments, one carrying the Group A sites and the other the sites of Groups B and C.

When the two sites involved in a cross are located within the same subsegment, all 6:2 asci result from conversion; conversely, when the two sites are not within the same subsegment, some 6:2 asci result from crossing-over, the remainder from conversion. The rare exceptions to this rule, shown in Tables 6 and 7, are not considered significant. Only 10 asci from the three A × B crosses 9, 11, and 12 (Table 6) where no crossing-over has been detected have been analyzed; among the crosses involving groups A and C (Table 7), crosses 10 and 12 (*1467* × *324* and *1510* × *324*) gave no detectable crossing-over, whereas crosses 9 and 11 (*1467* × *55* and *1510* × *55*), which involve a homoallele of *324* (Group C), gave both crossing-over and conversion.

TABLE 6
Genetic Structure of 6:2 Asci from Various A × B Crosses

No.	Cross	No. of asci analyzed	Crossing-over	Conversion	Majority parent	Minority parent
1	60 × 19	10	6	4	19(3), 60(1)	60(3), 19(1)
2	60 × 82	22	17	5	82	60
3	60 × 158	8	4	4	158	60
4	60 × 243	2	1	1	243	60
5	60 × 484	2	1	1	484	60
6	60 × 622	9	3	6	622(4), 60(2)	60(4), 622(2)
7	60 × 1844	1	0	1	1844	60
8	270 × 19	4	2	2	19	270
9	270 × 72	3	0	3	72	270
10	270 × 622	15	1	14	622(9), 270(5)	270(9), 622(5)
11	270 × 900	5	0	5	900	270
12	270 × 1292	2	0	2	1292	270
13	270 × 1844	4	1	3	1844	270
	Total	87	36	51		

In addition, the results given in Table 7 indicate that the relative frequency of the two types of events resulting in wild recombinants varies a good deal from cross to cross. This is more obvious from the data of Table 8, which gives the results of crosses carried out in a single experimental series, extracted from Table 7.

These results can be accounted for by assuming that, in the segment corresponding to Series 19, crossing-over is localized between the two subsegments within each of which only conversions occur. This intermediate zone will be tentatively referred to as the *linkage structure*.

No cross has ever been found which owed all the 6:2 asci to crossing-over. However, it is known that this can occur (Mme. N. Job-Engelmann: unpublished results concerning Series 12).

TABLE 7
Genetic Structure of 6:2 Asci from Various A × C Crosses

No.	Cross	No. of asci analyzed	Crossing-over	Conversion	Majority parent	Minority parent
1	49 × 55	8	6	2	49	55
2	60 × 55	134	88	46	60(41), 55(5)	55(41), 60(5)
3	60 × 218	6	3	3	60	218
4	60 × 324	24	15	9	60(7), 324(2)	324(7), 60(2)
5	65 × 55	2	1	1	65	55
6	270 × 55	2	1	1	270	55
7	1028 × 55	35	12	23	1028	55
8	1130 × 55	21	17	4	1130	55
9	1467 × 55	10	7	3	1467	55
10	1467 × 324	3	0	3	1467	324
11	1510 × 55	10	3	7	1510	55
12	1510 × 324	4	0	4	1510	324
13	1656 × 55	11	7	4	1656	55
14	1656 × 324	7	1	6	1656	324
15	1678 × 55	47	11	36	1678	55
16	1678 × 324	86	15	71	1678(67), 324(4)	324(67), 1678(4)
17	2073 × 55	25	7	18	2073	55
	Total	435	194	241		

TABLE 8
Relative Frequencies of Crossing-over and Conversion in the Progeny of Various A × C Crosses*

Cross	No. of asci analyzed	Crossing-over	Conversion
60 × 55	64	43	21
60 × 324	24	15	9
1678 × 55	47	11	36
1678 × 324	61	11	50

* Results of a single series of experiments at 21°C.

Further inferences can be made from Tables 5, 6, and 7, particularly concerning the majority mutants in the cases of conversion.

It can be seen that, with a few exceptions, the majority mutant is always the same in a given cross. This will be made use of below.

V. Analysis of Series 46; Evidence for a Polarized Genetic Unit: the Polaron

A. THE MUTANTS OF SERIES 46

This series has been the subject of several publications (Lissouba and Rizet, 1960; Rizet *et al.*, 1960a,b; Lissouba, 1961). In addition, over twenty recently isolated mutants have been studied. In what follows, a brief summary of all the results obtained is presented.

At the time of writing, thirty-four mutants are known in this series. Of these, thirty-one produce only white spores; three of them (*1546, 1604,* and *30*) produce pink ascospores in at least some of the asci; this phenotype is also exhibited in 4 spores of some of the asci formed following one cross of these mutants with other mutants of the same series: in the spores which have the genotype of one of the three mutants, the phenotype is not always expressed.

Several of these mutants are, no doubt, results of minute rearrangements. Such is the case of mutants *B, 188* (Lissouba, 1961), and *1180* (which behaves in some respects like *188*); this is also the case with both *1021* and *1026* which are homoallelic to several mutants involving relatively distant sites; they can be regarded as deficiencies.

Other mutants, when crossed with other members of the series, irrespective of their localization, always gave very few 6:2 asci; they have not been thoroughly investigated and their nature remains completely unknown.

Aside from these anomalies, seventeen mutants have been investigated in a somewhat more detailed manner. These are mutants *1604, 1546, 63, 665, 1066, 1330, 1425, 46, 1167, W, 686, 1416, 1590, 1216, 137, 138,* and *277*. No reversion has been detected among 20,000 to 100,000 tetrads. However, more recently, mutant *277* gave five 6:2 asci and eight 4:4 asci among a total of 262,000. Sites *665, 1066, 1330,* and *1425*, which so far have been little studied, are homoalleles of *63* or very close to it; sites *1167, 686, 1416,* and *1590* are homoalleles of *W* or very close to it. Mutants *137* and *138*, originally found side by side, behave identically; undoubtedly they involve the same site and resulted from a single mutational event.

B. THE POLARON

1. *Evidence from Two Independent Methods for the Order of Sites W, 46, 63, 137-138, 188, and 1216*

We shall first consider mutants *W, 46, 63, 137-138, 188,* and *1216*. All possible intercrosses have been made, except one. Mutant *188*, which has been shown previously to behave abnormally (Lissouba, 1961), is no longer used, and the cross *188* × *1216* has not been made.

TABLE 9

Number of 8:0 and 6:2 Asci and Frequency of 6:2 Asci in Crosses between Various Mutants of Series 46

Mutant	Asci	188	63	46	W	1216	137–138
188	8:0 6:2	>35,000 0	0.76 0.29 to 1.98	1.24 0.52 to 2.96	— 4.97 ± 0.78	— —	— 9.44 ± 1.04
63	8:0 6:2	5,264 4	>280,000 0	0.53 0.27 to 1.14	— 2.82 ± 0.68	— 2.90 ± 0.33	— 12.85 ± 2.47
46	8:0 6:2	4,008 5	11,278 6	>190,000 0	0.74 0.39 to 1.44	— 2.03 ± 0.27	— 11.51 ± 1.15
W	8:0 6:2	8,047 40	5,974 17	12,021 9	>186,000 0	0.88 0.64 to 1.36	— 8.30 ± 0.67
1216	8:0 6:2	— —	25,794 75	26,940 55	24,866 22	>120,000 0	— 8.35 ± 0.47
137–138	8:0 6:2	8,595 81	2,100 27	8,600 99	18,072 150	36,786 310	>222,000 0

The frequencies of 6:2 asci, with either the standard deviation or the confidence limits, are given in Table 9. No reverse mutation has been detected.

If the frequencies of 6:2 asci corresponding to the three pairwise crossings are compared, one of them is found never to differ significantly from the sum of the two others. A linear map of all the mutant sites can therefore provisionally be drawn (Fig. 2).

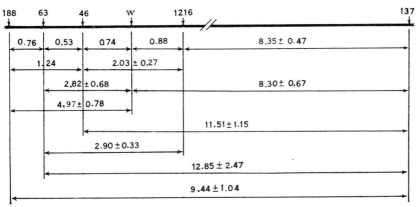

FIG. 2. Relative positions of sites *188, 63, 46, W, 1216,* and *137-138* (Series 46) determined by the frequency of 6:2 asci (average number per 1000 tetrads).

It would be of obvious interest to know the origin of the 6:2 asci in such crosses. If they all resulted from the same type of event, the distances given in Fig. 2 would have a more precise meaning. The analysis of 155 asci from these crosses has been made and the results obtained are summarized in Table 10.

It can be seen from this table that 154 out of the 155 asci analyzed were the result of conversion. The exceptional ascus (from the cross *63 × 1216*) contained a pair of spores which behaved as double mutants but several concordant observations suggest that they resulted from a chromosomal accident and not from crossing-over. We consider therefore (Lissouba, 1961) that, on this segment, only conversions are possible.

Table 10 shows that, in a given cross, the majority parent is always the same, and that the same parent can be the majority or the minority parent depending on the cross studied. If Fig. 2 and Table 10 are compared it appears clear that, for a given cross, the majority parent is always the one located on the left on the site map.

If one disregards the site map, and considers only the results concerning the majority-minority relationships given in Table 10, the simplest assumptions lead one to the order of the sites indicated in Fig. 2.

This order is also supported by other observations.

2. *Confirmation of the Validity of the Site Map*

   a. *Site 137 Is at One End of the Segment under Consideration.* When mutant *1021*, which behaves as a deficiency, is crossed with mutants *63*, *46*, *W*, *1216*, and *137*, only cross *1021* × *137* produces 6:2 asci.

TABLE 10
Genetic Structure of 6:2 Asci from Crosses between Mutants *188*, *63*, *46*, *W*, *1216*, and *137–138* of Series 46

Cross	No. of asci analyzed	Crossing-over	Conversion	Majority parent	Minority parent
188 × *63*	8	0	8	*188*	*63*
188 × *46*	2	0	2	*188*	*46*
188 × *W*	10	0	10	*188*	*W*
188 × *138*	11	0	11	*188*	*138*
63 × *46*	2	0	2	*63*	*46*
63 × *W*	18	0	18	*63*	*W*
63 × *1216*	5	1	4	*63*	*1216*
63 × *138*	35	0	35	*63*	*138*
46 × *W*	7	0	7	*46*	*W*
46 × *138*	22	0	22	*46*	*138*
W × *1216*	18	0	18	*W*	*1216*
W × *138*	15	0	15	*W*	*138*
1216 × *137*	2	0	2	*1216*	*137*
Total	155	1	154		

   b. *The Three Mutants W, 1216, and 137 Are in the Order Indicated on the Map.* When these mutants are crossed with mutants *665*, *1066*, and *1330*, all of which are very close to or are homoalleles of *63*, all the frequencies of the resulting 6:2 asci are compatible with this order and incompatible with any other. The same is true for crosses with mutants *1167*, *686*, *1416*, and *1590* which are very close to or homoalleles of *W*.

   c. *The Order of Sites 63, 46, W, and 1216 Is the Previously Described One.* This region could be confusing owing to the shortness of the "distances." Two recently isolated mutants (*1604* and *1546*), behaving as homoalleles of each other, allowed us to eliminate the last ambiguities, as shown by the results of crosses with the mutants represented on the map (Table 11).

Table 11 shows that these two sites behave as being very close to *63*, but when these two mutants are crossed with mutants *46*, *W*, and *1215* they give a large but significantly different number of 6:2 asci. The increasing order of "distances" is obvious.

TABLE 11
Number of 8:0 and 6:2 Asci in Crosses of Mutants *1604* and *1546* with Other Members of Series 46

Mutant	Asci	*63*	*46*	*W*	*1216*	*137*
1604	8:0	19,088	13,636	8,935	7,661	16,535
	6:2	3	14	52	95	352
1546	8:0	18,033	18,390	18,309	4,587	11,669
	6:2	2	14	58	71	216

d. *The Results of the Analysis of 6:2 Asci from Crosses Made with Certain New Mutants Agree with Previous Results* (Table 12). Added to the data of Table 11, the results given in Table 12 concerning sites *1604* and *1546* show that the two sites are not only very close to but probably on the left of *63* (Table 12, cross *1604* × *63*). These two mutants do raise several problems due to the high frequency of 6:2 tetrads observed and to the variability of their phenotype (pink spores). However, in our opinion, such unsolved questions do not detract from the value of the

TABLE 12
Genetic Structure of 6:2 Asci from Various Crosses between Mutants of Series 46

Cross	No. of Asci analyzed	Crossing-over	Conversion	Majority parent	Minority parent
1604 × *63*	1	0	1	*1604*	*63*
1604 × *46*	1	0	1	*1604*	*46*
1604 × *W*	7	0	7	*1604*	*W*
1604 × *1216*	9	0	9	*1604*	*1216*
1604 × *137*	15	0	15	*1604*	*137*
1546 × *46*	5	0	5	*1546*	*46*
1546 × *1216*	3	0	3	*1546*	*1216*
665 × *137*	2	0	2	*665*	*137*
1330 × *137*	3	0	3	*1330*	*137*
1167 × *137*	1	0	1	*1167*	*137*
1590 × *137*	1	0	1	*1590*	*137*
Total	48	0	48		

confirmatory evidence provided by these two mutants concerning the order of the sites *63*, *46*, *W*, and *1216*.

3. *Properties of the Polaron*

Be it as it may, the facts can be summarized by the following propositions. Sites *188*, *63*, *46*, *W*, *1216*, and *137-138* can be arranged in a linear order and define a chromosomal segment. No crossing-over occurs within this segment. All recombinations within it are of the conversion type: a tetrad never contains reciprocal recombinants. Finally, the rule which makes it possible for a given cross to predict the majority parent implies that the segment is polarized; in order to emphasize this remarkable property, we have named such a segment a *polaron* (Lissouba and Rizet, 1960).

C. Confirmation of the Polarity

Several questions can be asked concerning the preceding observations. (1) In addition to the 6:2 tetrads in which the recombinants are wild type and therefore easily detectable, are there tetrads among the 8:0 asci where the recombinants are double mutants and consequently undetectable? (2) When two mutants are crossed, the 6:2 asci resulting from conversion imply a segregation 1 mutant:3 wild type at one site; is it possible to detect such segregations when the wild strain is crossed with a mutant? (3) In the affirmative, do the "symmetrical" segregations (3 *wild type* : 1 *mutant*) occur in the same cross?

A priori the answers to the latter two questions should be the easiest to obtain because in the crosses *wild* × *mutant*, both the 3:1 and 1:3 segregations should be detectable (asci containing 6 white and 2 black spores and asci containing 6 black and 2 white spores). However, such an investigation is very laborious (at least in the case of the various mutants of this series), owing to the occurrence of false segregations [phenotypically white spores which are really wild type or vice versa (Lissouba, 1961)]. In any case two facts stand out. (1) Both 3:1 and 1:3 segregations occur in each cross and there is no indication that their frequencies are different; (2) the frequency of these aberrant segregations is much higher for *137* (right end of the polaron) than for *63*, *46*, or *W* (left end of the polaron).

Since both 3:1 and 1:3 segregations exist for a given couple of sites, it is difficult to see why, in the case of double heterozygotes, tetrads containing the double mutant would not also be produced. A simple calculation shows that 300 8:0 asci from the cross *63* × *137* should be analyzed in order to have a very high probability of finding one double mutant, if it is assumed that such tetrads occur with the same frequency as the

6:2 tetrads. At the time of writing no double mutant has been found among the 200 tetrads studied. This investigation is being continued.[1]

Be this as it may, the difference in the frequencies of 6:2 or 2:6 tetrads, depending on the position of the sites on the polaron, is further evidence of polarity.

VI. Structure of Series 19 and the Concept of the Polaron

We have seen that, on the chromosomal segment involved in Series 19, it is possible to distinguish two subsegments within which only conversion occurs, viz., the subsegment carrying the sites of Group A and the subsegment carrying the sites of Groups B and C. Can one consider these subsegments as polarons? As previously indicated, this question is difficult to answer owing to the very low frequency of 6:2 asci: both the calculation of frequencies and the tetrad analysis are very tedious.

Information concerning the first subsegment has been obtained from four mutants of Group A (*60, 270, 1656,* and *1678*) used both for counts and analysis of 6:2 asci. The data are summarized in Tables 2 and 5. Since mutants *60* and *1678* behave as homoalleles (Table 2), 6:2 asci from only four crosses: *60 × 270, 60 × 1656, 270 × 1656,* and *270 × 1678* have been analyzed (Table 5). All the analyzed asci correspond to conversions only, and, in each of the crosses, the majority parent is the same. Site *270* is always the minority parent; site *60* is the minority parent only in cross *60 × 1656*.

These results are consistent with the view that we are dealing with a polaron or a part of a polaron, oriented in the same way as the polaron originally studied and carrying the sites in the order: *1656 → 60* and *1678 → 270*. This order is compatible with the counts given in Table 2; it is also the only one compatible with a more complete map of Group A constructed on the assumption that two mutants which, when intercrossed, yield 6:2 asci, do not overlap (unpublished map).

Analyses of asci concerned with subsegment B-C are also given in Table 5. These asci are from (1) four crosses B × B (*865 × 622, 19 × 1844, 19 × 622,* and *1844 × 622*); (2) six crosses B × C (*865 × 55, 19 × 218, 19 × 55, 1844 × 218, 1844 × 55,* and *622 × 55*); and (3) one cross C × C (*218 × 55*). All the 63 tetrads analyzed result from conversion. Except for one case (a 6:2 ascus from cross *1844 × 55* to be discussed later), the majority parent was the same in each of the crosses. The origin of the 6:2 asci and the majority-minority relationships are thus established in eleven out of the fifteen possible crosses between these six mutants. The results obtained allow the order of the sites to be

[1] It has been found recently that there exist tetrads as a result of conversion in which the recombinants are double mutants.

arranged in such a way that, for a given cross, the left site is always the majority site. This order is as follows: *865* or *19* → *1844* → *622* or *218* → *55*. Although the relative positions of *865* and *19*, on the one hand, and *622* and *218*, on the other hand, are not defined by these experiments, the knowledge of the majority-minority relationships in only six crosses (*865* × *1844*, *19* × *1844*, *1844* × *622*, *1844* × *55*, *622* × *55*, and *218* × *55*) was sufficient to establish this order. The results given by the five additional crosses confirm this order.

If this segment is really a polaron, confirmatory evidence should be provided by the ratios of the frequencies of 6:2 asci in the various crosses. These frequencies being very low, the counts of 8:0 and 6:2 asci are given in Table 13. It is clear that these results agree with the previously established order. The latter becomes even more precise because the positions of *1844* and *622* between the extremely close sites *865* and *19*, on the one hand, and *218* and *55*, on the other, is now clear.

TABLE 13
Number of 8:0 and 6:2 Asci from Crosses between Various Mutants of Groups B and C of Series 19*

Mutant	Asci	*865*	*19*	*1844*	*622*	*218*	*55*
865	8:0	>72,000	41,000	27,000	35,300	12,500	20,208
	6:2	0	1	6	15	16	75
19	8:0		>69,500	23,960	16,000	6,480	10,760
	6:2		0	3	2	8	16
1844	8:0			>90,700	20,000	9,960	14,740
	6:2			0	3	3	4
622	8:0				>87,000	13,400	30,800
	6:2				0	3	7
218	8:0					>75,000	>29,500
	6:2					0	1
55	8:0						>77,500
	6:2						0

* For crosses with low frequency of 6:2 asci the counting technique described in Table 2 was used.

VII. Tentative Interpretation of the Properties of the Polaron

Any interpretation of the properties of the polaron must (1) account for the fact that conversion is the only mechanism of recombination; (2) be compatible with the additivity of conversion frequencies (i.e., of 6:2

asci) between closely linked sites; and (3) explain the increase of the conversion frequency from one end to the other of the segment concerned.

The observed properties can be accounted for by the two assumptions (a) that the 3:1 segregations result from copy-choice during chromosome replication at meiosis, and (b) that this replication proceeds progressively

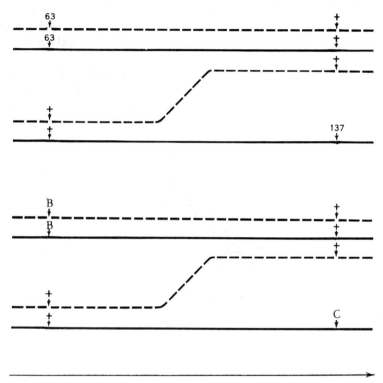

FIG. 3. Interpretation of the properties of the polaron as inferred from the analysis of segments *63-137* (Series 46) and B-C (Series 19).

from one end of the polaron toward the other. This interpretation, applied to the polaron studied in Series 46 and the one involving the segment B-C in Series 19, is schematically represented in Fig. 3.

This figure explains the origin of a 6:2 ascus and the fact that the left sites and the right sites are, respectively, the majority (segregation 2:2) and the minority (segregation 1:3) sites. If the direction of replication is taken into account, this figure also shows that 1:3 segregations are necessarily more frequent for the sites located on the right (*137*) than for those on the left (*63*). According to this interpretation, the absence of inversion of the majority parent means that the error of copying, once started, continues all the way down to the right end of the polaron; in other words, only a single event can occur in the polaron. The verification of this hypothesis would require a "three factor cross" within the polaron (for example, *63.137* × *W*), i.e., the availability of a double mutant that we do not have as yet. However, the fact that, in the course of the analysis of nearly 300 6:2 asci bearing on the three segments considered as polarons, only one tetrad involving an inversion of the majority parent in the cross B × C (Table 5) was found, possibly indicates that extremely rare exceptions do occur.

The following remarks may be added concerning this interpretation which, as a working hypothesis, has the advantage of simplicity. In the cross *63* × *wild type*, 1:3 and 3:1 segregations occur with a non-negligible frequency (Lissouba, 1961); is it to be inferred from this that *63* is at a certain distance from the left end of the polaron, or rather that "earlier" errors of copying continue along the length of the polaron? It seems that in both polarons *63-137* and B-C, the mutant sites are clustered at the ends, and that they are more numerous at the end from which the replication begins (region of Site *63* and Group B); however, it is as yet impossible to say whether this is really so or whether there is an unequal probability of copy errors along the chromosome segments concerned. Both these remarks lead to a consideration of the limits of the polaron and, more generally, of the relationship between polarons.

VIII. Linkage Structures and Relationships between Polarons

A. Results Concerning Mutant *277* of Series 46

Site *277* is close to site *137* since the frequency of 6:2 asci in cross *137* × *277* is equal to 0.8 per 10^3 tetrads at 25°C (seven 6:2 asci and 8747 8:0 asci). This new mutant at once showed two remarkable properites. When crossed with the other mutants of the same series, it gave not only 6:2 but also 7:1 asci, the structure of which will be discussed later. Although the frequency of the 7:1 asci is always lower than that of the 6:2 asci, it is not negligible, (for example, 38 7:1 asci were found with 177 6:2 asci in cross *63* × *277*, 24 with 48 6:2 asci in cross *46* × *277*, and 6 with 27 6:2 asci in cross *W* × *277*). However, they have not been used for the calculation of "distances" involving *277* (Lissouba, 1961). Among the 6:2 asci produced by crosses between *277* and the other mutants, some were found to result from conversions (this being thus far the general mechanism for this series), some others from crossing-over: 4 out of 11 in cross *63* × *277* and 2 out of 10 in cross *W* × *277* (Table 15).

The importance of the latter peculiarity is revealed by the comparison of the behavior of *277* and *137-138*. In the case of *137-138*, out of 107 tetrads from various crosses analyzed (Tables 10 and 12), none resulted from crossing-over. Moreover, all the previously known "distances" agreed with the localization of *277* to the right of *137-138*. Also the localization of crossing-over demonstrated in Series 19 had to be taken into account. These facts led one of us (Lissouba, 1961) to the conclusion that between *137-138* and *277* there exists a linkage structure. He discussed the role of such structures and analyzed certain aspects of negative interference.

The importance of the problem raised by this study, and the fact that the same problems are posed by Series 19, stimulated a fuller investigation of mutant *277*.

The "distances" between some new mutants and sites *137* and *277*, measured by the frequency of 6:2 asci, have been estimated. Table 14 lists most of the "distances" known at present between the following mutants: *1604* and *1546*; *665* and *1330*, both of which are homoalleles of *63*; the three mutants *1167*, *686*, and *1416*, which are very close to or homoalleles of *W*; and finally, *1021* and *1026*, which must have a certain length since they cover the whole *1604-1216* segment.

TABLE 14

Comparison of Frequencies of 6:2 Asci* Obtained from Crosses of Various Mutants with Mutants *137* and *277*

Mutant	137	277	Mutant	137	277
1604	20.84 ± 1.11	13.55 ± 1.12	1167	5.76 ± 0.86	4.87 ± 0.54
1546	18.17 ± 1.23	12.53 ± 1.02	686	7.56 ± 0.71	5.43 ± 1.21
188	9.33 ± 1.04	7.50 ± 2.47	1416	7.56 ± 1.06	4.23 ± 0.56
63	12.85 ± 2.47	14.50 ± 0.96	W	8.30 ± 0.67	9.88 ± 1.43
665	9.79 ± 1.41	13.53 ± 1.52	1216	8.35 ± 0.47	6.84 ± 1.10
1330	9.18 ± 1.59	8.26 ± 1.08	1021	2.25 ± 0.39	0.91 (0.62 to 1.33)
46	11.51 ± 1.15	10.06 ± 1.25	1026	2.13 ± 0.18	0.79 (0.60 to 1.03)

* Average number per 1000 tetrads.

For certain mutants and particularly those which have been used in the previously described experiments (Lissouba, 1961), the "distances" to *137* and to *277* are not statistically different. The distance from *1604*, *1546*, *1021*, or *1026* to *277* is on the contrary smaller than the "distance" to *137*. These results, taken as a whole, make it tempting to conclude that, contrary to what was previously assumed, the "distance" to *277* is smaller than that to *137*.

More 6:2 asci from crosses involving *277* were analyzed. Thus data from a total of fifty-four 6:2 asci are available. They are shown in Table 15.

This table shows the presence of anomalies in four of the five crosses studied. Some of the 6:2 asci from crosses 63×277 and $W \times 277$ result from conversion, others from crossing-over; inversions of the majority parent are observed in crosses 137×277 and 1216×277. Only six asci of crosses 46×277 have been analyzed and it is therefore difficult to conclude that neither crossing-over nor inversion of the majority parent take place.

TABLE 15
Genetic Structure of 6:2 Asci from Crosses with Mutant 277

Cross	No. of asci analyzed	Crossing-over	Conversion	Majority parent	Minority parent
63×277	11	4	7	63	277
46×277	6	0	6	46	277
$W \times 277$	12	2	10	W	277
1216×277	8	0	8	$1216(6)$ $277(2)$	$277(6)$ $1216(2)$
137×277	17	0	17	$277(16)$ $137(1)$	$137(16)$ $277(1)$
Total	54	6	48		

The results of three factor crosses involving the double mutant 63.277 and $137\text{-}138$ or W (Lissouba, 1961) should be recalled. They are given in Table 16. Ten 6:2 asci from crosses 63.277×138 have been analyzed: the eight strains from each ascus were individually crossed with strains 63, 277, and 138. Five of the eight asci were completely identified and found to have the same composition. The results concerning the other,

TABLE 16
Results of Crosses $63.277 \times W$ and $63.277 \times 137\text{-}138$

	Cross	
	$63.277 \times W$	63.277×137 63.277×138
Asci		
8:0	95,913	184,467
6:2	17	153
Number of 6:2 asci per 1000 tetrads	0.177 ± 0.042	0.828 ± 0.066

partially known, asci indicate no discrepancy with the former five. In each of these asci, two of the four products of meiosis are identical with parent *63.277*, one is identical with *138*, and the fourth is wild type. The only one 3:1 segregation detected occurred therefore at *138*.

B. Results Concerning Series 19

We were earlier led to believe (1) that on the chromosome segment corresponding to this series, the groups of sites were arranged in the order A → B → C; (2) that this segment comprised two polarons (sites A, on the one hand, and sites B and C, on the other); and (3) that crossing-over was localized in a linkage structure uniting these two polarons.

According to the hypothesis suggested, in order to account for the properties of the polaron, replication of the segment B-C proceeds in the direction B → C; so far there is nothing to indicate that the replication proceeds in the direction A → B because it is not known which site of polaron A is nearest to the linkage structure.

So far, only the occurrence of crossing-over in the crosses A × B and A × C has been discussed, but from the results shown in Tables 6 and 7 it can be seen that in these crosses many 6:2 asci resulted from conversion. The characteristics of the latter can be easily stated. In A × B crosses (Table 6), site B is generally the majority site since conversion occurs at the level of site A in 43 tetrads out of 51; only 8, 7 of which involve mutant *622*, show an inversion of majority parent. In A × C crosses (Table 7), site A is generally the majority site (230 out of 241 tetrads); the 11 inversions of majority parent have been found in crosses in which the number of asci analyzed is relatively large.

In addition to crosses A × B and A × C, 6:2 tetrads from the three factor cross A.C × B (*1678.324* × *484*) were analyzed, the eight strains from each ascus being crossed with the strains *1678*, *324*, and *484*. The five known asci have the same composition: one product of meiosis is wild type, a second one has the genotype of one of the parents (*1678.324*), and the remaining two have the genotype of the other parent (*484*). Therefore in these tetrads, a single 2:2 segregation was observed at site B (*484*) while two 1:3 segregations were observed at sites A (*1678*) and C (*324*).

C. Interpretation of the Results

1. *The Problem*

There are few problems concerning Series 19 that have not already been discussed. In the results of crosses made between mutants, the sites of which are localized on different polarons (A × B and A × C), it is necessary to account for site *A* being generally the majority one in cross *A* × *C* but the minority one in cross *A* × *B*. In the result of crosses *A.C* × *B*, we have to interpret the constitution of 6:2 tetrads where three products are parental (1 *A.C* and 2 *B*) and the fourth one wild type.

Series 46 raises apparently more complex problems due to mutant *277*. The first problem to solve is that of the localization of this site in relation to *137* which, we know, is very close (0.8×10^{-3}). A linkage structure is indicated by the occurrence of crossing-over in crosses *63* × *277* and *W* × *277*. Since no crossing-over was detected in crosses involving *137*, the linkage structure seems to lie between *137* and *277*. If this is so, crosses *137* × *277* can be regarded as similar to crosses *A* × *B* in Series 19.

However, several observations suggest that *277* is to the left of *137*. Certain sites, localized to the left of both, are more "distant" from *137* than from *277*, at least when the "distances" are measured by the frequency of 6:2 asci; for other sites, the "distances" are compatible with both possible positions (Table 14). All 6:2 asci from cross *137* × *277* result from conversion, and site *277* is the majority parent in 16 out of 17 tetrads (Table 15). Lastly, in cross *63.277* × *137* (Table 16), the frequency of 6:2 asci (per 1000 tetrads) is 0.8, i.e., equal to the corresponding frequency observed in cross *137* × *277* (page *364*).

What is the value of these arguments? Do the above notions provide an adequate interpretation of the facts revealed by the study of Series 46 and 19?

We know that the frequency of 6:2 asci is a complex value. In the case of crosses involving *137* it measures apparently only conversions while in crosses involving *277* it measures a mixture of conversion and crossing-over, the proportion of which is variable depending on the sites considered (Table 15). Furthermore, we know that crosses involving *277* produce 7:1 asci, but these have not been taken into account in calculating the "distances" for their origin is obscure. What then is the significance of comparisons of two distances, such as *63* → *137* and *63* → *277*, measured by the frequency of 6:2 asci? Finally, anticipating the following discussion and assuming that *277* is to the right of *137*, we are led to conclude that certain distances to *137* must be larger than the corresponding distances to *277*.

The fact that, in crosses *277* × *137*, *277* is majority parent may appear to be a more serious argument. It must be noted, however, (1) that *277* does not always behave in this way (Table 15); (2) that, in the crosses *1216* × *277* (see also Table 15) which involve sites that can be localized with certainty, either of the two parents can be the majority parent; (3) that if a linkage structure exists between *137* and *277* (postulated on the basis of the results obtained in crosses A × B of Series 19), *277* is expected to be the majority parent even if this site is on the right of *137* (see below, Fig. 4).

It is pertinent to consider whether the equality of the frequencies of 6:2 asci in crosses *137* × *277* and *63.277* × *137* is more than a coincidence and whether this equality indicates that *277* is on the left of *137*. If *277* is on the right, then each 6:2 ascus from cross *63.277* × *137* results from a double event. The distances $63 \rightarrow 137$ and $137 \rightarrow 277$ are, respectively, 12.85×10^{-3} and 0.8×10^{-3}. In the absence of interference, the frequency of 6:2 asci in such a "three factor cross" would be expected to be 0.010×10^{-3}; the observed frequency was found to be 0.8×10^{-3} which corresponds to a coincidence of 82.3. One may obviously consider this consequence as unlikely and, therefore, reject it with the conclusion that *277* is on the left of *137*. In our opinion such a conclusion is not justified for it accounts for only one part of the facts and does not agree with the others.

Whatever the relative positions of *277* and *137*, a very clear interference is observed also in crosses *63.277* × *W* where there is no doubt about the relative positions of the three sites: the frequency of 6:2 asci is 0.177 ± 0.04 (Table 16) while the frequency expected in the absence of interference is 0.028 (2.82 × 9.88, Fig. 2 and Table 14). If this observation is unrelated to that concerning cross *63.277* × *137*, it is difficult to understand its significance.

If, however, we assume that *277* is on the right of *137* and, consequently, that the two crosses *63.277* × *137* and *63.277* × *W* are very similar, an interesting and simple relationship emerges from these results. Assuming that each of the 6:2 asci produced in these crosses results from a double event (see Fig. 4), we can calculate the ratio between the frequency of single events ("two factor crosses") involving regions *63-137* and *63-W*, and the frequency of corresponding double events: in both cases the ratio is found to be close to 16 (Lissouba, 1961). Can this be a coincidence?

Irrespective of the relative positions of *277* and *137*, we face the following situation. Crosses involving *137* produce 6:2 asci, all of which result from conversion. In every one of these crosses, the majority parent is unique and it corresponds to the site on the left. Although investigations concerning *277* were much less extensive (Table 15), they revealed four cases of anomalies: in crosses *63* × *277* and *W* × *277* both crossing-over and conversion were found; in cross *1216* × *277* there was a tendency for *277* to be minority site; and in the cross *137* × *277*, *277* was almost always the majority site.

All these results can be accounted for by assuming that *277* is to the right of *137*. Moreover this interpretation also accounts for the observations concerning Series 19.

2. *The Bases of the Interpretation*

The interpretation already proposed concerning the polaron must be extended. It is necessary to take into account that crossing-over is localized on a structure of unknown nature which is situated between adjacent polarons; and also for the fact that, when a cross involves two sites located on the two sides of the linkage structure, the characteristics of the tetrads resulting from conversions are, on the one hand, relatively simple, and, on the other, related to the positions of the sites on the polaron. This scheme, which describes the results obtained in Series 19, seems to be the most suitable for Series 46, in which the second polaron is apparently marked only by site *277*.

It seems possible to account for the great majority of the observations if it is assumed (1) that replication proceeds not only from one end to the other of a polaron, but moreover, continues, without change of direction, through the linkage structure onto the adjacent polarons, and (2) that the switch postulated earlier and initiated on a polaron can also continue beyond the limits of the polaron (with a considerable tendency to switch back to the original model at the level of the linkage structure).

3. *Interpretation of the Observations Concerning Series* 46

This interpretation is diagrammatically represented in Fig. 4. Scheme a shows that site *63* and the nearby sites are majority sites with respect to *277*. Scheme b is analogous to the former and concerns site *1216* which is nearer to the linkage structure. Owing to the position of the latter, copy-errors occurring on the left side are more frequent, so that it is possible to detect switches back to the original model (Scheme e). This interruption brings about an inversion of the majority parent, an inversion which becomes the rule in cross *137* × *277* (Scheme d) where the probability of copy-error between *137* and the linkage structure (Scheme c) or between the latter and *277* is low but not nil.

Three factor crosses (e.g., *63.277* × *137*) are represented in Scheme f. The two events required for the production of a 6:2 ascus are two conversions involving the same chromatid; the second conversion implies a return to the original model at the level of the linkage structure and appears as a "half crossing-over." It will be noticed that in Schemes d and e also, each 6:2 ascus results from two events of which only the second (arrest of the aberrant replication and switch-back to the original chromatid) occurs in the marked region.

FINE STRUCTURE OF GENES IN *Ascobolus Immersus*

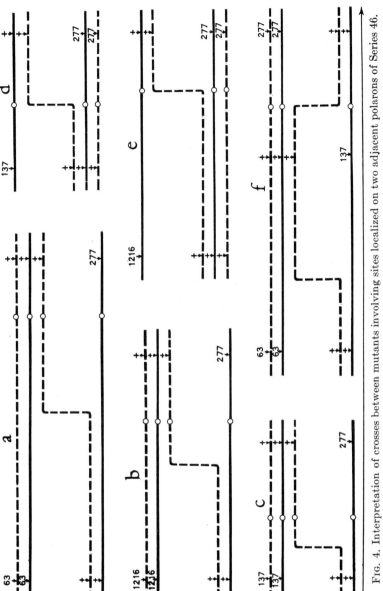

Fig. 4. Interpretation of crosses between mutants involving sites localized on two adjacent polarons of Series 46.

It should be emphasized that, in such an interpretation, the "distances" to the various sites of the series to *277* are complex, for they reflect a mixture of crossing-over and conversion for *63* and *W*, and a mixture of single and double events for *1216* and *137* (Table 15) with a predominancy of the latter when a site near to the linkage structure is involved. It is therefore understandable why, at least for certain sites, the "distance" to *137* is larger than the "distance" to *277*: in the latter case, the aberrant replications which have switched back to the original chromatids at the level of the linkage structure are not detected.

4. *Interpretation of the Facts Concerning Series 19*

This interpretation is shown diagrammatically in Fig. 5 in which most of the schemes are superimposable on those shown in Fig. 4. Scheme a shows that site A is the majority site in crosses A $\times$ C. Except in the cross A $\times$ *622* (this mutant being at the right in Group B and therefore located between B and C, see Tables 5 and 13), this mechanism (Scheme d) can only rarely occur in A $\times$ B crosses owing to the proximity of both sites to the linkage structure. In these crosses, the 6:2 asci resulting from conversions generally involve an inversion of the majority parent. Two events in succession occur, the second of which, involving an arrest of the aberrant replication, occurs in the marked region (Scheme c). As expected, A $\times$ C crosses also give some 6:2 asci resulting from double events (Scheme b).

Scheme e represents the mechanism resulting in the production of 6:2 tetrads in cross A.C $\times$ B; as in the other cases where several events are involved, a single chromatid is recombined. In this case the occurrence of three events must be postulated: the mechanism which produces 6:2 tetrads (conversion) in crosses A $\times$ B, and the one producing the same type of tetrads in A $\times$ C (Schemes c and a).

5. *Structure of the Genetic Material*

In Series 19 and 46, if only the better known sites are considered, the corresponding chromosome segment is apparently composed of two contiguous polarons separated by a linkage structure. In both cases we were able to account for the observations by assuming a single polarity reflecting the progressive replication along the entire length of the segment. We had also to postulate the intervention of events occurring outside the studied region but compatible with this conception.

The question arises as to whether this polarity can be extended to larger segments. The investigation of Series 75 was undertaken in order to answer this question. If crossing-over and inversion of the majority parent

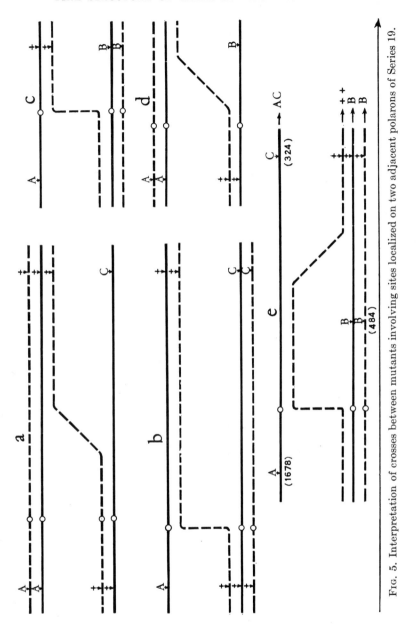

Fig. 5. Interpretation of crosses between mutants involving sites localized on two adjacent polarons of Series 19.

can be taken to reflect the existence of a linkage structure, then there are at least five polarons in this series (Table 1), a figure that is known to be an underestimate (unpublished results). This seems to hold also for Series 12 (Mme. N. Job-Engelmann, unpublished results). Table 1 and Fig. 1 (concerning Series 75) show that, in the tetrads resulting from conversions, the majority site is most often the one on the left; this means that the polarity does actually extend over the whole segment. The obviously required more precise investigations are in progress.

It can also be seen from Table 1 that, in the various crosses made, the relative frequencies of conversion and crossing-over are quite variable. This situation is observed also in A × B (Table 6) and A × C (Table 7) crosses of Series 19 where these variations do not appear to be due solely to the position of the sites. We are confronted here with a question to which no satisfactory answer can be given at present.

IX. Occurrence of Odd Segregations

Odd segregations (5:3), implying "impurity of the gametes," have already been observed in *Sordaria* (Olive, 1959). We have observed them in *Ascobolus*, in crosses of wild type with mutant, but their genetic analysis was not carried out.

In the cases reported here, these anomalies appeared as 7:1 segregants in crosses between two mutants of the same series, but only when particular mutants were involved. Several mutants of Series 75 can produce these odd segregations, whereas in Series 46, only *277* and *1216* appear to have this property, and in Series 19 it is exhibited by only mutant *60* and not even shared by its homoallele *65*.

These three mutants (i.e., *277*, *1216*, and *60*) yield 7:1 segregants when crossed with any of the well-known mutants of the same series, irrespective of the relative positions of the parents on the same or on adjacent polarons. However, when mutant *277* is crossed with *137*, no 7:1 asci are observed while there is a relatively high frequency of such segregations when *277* is crossed with the other mutants of the series.

The genetic analysis of some thirty 7:1 asci from crosses involving either mutant *60* or mutant *277* has been carried out. They all have the same general composition, viz., the black spore is wild type, three of the white spores have the genotype of *60* or *277*, and the four remaining white spores are identical to the second parent. In other words, the anomaly reflects a 5:3 segregation at the sites *277* or *60*.

Although it is clear that this type of anomaly is limited to certain mutants, it is as yet impossible to say if this property depends on the localization of these mutants or on their nature. The absence of an obvious relation between the frequencies of 6:2 and 7:1 asci in a given cross suggests that the mechanisms by which these two types of tetrads are produced are different.

X. Conclusions

Two essential problems were posed at the beginning of this paper. One concerns the fine structure of the gene, which is being thoroughly investigated in a number of organisms (Benzer, 1957; Demerec, 1956; De Serres, 1958; Forbes, 1956; Giles, 1958; Green and Green, 1956; Leupold, 1958; Pritchard, 1960; Strickland, 1958). The second problem can only be approached if the first one is, at least in part, solved, for it amounts to situating the gene on the structure as conceived by the biochemist and the biophysicist. The solution of these problems would obviously be facilitated if a genetic unit corresponding to the DNA molecule could be recognized and defined.

In each of the series studied, we have shown that, in a tetrad from a cross between two mutants, wild recombinants can result from two types of events, viz., crossing-over giving reciprocal recombinants, and conversion yielding non-reciprocal recombinants corresponding to only one of the four products of meiosis. There is no doubt as to the occurrence of tetrads containing only double mutant recombinants; the pink spore mutants of Series 46 will perhaps make their detection easier.

In each of these series, and even in Series 75, which involves a quite long chromosome segment, conversions are distinctly more frequent than crossing-over.

Crossing-over and conversion are differentiated by other characteristics as well. The variation of their frequency with temperature (Q_{10}) is different (Lissouba, 1961), but, above all, their localization is not the same. Evidence has been presented for a genetic unit termed *polaron* (Lissouba and Rizet, 1960); it is a linear structure on which mutant sites are located and within which only non-reciprocal exchanges (conversions) can occur. It was shown that these genetic units were polarized. They have been found to occur in both of the best investigated series.

In no case does the polaron represent the total genetic material of a series. In the simplest case (Series 46 and Series 19), the chromosomal segment concerned appears to comprise at least two contiguous polarons separated by a region tentatively designated as "linkage structure." Because only conversion can occur on polarons, "linkage structures" were postulated to account for the localization of crossing-over.

The term "conversion," as used here, has a more precise meaning than that defined by Lindegren (1953). All reported observations concerning conversion can be interpreted by the well-known scheme of copy-choice implying a double replication of one of the chromatids while the corresponding part of the other is not replicated. Conversion is therefore nothing but an accident in the replication of the chromatid, the characteristics of this replication being, of course, the main problem. It would appear therefore that replication starts at one end of the polaron and proceeds toward the other end; the "linkage structure" apparently represents a point of possible switch-back to the original chromatid; but replication can proceed beyond it and on to the contiguous polaron without change of direction. Observations on Series 75 suggest that there exists a single polarity on a sequence of several polarons; it is obviously necessary to look for this polarity on longer segments and on whole chromosome arms (Taylor, 1960). If this is so, it will be interesting to know if replication begins at one end of the chromosome or in the region of the centromere.

The view that a linkage structure constitutes a region of possible switch-back raises problems and brings interesting indications. The arrest of an aberrant replication and a switch-back at a linkage structure is expressed as a half cross-over and this again raises the question of relationship between conversion and crossing-over. The aberrant replication covers a certain distance but the latter corresponds only to a short segment; it is not known whether it can extend beyond the limits of a locus, but it may be thought that it cannot produce chromatids of a nonparental type over a larger part of their length; such a situation has recently been described in studies of phage recombinations (Kellenberger et al., 1961; Meselson and Weigle, 1961). Formally, therefore, a conversion implies the succession of two events within a very short distance, and this manifests itself experimentally as negative interference (three factor crosses of Series 46).

However we cannot account in as simple a manner for all the facts related to negative interference. In the case of crosses A.C $\times$ B in Series 19 (Fig. 5e) we had to imagine two successive copy-errors, and it appeared therefore that the occurrence of a first accident increases the probability of a similar second accident in the immediately adjacent region as the replication of the same chromatid continues. When the "distances" $A \to C$ and $A \to B + B \to C$ (Table 4) are compared, the previously noted anomaly (large size of the distance $A \to C$) is not easily explained, even if the origin of the 6:2 asci in crosses $B \times C$, $A \times B$, and $A \times C$ is taken into account (Tables 5, 6, and 7). In $A \times C$ crosses the absolute frequency of crossing-over is much higher than in $A \times B$ crosses even though sites B and C are on the same polaron and differ only by their position. In the same $A \times C$ crosses, the frequency of switches detectable on the segment B-C (6:2 asci from conversions in which A is the majority parent) is considerably higher than the frequency of the same events occurring on the same segment and detected in B × C crosses. This double increase in frequency is possibly due to a unique and at present unknown determinism.

This however does not detract anything from the essential conclusion, namely, that the chromosomal segment corresponding to each locus is made up of a sequence of polarons; these polarons are the zones of recombinations which appear "as miscopying of small segments of genetic material" (Beadle, 1957) and never as reciprocal exchanges; these units are linked by particular structures within which crossing-over is localized.

Is it possible to superimpose these results onto a biochemical or biophysical model? We shall not go beyond pointing out briefly the parallelism between the model arrived at here and the duplex linear model suggested by Freese (1958) and Taylor (1958): the latter was based not only on the intragenic recombinations known at that time, but also on the structure of the DNA molecule (two complementary chains), together with information obtained from autoradiographic studies (Taylor, 1958) and the mechanical requirements of the unwinding of helices during replication.

The authors visualize the chromosome as a sequence of "particles" which can be regarded as DNA molecules linked by "strong-bonds" (Freese) or "protein-links" (Taylor). No precise hypothesis is formulated concerning the structure of these bonds, except that only one helix of each adjacent DNA molecule is involved in the link (Fig. 6).

In the opinion of Taylor, crossing-over takes place at the level of the "protein-links." The replication is supposed to proceed along "DNA particles"; it can jump over the bonds, but at this point, it can also give rise to a reciprocal exchange, or no exchange at all. Within the DNA molecules themselves, only conversion would occur as results of switches between the models. Finally Taylor considers the genetic consequences of this model and in particular predicts 3:1 and 7:1 segregation. An extended discussion is not needed here to show the obvious analogies between the Freese and Taylor models and our observations (see Lissouba, 1961); our "polarons" can obviously be compared to the "particles" or to the DNA molecules of Taylor, the duplex structure of which is revealed in *Ascobolus* by the occurrence of the odd segregations of the 7:1 asci. Our linkage structures are the equivalent of the "protein-links" since they have the same properties. It must be emphasized that they were independently postulated from different considerations. As to 3:1 segregations in *Ascobolus*, most of them correspond to those foreseen by Taylor.

It is impossible to say that all the predictions have been verified, nor that they will be verified in the future. We have seen, for instance, that contrary to the predictions, 7:1 segregations are not always very rare and that they are characteristic of certain mutants.

In this paper an attempt was made to summarize all the results at present available at the risk of making the reading difficult. Interpretations of the results have been offered but not all have the same degree of solidity: some have to be completed, others may have to be revised as a result of investigations under way. Some very important problems were

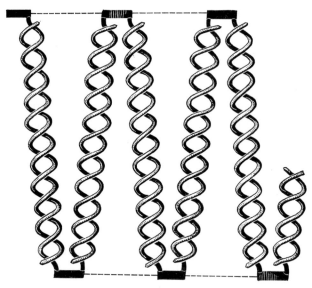

Fig. 6. Schematic representation of the linear duplex structure of a chromosome (after Taylor, 1958).

not even discussed in this study. We shall only mention that of the relation between our units, the *polarons*, and the functional units or *cistrons*. Such an analogy, however, does not seem unlikely if it is recalled that both white and pink spore mutants occur within a single series (Series 46 and Series 19); and that polynucleate spores are sometimes produced, the visible characters of which may provide evidence concerning the presence or absence of complementation. Such questions can now be approached without awaiting the definitive solution to the problems already posed.

Acknowledgments

Professors B. Ephrussi, P. L'Heritier, and M. Delbrück kindly read and discussed the manuscript of the present paper and made many helpful suggestions. The translation of the French manuscript is due to the cooperative effort of Dr. L. O. Butler,

M. M. Sicard, and Prof. B. Ephrussi. To all these colleagues we wish to express our sincere appreciation.

REFERENCES

Beadle, G. W. 1957. The role of the nucleus in heredity. *In* "The Chemical Basis of Heredity" (W. D. McElroy and B. Glass, eds.), pp. 2–22. Johns Hopkins Press, Baltimore, Maryland.

Benzer, S. 1957. The elementary units of heredity. *In* "The Chemical Basis of Heredity" (W. D. McElroy and B. Glass, eds.), pp. 70–93. Johns Hopkins Press, Baltimore, Maryland.

Benzer, S. 1959. On the topology of the genetic fine structure. *Proc. Natl. Acad. Sci. U.S.* **45**, 1607–1620.

Bistis, G. 1956. Studies on the genetics of *Ascobolus stercorarius* Schrot. *Bull. Torrey Botan. Club* **83**, 35–61.

Case, M. E., and Giles, N. H. 1958. Evidence from tetrad analysis for both normal and aberrant recombination between allelic mutants in *Neurospora crassa*. *Proc. Natl. Acad. Sci. U.S.* **44**, 378–390.

Demerec, M. 1956. A comparative study of certain gene loci in *Salmonella*. *Cold Spring Harbor Symposia Quant. Biol.* **21**, 113–122.

De Serres, F. J. 1958. Recombination and negative interference in the *ad-3* region of *Neurospora crassa*. *Cold Spring Harbor Symposia Quant. Biol.* **23**, 111–118.

Forbes, E. C. 1956. Recombination in the *pro* region of *Aspergillus nidulans*. *Microbial Genet. Bull.* **13**, 9–11.

Freese, E. 1958. The arrangement of DNA in the chromosome. *Cold Spring Harbor Symposia Quant. Biol.* **23**, 13–18.

Giles, N. M. 1958. Mutations at specific loci in *Neurospora*. *Proc. Intern. Congr. Genet. 10th Congr. Montreal, 1958* **2**, 261–279.

Green, M. M., and Green, K. C. 1956. A cytogenetic analysis of the lozenge pseudoallele. *Z. Induktive Abstammungs-u. Vererbungslehre* **87**, 708–721.

Kellenberger, G., Zichichi, M. L., and Weigle, J. J. 1961. Exchange of DNA in the recombination of bacteriophage. *Proc. Natl. Acad. Sci. U.S.* **47**, 869–878.

Lederberg, J. 1955. Recombination mechanism in bacteria. *J. Cellular Comp. Physiol.* **45** (Suppl. 2), 75–107.

Leupold, U. 1958. Studies on recombination in *Schizosaccharomyces pombe*. *Cold Spring Harbor Symposia Quant. Biol.* **23**, 161–169.

Levinthal, C. 1956. The mechanism of DNA replication and genetic recombination in phage. *Proc. Natl. Acad. Sci. U.S.* **42**, 394–402.

Lindegren, C. C. 1953. Gene conversion in *Saccharomyces*. *J. Genet.* **51**, 625–647.

Lissouba, P. 1961. Mise en évidence d'une unité génétique polarisée et essai d'analyse d'un cas d'interférence négative. Thèse Sciences, Paris; 1960. *Ann. sci. nat. Botan. et Biol. végétale* **44**, 641–720.

Lissouba, P., and Rizet, G. 1960. Sur l'existence d'une unité génétique polarisée ne subissant que des échanges non réciproques. *Compt. rend. acad. sci.* **250**, 3408–3410.

Meselson, M., and Weigle, J. J. 1961. Chromosome breakage accompanying genetic recombination in bacteriophage. *Proc. Natl. Acad. Sci. U.S.* **47**, 857–868.

Olive, L. S. 1959. Aberrant tetrads in *Sordaria fimicola*. *Proc. Natl. Acad. Sci. U.S.* **45**, 727–732.

Pritchard, R. M. 1960. The bearing of recombination analysis at high resolution on genetic fine structure in *Aspergillus nidulans* and the mechanism of recombination in higher organisms. *Symposium Soc. Gen. Microbiol.* **10**, 155–180.

Rizet, G. 1939. Sur les spores dimorphes et l'hérédité de leur caractère chez un nouvel *Ascobolus* hétérothallique. *Compt. rend. acad. sci.* **208**, 1669–1671.

Rizet, G., Lissouba, P., and Mousseau, J. 1960a. Sur l'interférence négative au sein d'une série d'allèles chez *Ascobolus immersus*. *Compt. rend. soc. biol.* **11**, 1967–1970.

Rizet, G., Lissouba, P., and Mousseau, J. 1960b. Les mutations d'ascospores chez l'Ascomycète *Ascobolus immersus* et l'analyse de la structure fine des gènes. *Bull. soc. franç. physiol. végétale* **6**, 175–193.

Rizet, G., Engelmann, N., Lefort, C., Lissouba, P., and Mousseau, J. 1960c. Sur un Ascomycète intéressant pour l'étude de certains aspects du problème de la structure du gène. *Compt. rend. acad. sci.* **250**, 2050–2052.

Roman, H. 1958. Sur les recombinaisons non réciproques chez *Saccharomyces cerevisiae* et sur les problémes posés par ces phénomènes. *Ann. Génét.* **1**, 11–17.

Stadler, D. R. 1958. Gene conversion of cysteine mutants in *Neurospora*. *Proc. Natl. Acad. Sci. U.S.* **44**, 647–655.

Strickland, W. N. 1958. An analysis of interference in *Aspergillus nidulans*. *Proc. Roy. Soc.* **B149**, 82–101.

Taylor, J. M. 1958. The organisation and duplication of genetic material. *Proc. Intern. Congr. Genet. 10th Congr. Montreal, 1958* **2**, 63–78.

Taylor, J. M. 1960. Asynchronous duplication of chromosomes in cultured cells of Chinese hamster. *J. Biophys. Biochem. Cytol.* **7**, 455–465.

Watson, J. D., and Crick, F. M. C. 1953. The structure of DNA. *Cold Spring Harbor Symposia Quant. Biol.* **18**, 123–131.

Yu, C. C. 1954. The culture and spore germination of *Ascobolus* with emphasis on *Ascobolus magnificus*. *Am. J. Botany* **41**, 21–30.

AUTHOR INDEX

Numbers in italics indicate the page on which the references are listed.

A

Acton, A. B., 26, 52, *87*
Adams, J., 109, 115, 121, 122, 124, 128, 129, 133, *137, 142*
Adelberg, E., 103, 124, 129, 134, *137, 139, 140, 144*
Aird, I., 38, *87*
Akdik, S., 41, *96*
Akiba, T., 106, *137*
Alexander, H. E., 274, *341*
Alfoldi, L., 104, 125, *137*
Alikhanian, S. I., 275, *339*
Allan, T. M., *270*
Allen, G., 244, *265*
Allen, S. L., 56, *87*
Alpinus, P., 148, *229*
Ames, B. N., 62, 65, *87*, 334, *340*
Anagnostopoulos, C., 59, 62, *87*
Anderson, D. L., 277, 303, 331, 332, *340*
Anderson, E. G., 51, *87*
Anderson, T. F., 103, 105, *137*, 339, *340*
Appleyard, R., 109, 112, 113, *138, 145*
Arber, W., 113, 114, 125, 127, 128, 134, *138*
Arditti, R. R., 306, *340*
Aristotle, 241, *265*

B

Baerecke, M., 174, *229*
Bailey, N. T. J., *340*
Balbinder, E., 62, *90*, 104, 123, 134, *139*
Banič, S., 62, *90*, 134, *139*
Barker, J. F., 44, *87*
Barker, J. S. F., 263, *266*
Barksdale, L., 109, *142*
Baron, L., 104, 129, 130, 133, *138, 143, 139*
Barr, M. L., 243, *266, 268*
Barratt, R. W., 62, *87*

Bartlett, H. H., 148, 189, *229, 230*
Bateman, N., 37, *87*
Bateson, W., 242, *266*
Baylor, M. B., 56, *87*
Beadle, G. W., 41, 42, 53, *88*, 377, *379*
Beasley, A. B., 37, *91*
Beck, S. L., 258, 259, *270*
Belling, J., 156, *230*
Belser, W., 104, *138*
Bennett, J. H., 66, *88*, 263, *266, 270*
Bentall, H. H., 38, *87*
Benzer, S., 40, 65, *88*, 343, 346, 375, *379*
Bernstein, H., 103, *138*
Bertani, G., 102, 104, 109, 110, 111, 115, 119, 120, 121, 124, 125, *138*
Bertani, L. E., 56, *87*, 121, *138*
Bertram, E. G., 243, *266*
Betz-Bareau, M., 104, 105, *140*
Bhaduri, P. N., 160, *230*
Bhaskharan, K., 274, *340*
Binet, F. E., 66, *88*
Bishop, D. W., 245, *269*
Bistis, G., 345, *379*
Blakeslee, A. F., 156, *230, 231*
Boam, T. B., 10, 11, 12, 17, 18, 19, 21, 27, 29, 50, *98, 99*
Bodmer, W. F., 6, 20, 24, 25, 34, 40, 48, 71, 72, 74, 75, 85, *88, 96*, 261, *266*
Boedijn, K., 190, 191, 199, 200, 201, *230, 232*
Boice, L., 110, *138*
Bonner, D. M., 55, *88*
Borisova, L. N., 275, *339*
Boyd, J., 121, *138*
Bradley, S. G., 275, 277, 279, 291, 303, 330, 331, 332, *340*
Braendle, D. H., 275, 276, 277, 279, 330, *340*
Breed, R. S., 274, 276, *340*
Breese, E. L., 14, 15, 16, 17, 20, 29, *88*

Bridges, C. B., 4, 48, 49, 51, *88*, 243, 258, 263, *266*, *268*
Brittingham, W. H., 156, 202, *230*, *231*
Brncic, D., 26, *88*
Brown, S. W., 40, *88*
Bunak, V. V., 246, *266*
Bunting, M., 104, *138*
Burla, H., 22, *91*
Burns, S., 124, 129, *137*
Burrous, J., 104, 109, 115, 121, 122, 124, 133, *142*
Buttin, G., 134, 135, *138*

C

Cain, A. J., 35, *88*
Calef, E., *138*
Campbell, A., 113, 115, 119, 123, 127, *138*, *141*
Carey, F., 104, 129, 133, *138*
Carlson, E. A., 40, *88*
Carson, H. L., 22, *88*
Carvajal, F., 274, *340*
Case, M. E., 344, *379*
Casinovi, C., 276, *342*
Castle, W. E., 48, *88*
Catcheside, D. G., 46, 58, *89*, 157, 165, 166, 174, 176, 181, 182, 191, 192, 196, 197, 198, 202, *230*, 280, *340*
Cavalli, L. L., 12, 56, *89*, 102, 104, *139*
Chimênes, A., 103, *139*
Chung, C. S., 38, *89*
Ciocco, A., *270*
Clark, A., 130, *139*
Clark, R. D., 258, *270*
Clarke, C. A., 35, 36, *89*
Cleland, R. E., 152, 153, 154, 155, 156, 157, 165, 190, 191, 199, 205, 217, 221, 227, *230*, *231*, *236*
Clowes, R. C., 62, *90*, 105, 109, 132, 135, *143*
Cohen, B. H., *270*
Cohen, D., 115, 117, 125, *139*
Colombo, B., 246, *266*
Coonradt, V. L., 53, *88*
Cooper, J. P., 20, *89*
Correns, C., 169, *231*
Cox, E., 125, *142*
Crawford, I. P., 59, 62, *87*
Crew, F. A. E., 48, *89*, 242, 261, *266*

Crick, F. M. C., 343, *380*
Crosby, J. L., 33, *89*
Crowe, L. K., 35, 40, *89*, *94*, 204, *231*
Curtiss, R., III, 134, *139*

D

da Cunha, A. B., 22, 23, *89*, *91*
Dahlberg, G., 246, *266*, *270*
Darlington, C. D., 2, 3, 32, 40, 41, 42, 44, 45, 46, 64, 84, *89*, 156, 160, 163, 165, 166, 175, 179, 198, *231*, 240, *266*
Darwin, C., 259, *266*
Datta, N., 129, *144*
Davidson, W. M., 243, 244, *266*
Davis, B. M., 189, *231*
Dawson, G. W. P., 62, *89*
Day, P. R., 35, *89*
de Haan, P., 104, *139*
De Lamater, E. D., 286, *340*
Delbrück, M., 131, *139*, 329, *342*
de Margerie, H., 125, *143*
Demerc, M., 56, 58, 62, 63, *90*, 104, 134, *139*, *143*, 281, 335, *340*, 375, *379*
Demerec, Z., 63, *90*
De Serres, F. J., 375, *379*
Detlefsen, J. A., 51, *90*
de Vries, H., 148, 176, 183, 188, 189, 191, 192, 198, 199, 200, 201, 208, *231*, *232*
De Winton, D., 33, 34, *95*
Digby, L., 153, *232*
Diver, C., 35, *90*, *92*
Djordjevic, B., 62, *90*
Dobzhansky, T., 22, 23, 24, 27, 28, 29, 30, 31, 51, 70, *89*, *90*, *91*, *98*, *100*, 258, 263, *269*
Dodge, B. O., 53, *91*
Doncaster, L., 242, 243, *266*
Doty, P., 130, *143*
Dowrick, G. J., 48, *91*
Dowrick, V. P. J., 32, *91*
Driskell, P., 103, 134, *139*
Dubinin, N. P., 23, *91*
Dunn, L. C., 37, *91*, 263, *266*, *268*
Düsing, C., 248, *266*

E

East, E. M., 204, *232*
Eberhart, B. M., 55, *91*

AUTHOR INDEX

Edwards, A. W. F., 247, 251, 253, 254, 257, 261, 262, 263, *266*, *267*, *270*
Edwards, J. H., *270*
Egawa, R., 105, *143*
Elliott, C. G., 48, *91*
Emerson, S. H., 156, 158, 171, 173, 174, 177, 179, 191, 192, 198, 201, 204, 205, *232*, *233*
Emsweller, S. L., 41, *91*
Engelmann, N., 346, *380*
Englesberg, E., 62, *92*
Ephrati-Elizur, E., 62, *91*
Ephrussi, B., 103, *139*
Epling, C., 22, *90*
Erikson, D., 276, *340*
Ernst, A., 32, 33, *91*
Eversole, R. A., 48, *91*

F

Falconer, D. S., 21, *91*, 258, *267*
Falkow, S., 129, 130, *139*, *143*
Fancher, H. L., *271*
Fein, A., 59, 62, 65, *92*
Fincham, J. R. S., 50, *91*
Fisher, R. A., 2, 4, 16, 21, 32, 35, 36, 38, 39, 48, 66, 70, 75, 76, 84, *91*, *92*, 240, 246, 250, 259, *267*, *271*
Fling, M., 55, *93*
Forbes, E., 62, *92*
Forbes, E. C., 375, *379*
Ford, C. E., 45, *92*, 243, 244, *267*
Ford, E. B., 35, 38, *92*
Fox, A. S., 55, *92*
Fraccaro, M., 253, *267*
Fraser, D., 109, 115, 121, 122, 124, *142*
Fraser, F. C., 49, *98*
Fredericq, P., 104, 105, 120, 125, 134, *139*
Freese, E., 377, *379*
Frost, L. C., 50, *92*
Fuerst, C., 123, *141*
Fukasawa, T., 106, 126, 131, 133, 134, *144*
Fukushima, T., 106, *137*
Fung, S. C., 258, *267*

G

Gardella, C., 161, 164, *233*
Garen, A., 103, *144*
Garmise, L., 109, *142*

Garnjobst, L., 62, *87*
Garry, B., 62, 65, *87*, 334, *340*
Gates, R. R., 152, 172, 189, 199, 204, *232*, *233*
Geckler, L. H., *231*
Geerts, J. M., 152, 167, *233*
Geiringer, H., 66, *92*
Geissler, A., 249, *267*
Gershenson, S., 258, 263, *267*
Gibson, J. B., 17, 18, 20, 29, 73, 84, 85, *92*
Giles, N. H., 344, 375, *379*
Gini, C., 248, 249, 251, 256, *267*
Glanville, E. V., 62, *90*, 134, *139*
Glass, B., *270*
Glavert, A. M., 274, 276, 285, 287, 289, *340*, *341*
Gluecksohn-Waelsch, S., 37, *92*
Goldman, I., 62, *90*, 134, *139*
Goldschmidt, E., 133, *143*
Goldschmidt, R. B., 187, *233*, 243, *267*
Goodale, H. D., 4, 8, 10, 20, *92*
Goodgal, S., 134, *140*
Goodman, L. A., 255, *267*
Gordon, M. J., 245, *267*
Gowen, J. W., 42, 51, *92*, 258, *267*
Graham, M. A., *268*
Grant, V., 46, *92*
Graubard, M. A., 49, *92*
Gray, W. D., 55, *92*
Green, K. C., 375, *379*
Green, M. M., 50, *92*, 375, *379*
Gross, J. D., 62, *90*, *92*, 134, *139*
Gross, S. R., 59, 62, 65, *92*
Grüneberg, H., 58, *92*
Gunstad, B., 250, *267*

H

Hagen, C. W., Jr., 204, *233*
Hahn, E., 274, *341*
Håkansson, A., 166, 191, 192, 198, 199, 201, *233*
Haldane, J. B. S., 24, 72, *92*
Hamerton, J. L., 45, *92*
Hammond, B. L., 217, *231*
Hamon, Y., 105, 120, *140*
Hansen, H. N., 54, *92*
Harada, K., 105, *140*, *143*
Harris, H., 256, *267*

Harris, J. A., 250, *267*
Harrison, B. J., 9, 13, 21, *93*, *95*
Hart, D., 246, *267*
Harte, C., 173, 174, *233*
Hartman, P. E., 58, 62, *90*, *93*, 129, *140*, 334, 335, *340*, *341*
Hartman, Z., 62, *93*
Haselwarter, A., 203, *233*
Hashimoto, K., 62, *90*, 105, 134, *139*, *143*
Haskell, G., 14, *93*
Haskins, F. A., 55, *93*
Haubenstock, H., 258, 259, *270*
Hayes, W., 102, *140*, 274, 339, *341*, *342*
Hayman, D. L., 47, 49, *93*
Heath, C. W., *271*
Hecht, A., 204, 205, *233*
Hemphill, R. E., 257, *267*
Heribert-Nilsson, N., 170, 189, *234*
Hershey, A., 108, *140*
Herzenberg, L. A., 62, 65, *87*, 334, *340*
Herzog, G., 196, 197, *234*
Heslot, H., 104, *139*
Hewitt, D., 252, *267*
Hildreth, P. E., 29, *93*
Hippocrates, 242, *267*
Hirota, Y., 103, 125, 126, 129, 130, 134, *140*
Hirsch, H. M., 55, *93*
Hoeppener, E., 168, 191, 216, *234*
Holloway, B. W., 104, *140*, 274, *341*
Holt, H. A., *271*
Holt, S. B., 21, *93*
Hopwood, D. A., 62, *93*, 274, 275, 276, 277, 278, 285, 287, 289, 290, 291, 297, 299, 300, 305, 310, *340*, *341*
Horiuchi, T., 133, 134, *140*, *143*
Horowitz, N. H., 55, *93*
Hotchkiss, R. D., 274, *341*
Hottinguer, H., 103, *139*
Howard, A., 258, *267*
Howarth, S., 62, *90*, 105, 109, 132, 135, *143*
Hurst, D. D., 56, *87*

I

Ikeda, Y., 275, 281, *342*
Ishidsu, J., 62, *90*, 104, *139*
Isshiki, Y., 106, *137*

J

Jacob, F., 48, 58, 62, 64, 65, *93*, *96*, *97*, 101, 102, 103, 104, 105, 106, 109, 110, 112, 114, 115, 116, 117, 118, 119, 120, 123, 124, 125, 128, 129, 134, 135, *137*, *138*, *140*, *141*, *142*, *145*, 274, 339, *342*
Jacobowicz, R., *272*
Japha, B., 163, 165, 166, 175, *234*
Jennings, H. S., 65, *93*
Jennings, P., 104, *140*
Jinks, J. L., 54, 56, *93*
John, B., 45, *93*
Johnsson, H., 41, *94*
Johnstone, J. M., *271*
Jones, H. A., 41, *91*
Jones, L. A., 331, 332, *340*

K

Kaiser, A., 113, 116, 118, 119, 128, *141*, *142*
Kajima, K., 3, 67, 69, 70, 79, *94*
Kalmus, H., 261, *267*
Kameda, M., 105, *140*
Kang, Y. S., 257, *268*
Kauffmann, F., 104, *143*
Kaufmann, B., 29, *99*
Kellenberger, G., 104, 117, 119, *141*, *142*, 376, *379*
Kelner, A., 275, *341*
Kemeda, M., *143*
Kimura, K., 106, *143*
Kimura, M., 3, 36, 66, 67, 69, 71, 72, 76, 77, 78, *94*
Kimura, S., 106, *137*
King, H. D., 258, 264, *268*
King, J. C., 29, *99*
Kirk, R. L., *272*
Kisch, R., 203, *234*
Kloepfer, H. W., *271*
Kohn, H. I., *271*
Koller, P. C., 48, *89*
Kolman, W. A., 261, *268*
Komai, T., 58, *94*
Koyama, T., 106, *137*
Kozloff, L., 107, *142*
Kutzner, H. J., 276, 337, *341*

L

LaCour, L. F., 45, *89*
Lahr, E. L., 62, *90*, 104, 134, *139*
Lamm, R., 41, *94*
Lancaster, H. O., 250, *268*
Langdendorf, J., 168, *234*
Lanni, F., 129, *142*
LaPlace, Marquis de, 248, *268*
Lavalle, R., 103, 105, *142*
Lawrence, M. J., 50, *94*
Lederberg, E. M., 62, 65, *94*, 102, 103, 104, 109, 125, 126, 133, *142*, *143*, 281, *341*
Lederberg, J., 40, *94*, 102, 103, 104, 105, 109, 125, 126, 132, 133, *142*, *143*, *144*, 274, 275, 281, 306, 328, 330, 339, *340*, *341*, *342*, *344*, *379*
Lee, B. T. O., 57, *94*, *96*
Lefort, C., 346, *380*
Leidy, G., 274, *341*
Leitch, A., 257, *267*
Lejeune, J., *271*, *272*
Lennox, E., 134, *142*
Leupold, U., 34, *94*, 375, *379*
Levan, A., 41, *94*
Levene, H., 24, 29, 30, 31, *90*, *91*, *94*, *98*
Levine, E. E., 50, *94*
Levine, M., 119, 125, *142*
Levine, R. P., 48, 50, *94*
Levinthal, C., 107, *142*, 344, *379*
Levitan, M., 25, 26, *94*
Lewis, B. M., 48, *99*
Lewis, D., 32, 34, 40, *94*, 206, *234*
Lewis, K. R., 45, *93*
Lewitsky, G. A., 160, *234*
Lewontin, R. C., 3, 26, 67, 69, 70, 79, *94*, 263, *268*
Li, K., 109, *142*
Licciardello, G., 112, *138*
Lieb, M., 104, 119, 121, 122, 126, *142*
Lienhart, R., 256, *268*
Lindahl, P. E., 259, *268*
Lindegren, C. C., 375, *379*
Linnert, G., 203, *235*
Lissouba, P., 346, 351, 355, 357, 360, 364, 365, 366, 375, *379*, *380*
Lively, E., 133, *142*
Loper, J. C., 62, *93*, 334, *341*
Lord, E. M., 259, *268*

Lowe, C. R., 245, *268*, *271*
Luria, S. E., 104, 109, 110, 115, 121, 122, 124, 128, 129, 133, *137*, *138*, *142*
Lutz, A. M., 152, 190, *234*
Lwoff, A., 102, 121, *142*

M

McClintock, B., 107, 136, *143*
McClung, C. E., 242, *268*
McDonald, W., 133, *143*
MacDowell, E. C., 259, *268*
McGunnigle, E. C., 29, *99*
McKeown, T., 245, *268*, *271*
McLaren, A., 258, *267*
MacLeod, H. L., 55, *93*
MacMahon, B., *271*
McWhirter, K. G., 258, *268*
Maccacaro, G. A., 12, 56, *89*
Madden, C. V., 29, *99*
Maeda, T., 41, *95*
Mahler, B., 62, *90*, 104, *139*
Mallyon, S. A., 48, *95*
Malogolowkin, C., 258, *268*
Mancinelli, A., 304, 305, 306, 310, 330, *342*
Marini-Bettolo, G. B., 276, *342*
Markert, C. L., 55, *95*
Marmur, J., 130, *143*
Marquardt, H., 160, 161, 163, 165, 166, 167, 202, *234*
Martin, W. O., *271*
Mather, K., 2, 3, 4, 6, 7, 8, 9, 10, 12, 13, 14, 15, 16, 17, 18, 20, 21, 22, 23, 27, 29, 30, 32, 33, 34, 35, 40, 41, 44, 45, 47, 49, 51, 52, 56, 64, 84, *88*, *93*, *95*, *98*, *99*
Mathew, N. T., *271*
Matney, T., 133, *143*
Mazé, R., 103, 105, 125, *137*
Melechen, N., 108, *140*
Mendel, G. J., 242, *268*
Meselson, M., 127, 131, *143*, *145*, 376, *379*
Michaelis, P., 202, *234*
Michie, D., 48, *95*, 258, *267*
Millicent, E., 17, 18, *95*
Mindlin, S. V., 275, *339*
Misro, B., 27, 29, *95*
Mitchell, H. K., 55, *93*, 334, *340*

Mitsuhashi, S., 105, *140*, *142*
Miyake, T., 62, *90*, 104, 134, *139*, *143*
Mizobuchi, K., 62, *90*, 104, *139*
Moffett, A. A., 48, *95*
Monod, J., 58, 62, 64, 65, *93*, *96*, 109, 115, 116, 117, 134, 135, *138*, *141*
Moody, J. D., 246, *267*
Moore, K. L., *268*
Morgan, T. H., 243, 258, 263, *268*
Morse, M., 126, *143*
Morton, N. E., 38, *89*
Moser, H., 62, *90*
Mousseau, J., 346, 355, *380*
Mukherjee, A. S., 51, *96*
Muller, H. J., 48, *96*, 148, *234*
Müntzing, A., 41, 44, *95*, *96*, *97*
Munz, P. A., 147, *234*
Murray, E. G. D., 274, 276, *340*
Murray, N. E., 62, 63, *96*
Myers, R. J., 249, *268*

N

Nanney, D., 137, *143*
Naylor, B., 43, *97*
Neel, J. V., *271*
Nelson, T., 133, *143*
Nevill, A., 275, *342*
Newcomb, S., 248, *268*
Newcombe, H. B., 275, *341*
Newmeyer, D., 62, *87*
Nice, S., 121, *138*
Novick, A., 133, 134, *140*, *143*
Novitski, E., 258, 264, *268*, *269*, *271*

O

Ochiai, K., 106, *143*
Oehlkers, F., 154, 155, 171, 177, 179, 183, 184, 185, 203, 204, *231*, *234*, *235*
Ørskov, F., 104, *143*
Ørskov, I., 104, *143*
Ogata, W. H., 62, *87*
Olive, L. S., 374, *379*
Ounsted, C., 252, *268*
Owen, A. R. G., 263, *269*
Owen, R. D., 55, *95*
Ozeki, H., 62, *90*, 105, 109, 125, 132, 134, 135, *143*

P

Paigen, K., 127, *145*
Painter, T. S., 243, *269*
Pandey, K. K., 40, *96*
Papavassiliou, J., 105, *140*
Parag, Y., 35, *96*
Pardee, A. B., 58, 62, 64, *96*
Parker Hitchens, A., 274, 276, *340*
Parkes, A. S., 249, 259, *269*, *271*
Parsons, P. A., 6, 20, 24, 25, 40, 47, 48, 49, 51, 71, 72, 74, 75, 85, *88*, *92*, *93*, *96*
Partridge, C. W. H., 55, *88*
Pateman, J. A., 57, *94*, *96*
Patterson, J. T., 48, *96*
Pavlovsky, O., 24, *90*
Payne, F., 8, 9, *96*
Penrose, L. S., 250, *269*
Perkins, D. D., 62, *87*, *96*
Perrin, D., 62, 64, 65, *93*, 117, *141*
Peto, F. H., 41, *96*
Piéchaud, M., 289, *341*
Pittenger, T., 306, *341*
Pitt-Rivers, G. H. L-F, 262, *269*
Plough, H. H., 47, 48, 51, *96*
Pohl, J., 199, *235*
Poisson, S. D., 248, *269*
Pontecorvo, G., 52, 53, 54, 58, 62, 63, *96*, *97*, 278, 281, 332, *342*
Prakken, R., 41, *97*
Preer, L. B., *231*
Pritchard, R. H., 306, *342*
Pritchard, R. M., 375, *379*
Pugh, T. F., *271*

R

Race, R. R., 39, 40, *97*, *271*
Raper, J. R., 35, *96*
Record, R. G., *271*
Redfield, H., 49, 51, *97*
Rees, H., 42, 43, 47, *97*
Rendel, J. M., 49, *97*
Renkonen, K. O., 253, *269*
Renner, O., 148, 151, 155, 158, 167, 168, 169, 170, 171, 173, 174, 175, 176, 177, 179, 186, 191, 192, 196, 197, 198, 208, 209, 210, 214, 216, *235*, *236*
Rethoré, M. O., *271*, *272*
Richter, A., 109, 110, *143*, *144*

AUTHOR INDEX

Rifaat, O. M., 50, *97*
Rife, D. C., 249, *269*
Rizet, G., 345, 346, 355, 360, 375, *379, 380*
Roberts, C. F., 62, *97*
Roberts, E., 51, *90*
Roberts, J. A. F., 38, *87, 97, 271*
Robertson, A., 251, *269*
Robinson, C. F., 287, *342*
Roman, H., 48, 62, *97*, 346, *380*
Roper, J., 106, *144*
Rothschild, 245, *269*
Rownd, R., 130, *143*
Rudloff, C. F., 168, 169, 202, *236*
Russell, L. B., 243, *270*
Russell, W. T., *271*

S

Saito, H., 275, 281, 330, *342*
Salzano, F. M., 25, 26, *94*
Sanchez, C., 62, 64, 65, *93*, 117, *141*
Sander, G., 258, *267*
Sandler, I., *268*
Sandler, L., 264, *269, 271*
Sanger, R., 39, 40, *97, 271*
Sanghul, L. D., *272*
Saunders, E. R., 38, *97*
Sawada, O., 106, *143*
Schaeffer, P., 106, 120, 124, *141*
Schötz, F., 211, 212, *236*
Schull, W. F., *271, 272*
Schultz, J., 49, 51, *97*
Schützenberger, M. P., 253, 254, *269, 270, 272*
Seitz, F. W., 163, *236*
Serman, D., 62, *93*, 334, *341*
Sermonti, G., 275, 276, 277, 278, 290, 304, 305, 306, 310, 330, *342*
Sharman, G. B., 45, *92, 97*
Shaw, R. F., 263, *269*
Sheppard, P. M., 35, 36, 74, 75, *88, 89, 97*
Shettler, L. B., 245, *269*
Shield, J. W., *272*
Shildkraut, C., 130, *143*
Shull, A. F., 48, *97*
Shull, G. H., 151, 172, 173, 177, 199, *236*
Siminovitch, L., 125, *141*
Sismanidis, A., 10, *98*
Six, E., 104, 110, 113, 115, 117, 123, 124, *144*

Skaar, P., 103, 109, *144*
Skellam, J. G., 251, *269*
Slater, E., 250, 257, *269*
Smith, C. A. B., 261, *267*
Smith, D. A., 62, *98*
Smith, D. R., 243, *266*
Smith-Keary, P. F., 62, *89*
Sneath, P., 129, 130, 134, *140*
Snyder, L. H., 249, *269*
Snyder, R. G., *272*
Snyder, W. C., 54, *92*
Sobels, F. H., 48, *98*
Spada-Sermonti, I., 275, 276, 277, 278, 290, 304, 305, 306, 310, 330, *342*
Spassky, B., 29, 30. *91, 98*
Spassky, N., 29, 30, *91, 98*
Spicer, C., 129, *144*
Spickett, S. G., 20, *92*
Spiess, E. B., 29, *98*
Spilman, W., 104, 129, 133, *138*
Spizizen, J., 274, *342*
Srinivasan, P. R., 62, *91*
Stadler, D. R., 344, *380*
Stadler, L. J., 48, *98*
Stahl, F., 112, 128, *144*
Stanier, R. Y., 276, *342*
Stebbins, G. L., 32, 45, 46, *98*
Steinberg, A. G., 49, *98*
Steiner, E. E., 207, 222, *236*
Stent, G. S., 131, *139*, 144
Stern, C., 47, *98*, 250, *269*
Stevens, N. M., 242, *269*
Stewart, A. M., 252,
Stinson, H. T., 220, *236*
Stocker, B., 105, 109, 120, 124, 125, 134, *143, 144*
Stomps, T. J., 199, *236*
Strickland, W. N., 62, *87*, 375, *380*
Strigini, P., 306, *340*
Stuart, J. R., 257, *267*
Stubbe, W., 202, 210, 212, 213, 214, 215, *236, 237*
Sturtevant, A. H., 7, 8, 22, 40, 48, 49, *90, 98*, 156, 158, 171, 173, 174, 177, *233, 258, 263, 268, 269*
Suche, M. L., 48, *96*
Sueoka, N., 55, *93*
Sundell, G., 259, *268*
Sutton, W. S., 242, *269*
Suzuki, M., 105, *140, 143*

Szilard, L., *272*
Szybalski, W., 275, 276, 277, 279, 330, *340*, *342*

T

Tatum, E. L., 48, 55, *91*, 102, *142*, 274, *341*
Taylor, A., 103, *144*
Taylor, J. M., 376, 377, *380*
Taylor, W., 249, *269*
Tebb, G., 29, *98*
Thoday, J. M., 10, 11, 12, 17, 18, 19, 20, 21, 26, 29, 45, 47, 50, 73, 84, 85, *92*, *95*, *98*, *99*
Thomas, M. H., 256, *269*
Thomas, R., 108, *144*
Thompson, J. B., 42, 43, 47, *97*, *99*
Thompson, J. S., *271*
Thomson, J. A., 242, *269*
Ting, R., 126, 129, 133, *142*, *144*
Tiniakov, G. G., 23, *91*
Tinker, H., 37, *91*
Tomizawa, J., 108, 133, *140*, *144*
Tonolo, A., 276, *342*
Tschetwerikoff, S. S., 27, 53, *99*
Tulasne, R., 289, *342*
Turpin, R., 253, 254, *270*, *271*, *272*

V

Van Overeem, C., 160, *237*
Vendrely, R., 289, *342*
Vermelin, H., 256, *268*
Vetukhiv, M., 30, 31, *99*
Vielmetter, W., 62, *90*
Vines, A., 32, *95*
Visconti, N., 329, *342*

W

Waaler, G. H. M., 249, *270*
Wahrman, J., 45, *99*
Wainwright, S. D., 275, *342*
Waksman, S. A., 274, 276, 277, 285, 337, *341*, *342*
Walker, C. B. V., *270*
Wallace, B., 24, 29, 31, *90*, *99*, 258, 263, *270*

Wallace, M. E., 49, *99*
Watanabe, T., 106, 125, 131, 133, 134, *144*
Watson, J. D., 343, *380*
Webb, J. W., 252, *267*
Weier, T. E., 165, 166, *237*
Weigle, J. J., 104, 117, 119, 127, 131, *141*, *142*, *143*, *145*, 376, *379*
Weir, J. A., 258, 259, *270*
Welshons, W. J., 243, *270*
Wexler, I. B., 39, *99*
White, M. J. D., 26, 45, 48, *94*, *99*
Whitfield, J., 112, *145*
Whittinghill, M., 48, *97*, *99*
Wiener, A. S., 39, *99*
Wiesmeyer, H., 135, *145*
Wigan, L. G., 10, 12, 13, 14, 29, 31, 56, *95*, *99*
Wilson, D., 129, *145*
Wilson, E. B., 242, *270*
Wilson, J. Y., 48, *99*
Winge, Ö., 37, *99*
Winkler, H., 186, *237*
Wisniewska, E., 160, 161, 165, *237*
Wolfe, H. G., 258, *270*
Wollman, E. L., 48, *93*, 101, 102, 103, 104, 106, 109, 110, 112, 114, 116, 118, 120, 123, 124, 125, *137*, *141*, *145*, 274, 339, *342*
Worms, R., 248, *270*
Wright, S., 3, 23, 66, 67, 70, 79, 80, *99*, *100*

Y

Yamanaka, T., 106, *143*
Yanofsky, C., 55, 59, 62, *88*, *100*
Yarmolinsky, M., 135, *145*
Yu, C. C., 345, *380*

Z

Zahavi, A., 45, *99*
Zamenhof, S., 62, *91*
Zichichi, M. L., 117, 119, *141*, *142*, 376, *379*
Zinder, N. D., 104, 109, 115, 117, 119, 121, 122, 128, 133, *145*, 274, *342*
Zohary, D., 40, *88*
Zürn, K., 203, *237*

SUBJECT INDEX

A

ABO blood groups, linkage and, 38–39
Alopecurus myosuroides, chiasma frequency in, 41
Apotettix, balanced polymorphism in, 32
Artificial selection, sex ratio and, 264–265
Ascobolus immersus,
 conversion,
  grouped mutants and, 349–352
  6:2 tetrad analysis and, 352–354
 crossing-over,
  grouped mutants and, 349–352
  6:2 tetrad analysis and, 352–354
 favorability of, 345–346
 genetic polarity in, 360–361
 odd segregations in, 374–375
 recombinations in, 346–349
 polaron of, 355–360, 361–374
 series 19 mutants of, 361–362, 367, 372–374
 series 46 mutants of, 355, 364–367, 370–372
Aspergillus, mycelial factor in, 106–107
Aspergillus nidulans,
 biochemical loci linkage in, 59, 60, 61, 62
 parasexual cycle in, 52, 53, 54–55
Auxotrophs, *Streptomyces coelicolor* and, 281

B

Bacillus subtilis, biochemical loci linkage in, 59, 60, 63
Bacteriophage, *see also* Lambda
 episomes in, 107–109, 110–120, 121–124, 125–135
 host range mutants and, 56–57
Balanced complex, evolution of, 81–84
Birth control, sex ratio and, 255–256
Blaberus discoidalis, recombination in, 45

C

Cepaea, balanced polymorphism in, 32
Cepaea nemoralis, linkage and, 35, 36
Chiasmata, recombination frequency and, 40–47
 recombination index and, 44–47
Chironomus, polygenic balance in, 26
Chlamydomonas reinhardi, recombination in, 48
Chloramphenicol, *Shigella*, resistance transfer factor and, 105–106
Chorthippus parallelus, chiasma frequency in, 44
Chromosomes,
 prophage, mode of attachment and, 111–114
 Streptomyces coelicolor and, 332–335
 structure, *Oenothera* and, 160–164
Chromosome theory, sex ratio and, 242–243
Cichorieae, recombination in, 45–46
Circles, chromosomal, *Oenothera* and, 156–160
Co-adaptation, polygenic balance and, 23
Colicin, episomes and, 104–105
Colonies, *Streptomyces coelicolor* and, 285–287
Competition,
 megaspore, *Oenothera* and, 167–169
 pollen tube, *Oenothera* and, 169–170
Crepis capilaris, chiasma frequency in, 42
Crossing-over,
 Ascobolus immersus,
  grouped mutants and, 349–352
  6:2 tetrad analysis and, 352–354
 Oenothera and, 172–180
Culture, *Streptomyces coelicolor* and, 277–279
Cytology,
 nuclear, *Streptomyces coelicolor* and, 287–290

Curing, episomes and, 125-126
  Oenothera and, 152-160

D

Datura, chromosomal circle formation and, 156
Diploidy, partial, Escherichia coli and, 132-134
Diplophase, Oenothera and, 165-167
Drosophila, sex determination and, 243
  sex ratio and, 258
Drosophila ananassae, crossing-over, selection and, 51-52
Drosophila funebris, polygenic balance in, 23
Drosophila guaramunu, polygenic balance in, 25-26
Drosophila melanogaster, 3, 4, 7, 8, 10-12, 13-20, 27, 29, 31
  chiasma frequency in, 42
  crossing-over, selection and, 51, 52
  recombination, variables and, 48-50
Drosophila pavani, polygenic balance in, 26
Drosophila persimilis, polygenic balance in, 29-30
Drosophila prosaltans, polygenic balance in, 29-30
Drosophila pseudoobscura, crossing-over in, 50
  polygenic balance in, 22, 28, 29-31
Drosophila robusta, polygenic balance in, 22, 25
Drosophila spp., recombination index and, 46-47
Drosophila willistoni, polygenic balance in, 22

E

Episomes,
  autonomous state,
    non-bacteriophage and, 109
    bacteriophage carrier state and, 108-109
  vegetative replication and, 107-108
  cellular regulatory mechanisms and, 135
  curing and, 125-126
  Aspergillus mycelial factor and, 106-107
  colicin factor(s) and, 104-105
  F agent and, 102-104
  maize controlling elements and, 107
  resistance-transfer factor and, 105-106
  sporulation factor and, 106
  fertility factors and, 102-104
  "gene pick-up," 136,
  F-mediated transduction and, 129-131
  phage-mediated transduction and, 126-129
  products and, 131-132
  homopolylysogens and, 110
  immunity,
    cytoplasmic nature and, 115-116
    genetic determination and, 116-120
    non-bacteriophage and, 120-121
    superinfection and, 114-115
  integrated-autonomous transition and, 124-125
  integrated state,
    attachment mode and, 111-114
    chromosomal localization and, 109-110
    lambda coliphage and, 108, 109, 110, 111, 112, 117, 118-119
  lysogenization,
    genetic incorporation and, 123-124
    infection and, 121-122
    non-bacteriophage and, 124
    not-to-lyse decision and, 122
    partial diploidy and, 132-134
    prophage and, 108-109, 110, 111-114
    repression and, 114-115, 116
    superinfection and, 108, 114-115
    transfer and, 134-135
Escherichia coli, 4, 12,
  biochemical loci linkage in, 59, 60, 61, 62, 64, 65
  chloromycetin-resistance in, 56
  colicin factor of, 104-105
  fertility factor of, 102-104
  partial diploidy in, 132-134
  phage lambda, episomes and, 108, 109, 110, 111, 112, 117, 118-119
  resistance transfer factor and, 106
Euoenothera, see Oenothera

SUBJECT INDEX 391

Evolution,
   Oenothera, 215–229
   supergenes, 35–36

F

Fertility,
   episomes and, 102–104
   Streptomyces coelicolor and, 303–304
Fritillaria imperialis, chiasma frequency and, 41

G

Gametophyte, *Oenothera* and, 167–169
Gene conversion,
   Ascobolus immersus,
      grouped mutants and, 349–352
      6:2 tetrad analysis and, 352–354
   Oenothera and, 183–187
Gene location, *Oenothera* and, 170–172
"Gene Pick-Up," episomes and, 126–132, 136
Genes,
   new combinations,
      polymorphic locus and, 71–73
      two loci and, 73–78
Genetics,
   Oenothera and, 148–152
   sex ratio and, 244–246
   study strains, *Streptomyces coelicolor* and, 276–277
Gerbillus, recombination in, 45

H

Heteroclones,
   Streptomyces coelicolor,
      analysis of, 308–310
      discovery of, 304–305
      linkage and, 310–321
      nature of, 321–332
      selection and, 305–310
Heterokaryons, *Streptomyces coelicolor* and, 330–331, 332
Heritability,
   human sex ratio and, 246–258
   mammalian sex ratio and, 259

Hfr,
   colicinogeny and, 104–105
   properties of, 103
Homopolylysogens, episomes and, 110
Human, blood groups, linkage and, 38–40
   sex ratio variability and, 246–258
Hydroxymethylcytosine, bacteriophage and, 107–108

I

Immunity,
   episomes,
      cytoplasmic nature and, 115–116
      genetic determination and, 116–120
      non-bacteriophage and, 120–121
      superinfection and, 114–115
Incompatibility,
   balanced polymorphisms and, 32–35
   Oenothera and, 204–208
Incubation, *Streptomyces coelicolor* and, 278–279
Interference, crossing-over frequency and, 49
Inversion, polymorphism and, 21–25
Irradiation, *Streptomyces coelicolor* and, 279–281

L

Lambda, episomes and, 108, 109, 110, 111, 112, 117, 118–119
Lebistes, balanced polymorphism in, 32
Leptophase, *Oenothera* and, 165
Linkage,
   balanced polymorphisms,
      artificial selection and, 37–38
      incompatibility systems and, 32–35
      linked complexes and, 35–38
      in man, 38–40
   biochemical loci and, 58–65
   blood groups and, 38–40
   Cepaea nemoralis and, 35, 36
   Matthiola annua and, 37–38
   MNSs blood groups and, 40
   mouse and, 37
   Papilio dardanus and, 35–36

polygenic balance, 4–7,
    artificial selection and, 7–21
    Rh blood groups and, 39
    Streptomyces coelicolor and, 296–299, 300–303, 310–321
    theoretical genetics of, 65–84
Lolium, balanced polygenes in, 20
Lolium-Festuca hybrids, chromosome behavior in, 41
Lymantria, sex determination and, 243
Lysogenization,
    episomes,
        genetic incorporation and, 123–124
        infection and, 121–122
        non-bacteriophage and, 124
        not-to-lyse decision and, 122

M

Maize, *see also Zea mays*, controlling elements in, 107
Mammals, sex ratio and, 259
Matthiola annua, linkage and, 37–38
Meiosis, early, *Oenothera* and, 164–167
Melandrium, pollen tube competition in, 169
Microorganisms, *see also* specific species
    linkage and recombination,
        biochemical loci and, 58–65
        genetic systems and, 52–55
        polygenic variation and, 55–58
Moraba scurra, polygenic balance in, 26
MNSs blood groups, linkage and, 40
Morphology,
    Streptomyces coelicolor,
        colony structure and, 285–287
        nuclear cytology and, 287–290
Mortality rates, sex ratio and, 245–246
Mouse,
    balanced polygenes in, 20–21
    linkage and, 37
Mutants, *Oenothera lamarckiana* and, 199–201
Mutations,
    de Vries', *Oenothera*, 188–202
    induced, *Oenothera*, 202–204
Mycelium, episomes and, 106–107
Myrmeleotettix maculatus, chiasma frequency in, 44

N

Natural selection, sex ratio and, 259–265
Neurospora, polygenic system in, 20
Neurospora crassa,
    biochemical loci, linkage in, 59, 60, 61, 62, 63, 65
    recombination, 50,
    temperature and, 48
Nomenclature, *Streptomyces coelicolor* and, 276–277
Nucleus,
    behavior, *Streptomyces coelicolor* and, 335–336
    cytology, *Streptomyces coelicolor* and, 287–290

O

Oenothera,
    chromosome structure,
        morphology and, 163–164
        size and, 160–163
    circle formation in, 156–160
    crossing-over in, 172–180
    cytological-genetic behavior of, 154–156
    cytology of, 152–160
    de Vries' mutants of, 188–202
    early meiotic stages of, 164–167
    evolution in, 215–229
    gene conversion in, 183–187
    gene location in, 170–172
    genetic analyses,
        female gametophyte development and, 167–169
        megaspore competition and, 167–169
        pollen tube competition and, 169–170
    genetic behavior, outline of, 148–152
    incompatibility in, 204–208
    induced mutations in, 202–204
    loci in, 173–174
    plastid behavior in, 208–215
    polygenic balance in, 31
    position effect in, 180–183
    recombination index and, 46
    trisomy in, 189–202
Oenothera lamarckiana, mutants in, 199–201
Onagra, see Oenothera

P

Pachyphase, Oenothera and, 165–167
Papilio dardanus, linkage and, 35–36
Periplaneta americana, recombination in, 45
Petunia, balanced polymorphism in, 34
Plastids, *Oenothera*, 208–215
Polarity, *Ascobolus immersus* and, 360–361
Polaron, *Ascobolus immersus* and, 355–360, 361–374
Polygenes, interactions and, 25–27
  location of, 14–19
Polygenic balance, linkage, 4–7, artificial selection and, 7–21
Polymorphism, inversion and, 21–25
  linkage,
    artificial selection and, 37–38
    incompatibility systems and, 32–35
    linked complexes and, 35–38
    man and, 38–40
Position effect, *Oenothera* and, 180–183
Primula obconica, balanced polymorphism in, 32–35
Primula vulgaris, balanced polymorphism in, 33–34
Prophage, bacterial chromosome, mode of attachment and, 111–114
  episomes and, 108–109, 110, 111–114
Pseudomonas, conjugation in, 104

R

Rat, Norway, sex ratio and, 258–259
Recombination,
  Ascobolus immersus and, 346–349
  chiasmata frequency and, 40–47
  selection and, 50–52
  Streptomyces coelicolor,
    analysis of, 293–301
    crossing techniques and, 291–293
    discovery of, 290–291
    fertility and, 303–304
    unequal frequencies and, 301–303
    temperature and, 47–48
Recombination index, chiasma frequency and, 44–47
Repression, episomes and, 114–115, 116
Resistance, episomes and, 105–106

Rh blood groups, linkage and, 39
Rye, chiasma frequency in, 41, 42–44

S

Salmonella spp. resistance transfer factor in, 106
Salmonella typhimurium, biochemical loci linkage in, 58–59, 60, 61, 62, 63, 64, 65
Schizosaccharomyces pombe, biochemical loci linkage in, 60
  mating type loci in, 34–35
  natural selection, 259–265
  sex determination,
    chromosome theory and, 242–243
    early theories and, 241–242
    human sex chromosomes and, 243–244
  sexual differentiation and, 240–241
Segregation, *Ascobolus immersus* and, 374–375
Selection,
  artificial,
    balanced polygenic complexes and, 7–21
    balanced polymorphisms and, 37–38
  recombination and, 50–52
Series 19, *Ascobolus immersus* and, 361–362, 367, 372–374
Series 46, *Ascobolus immersus* and, 355, 364–367, 370–372
Serratia, conjugation in, 104
Sex chromosomes, human, sex ratio and, 243–244
Sex determination,
  sex ratio,
    chromosome theory and, 242–243
    early theories and, 241–242
    human sex chromosomes and, 243–244
Sex differentiation, sex ratio and, 240–241
Sex ratio,
  birth control and, 255–256
  Drosophila, variability and, 258
  genetic problem and, 244–246
  human, variability and, 246–258
  mammals, variability and, 259
  mortality rates and, 245–246

Shigella, resistance transfer factor of, 105–106
Sordaria, odd segregations in, 374
Sorex araneus, recombination in, 45
Streptomyces coelicolor and, 219
Sporulation, episomes and, 106
Staphylococcus, penicillin-resistance in, 55–56
Strains,
  Streptomyces coelicolor,
    generic study strains and, 276–277
    nomenclature and, 276
Streptomyces albus, 275, 330
Streptomyces aureofaciens, recombination and, 275
Streptomyces coelicolor,
  biochemical loci linkage in, 60, 63
  culture media and, 277–278
  heteroclones,
    analysis of, 308–310
    discovery of, 304–305
    linkage and, 310–321
    nature of, 321–332
    selection and, 305–310
  heterokaryons and, 330–331, 332
  incubation and, 278–279
  morphology,
    colony structure and, 285–287
    nuclear cytology and, 287–290
  mutant characterization,
    auxotroph isolation and, 281
    irradiation and, 279–281
    mutant list and, 281–285
    streptomycin-resistant mutants and 281
  mutations in, 283–284
  nuclear behavior in, 335–336
  recombination,
    analysis of, 293–301
    crossing techniques and, 291–293
    discovery of, 290–291
    fertility and, 303–304
    unequal frequencies and, 301–303
  short chromosome regions and, 332–335
  spore suspensions and, 279
  stock culture maintenance and, 278–279
  strains,
    genetic study strains, 276–277
    nomenclature, 276
Streptomyces cyaneus, 275, 330
Streptomyces fradiae, recombination and, 275, 279
Streptomyces griseoflavus, recombination and, 275, 330
Streptomyces griseus, recombination and, 275, 279, 330
Streptomyces rimosus, recombination and, 275
Streptomyces scabies, recombination and, 275
Streptomyces venezuelae, 275, 330
Streptomyces violaceoruber, 276
Streptomycin,
  Shigella, resistance transfer factor and, 105–106
  Streptomyces coelicolor and, 281
Sulfonamide, *Shigella*, resistance transfer factor and, 105–106
Supergene, evolution and, 35–36
Superinfection, episomes and, 108, 114–115

T

Temperature, recombination and, 47–48
Tetracycline, *Shigella*, resistance transfer factor and, 105–106
Tetrads,
  non-reciprocal recombinants, *Ascobolus immersus* and, 346–349
  reciprocal recombinants, *Ascobolus immersus* and, 346–349
Transduction, episomes and, 126–132
Trisomy, *Oenothera* and, 189–202
Two locus theory, formulation of, 67–70

V

Variability, human sex ratio and, 246–258
Vhf, properties of, 103–104

Z

Zea mays, chiasma frequency in, 42
Zygophase, *Oenothera* and, 165